T0141840

Intelligent Systems Reference Library

Volume 175

Series Editors

Janusz Kacprzyk, Polish Academy of Sciences, Warsaw, Poland

Lakhmi C. Jain, Faculty of Engineering and Information Technology, Centre for Artificial Intelligence, University of Technology, Sydney, NSW, Australia;
Faculty of Science, Technology and Mathematics, University of Canberra, Canberra, ACT, Australia;
KES International, Shoreham-by-Sea, UK;
Liverpool Hope University, Liverpool, UK

The aim of this series is to publish a Reference Library, including novel advances and developments in all aspects of Intelligent Systems in an easily accessible and well structured form. The series includes reference works, handbooks, compendia, textbooks, well-structured monographs, dictionaries, and encyclopedias. It contains well integrated knowledge and current information in the field of Intelligent Systems. The series covers the theory, applications, and design methods of Intelligent Systems. Virtually all disciplines such as engineering, computer science, avionics, business, e-commerce, environment, healthcare, physics and life science are included. The list of topics spans all the areas of modern intelligent systems such as: Ambient intelligence, Computational intelligence, Social intelligence, Computational neuroscience, Artificial life, Virtual society, Cognitive systems, DNA and immunity-based systems, e-Learning and teaching, Human-centred computing and Machine ethics, Intelligent control, Intelligent data analysis, Knowledge-based paradigms, Knowledge management, Intelligent agents, Intelligent decision making, Intelligent network security, Interactive entertainment, Learning paradigms, Recommender systems, Robotics and Mechatronics including human-machine teaming, Self-organizing and adaptive systems, Soft computing including Neural systems, Fuzzy systems, Evolutionary computing and the Fusion of these paradigms, Perception and Vision, Web intelligence and Multimedia.

** Indexing: The books of this series are submitted to ISI Web of Science, SCOPUS, DBLP and Springerlink.

More information about this series at http://www.springer.com/series/8578

Margarita N. Favorskaya · Lakhmi C. Jain
Editors

Computer Vision in Advanced Control Systems-5

Advanced Decisions in Technical and Medical
Applications

 Springer

Editors
Margarita N. Favorskaya
Department of Informatics and Computer
Techniques
Institute of Informatics
and Telecommunications
Reshetnev Siberian State University
of Science and Technology
Krasnoyarsk, Russia

Lakhmi C. Jain
University of Canberra
Canberra, SA, Australia

University of Technology Sydney
Sydney, Australia

Liverpool Hope University
Liverpool, UK

KES International
UK

ISSN 1868-4394 ISSN 1868-4408 (electronic)
Intelligent Systems Reference Library
ISBN 978-3-030-33797-1 ISBN 978-3-030-33795-7 (eBook)
https://doi.org/10.1007/978-3-030-33795-7

This Springer imprint is published by the registered company Springer Nature Switzerland AG
The registered company address is: Gewerbestrasse 11, 6330 Cham, Switzerland

Preface

The research book is a continuation of our previous books which are focused on the recent advances in computer vision methodologies and technical solutions using conventional and intelligent paradigms.

- Computer Vision in Control Systems—1, Mathematical Theory, ISRL Series, Volume 73, Springer-Verlag, 2015
- Computer Vision in Control Systems—2, Innovations in Practice, ISRL Series, Volume 75, Springer-Verlag, 2015
- Computer Vision in Control Systems—3, Aerial and Satellite Image Processing, ISRL Series, Volume 135, Springer-Verlag, 2018
- Computer Vision in Control Systems—4, Real Life Applications, ISRL Series, Volume 136, Springer-Verlag, 2018.

Image processing and analysis remain a vital part of numerous real-time applications in every discipline. The main contribution of this book is an attempt to improve algorithms by novel theories and complex data analysis in different scopes including object detection, remote sensing, data transmission, data fusion, gesture recognition, and medical image processing and analysis.

Recent research results in the image and video processing, transmission, and image analysis are included in Part I, while a wide spectrum of algorithms for medical image processing are included in Part II of the book.

The book is directed to the Ph.D. students, professors, researchers, and software developers working in the areas of digital video processing and computer vision technologies.

We wish to express our gratitude to the authors and reviewers for their contribution. The assistance provided by Springer-Verlag is acknowledged.

Krasnoyarsk, Russian Federation

Canberra, Australia

Margarita N. Favorskaya

Lakhmi C. Jain

Contents

About the Editors

Dr. Margarita N. Favorskaya is a Professor and Head of Department of Informatics and Computer Techniques at Reshetnev Siberian State University of Science and Technology, Russian Federation.

Professor Favorskaya is a member of KES organization since 2010, the IPC member, and the Chair of invited sessions of over 30 international conferences. She serves as an associate editor of *Intelligent Decision Technologies Journal, International Journal of Knowledge-Based and Intelligent Engineering Systems, International Journal of Reasoning-Based Intelligent Systems*, a Honorary Editor of the *International Journal of Knowledge Engineering and Soft Data Paradigms*, Guest Editor, and Book Editor (Springer). She is the author or the co-author of 200 publications and 20 educational manuals in computer science. She co-edited several books for Springer. She supervised nine Ph.D. candidates to completion and presently supervising four Ph.D. students.

Her main research interests are digital image and video processing, remote sensing, pattern recognition, fractal image processing, artificial intelligence, and information technologies.

Dr. Lakhmi C. Jain, Ph.D., ME, BE(Hons), Fellow (Engineers Australia) is with the University of Technology Sydney, Australia, and Liverpool Hope University, UK.

Professor Jain founded the KES International for providing a professional community the opportunities for publications, knowledge exchange, cooperation, and teaming. Involving around 5000 researchers drawn from universities and companies worldwide, KES facilitates international cooperation and generates synergy in teaching and research. KES regularly provides networking opportunities for professional community through one of the largest conferences of its kind in the area of KES.

http://www.kesinternational.org/organisation.php

Chapter 1
Advanced Decisions in Technical and Medical Applications: An Introduction

Margarita N. Favorskaya and Lakhmi C. Jain

Abstract This chapter presents a brief description of chapters pertaining to advanced decisions for technical and medical systems. Recent research results in the image and videos processing, transmission, and image analysis are included in Part I, while a wide spectrum of algorithms for medical image processing are included in Part II of this book. Each chapter involves detail practical implementations and explanations.

Keywords Autoregressive models · Random fields · Image analysis · Multidimensional image processing · Pseudo-inverse matrix singular-value · Copyright protection · Sign language · Clinical decision support system · Histological image processing · Convolutional neural network

1.1 Introduction

At present, image processing and analysis remain a vital part of numerous applications in many fields. In spite of numerous methods and algorithms developed in the past, this topic is of current interest in various control systems based on computer vision paradigms. Researchers are looking for more effective and accurate

M. N. Favorskaya (✉)
Institute of Informatics and Telecommunications, Reshetnev Siberian State University of Science and Technology, 31, Krasnoyarsky Rabochy ave., Krasnoyarsk 660037, Russian Federation
e-mail: favorskaya@sibsau.ru

L. C. Jain
University of Technology Sydney, Ultimo, Australia
e-mail: jainlakhmi@gmail.com; jainlc2002@yahoo.co.uk

University of Canberra, Canberra, Australia

Liverpool Hope University, Liverpool, UK

© Springer Nature Switzerland AG 2020
M. N. Favorskaya and L. C. Jain (eds.), *Computer Vision in Advanced Control Systems-5*, Intelligent Systems Reference Library 175,
https://doi.org/10.1007/978-3-030-33795-7_1

algorithms to process the high resolution images in real-time mode with low computational costs. However, this requirement cannot be achieved fully at current stage of technical development. The main contribution of this book is the attempts to improve algorithms by novel theories and complex data analysis in different scopes including object detection, remote sensing, data transmission, data fusion, gesture recognition, and medical image processing and analysis. Part I includes the Chaps. 2, 3, 4, 5 and 6 and Part II of the book contains Chaps. 7, 8, 9 and 10.

1.2 Chapters Including in the Book

Chapter 2 explores the processing of multidimensional images, referring to aerospace images and remote sensing multispectral and hyperspectral images [1, 2]. New mathematical models of multidimensional images are proposed. Based on the procedures of vector Kalman filtering, optimal recurrent estimates of the autoregressive sequences with multiple roots of characteristic equations are constructed. Additionally to Kalman algorithm, Wiener filter was synthesized and investigated when processing random sequences generated by autoregressive model with multiple roots of characteristic equations. Compact analytical relations have been obtained for analyzing the effectiveness of random fields filtering algorithms with multiple roots of characteristic equations [3]. The behavior of filters was studied at various correlation intervals and in the processing of random fields with the roots of characteristic equations of various multiplicities. Thus, for large correlation intervals for the model with multiple roots, the filtering error variance is 5–10 times less than that for the ordinary autoregressive model. An algorithm for identification of the parameters and orders of the autoregressive models with multiple roots of characteristic equations based on solving Yule-Walker system of equations is proposed. Quasi-optimal and optimal filtering algorithms for random fields based on autoregressive models with multiple roots of characteristic equations have been developed and investigated. In particular, on the basis of Kalman filter, a solution of sequential row-by-row and column-by-column estimation was obtained, which makes it possible to reduce computational costs in comparison with optimal vector filtering. For models of the 1st and 2nd orders, the proposed algorithm loses no more than 10% in terms of the variance of the filtering error. The obtained algorithms are tested on real images. Experiments show that the use of autoregressive models with multiple roots provides a significantly lower filtering error variance than traditional approaches based on first-order models.

Chapter 3 studies the problems of multidimensional images and image sequences representation and processing within the framework of the Earth remote sensing [1]. Correlated data multidimensional arrays description and optimal and suboptimal processing are based on the proposed doubly stochastic autoregressive models. Application of doubly stochastic autoregressive models is twofold. First, the doubly stochastic models are used for description of multidimensional random fields and time sequences. A time sequence of multidimensional satellite images have four-dimensional random field with one-dimension corresponding to discrete time [4]. Second, doubly stochastic models allow to estimate the spatially hetero-geneous images with probabilistic properties, such as rivers, forests, fields, etc. Doubly stochastic models are based on autoregressive with multiple roots models of characteristic equation. For two-dimensional case, random field model based on doubly stochastic model ought to have the characteristics of a real image [5]. Identification technique based on a combination of on doubly stochastic model varying parameters estimation by means of a sliding window and by means of pseudogradient procedures is developed. The proposed algorithms are available to synthesize multidimensional images filtering algorithms and even several classes of such algorithms can process the multispectral satellite images as the real-time sequences. Actual satellite observations obtained in 2001–2017 are used for experiments.

Chapter 4 covers the analysis of inverse problems related to the applied elec-trodynamics, and radio, acoustic and optical wave physics [6]. For this purpose, a concept of matrix or tensor equations technique for direct problem derivation, as well as, the inverse problem resolving, which deals with determination and local-ization of the radiated sources' distribution a few limited cases of canonical objects and media, is developed. Matrix-vector systems of linear equations allows to find a solution of the inverse problems occurring in the optic and radio communication, wired and wireless, presented in 1D and 2D forms [7]. It is shown that the most of the inverse problems can be declared via system of linear algebraic equations with singular-value decomposition based on Moore-Penrose matrix. For non-linear equations solutions, Levenberg–Marquardt algorithm is applied. Also, Wiener's filtering with regularization is investigated in order to increase the accuracy of solution of the inverse problem, for example, reconstruction of blurred images. Some practical examples regarding to inverse problems, viz. source localization, micro-strip sensors reconstruction, and signal analysis, are discussed.

Chapter 5 contains a detailed overview of the transmitting specifications of video content, such as H.264/AVC, H.264/SVC, and H.265/HEVC in the sense of authentication and copyright protection. Internet attacks against video sequence are classified as the intentional and accidental attacks. Intentional attacks are directed on the distortions of a part of video or a single frame and categorized into common image processing and geometric attacks. Accidental attacks are concerned to the common processing attacks of video. Authentication and copyright protection of videos represented in some formats are developed in this chapter. The proposed video watermarking method is robust to several types of typical Internet attacks, e.g. the common image processing attacks, global and local geometric attacks, and permutation attacks. Proposed method supports the detection of I-frames and selection the best regions for embedding using the joint map, which excludes moving and salient regions [8] and involves high textural regions [9] with prevailed blue component [10]. Invariance to the main types of attacks regarding the com-pressed videos is provided by a feature-based approach for embedding with the original procedures. The novelty is that the coordinate values of speeded up robust features as they were in the host frame are embedded in the stable regions.

This allows to avoid the corresponding matches between SURF descriptors in the host and watermarked frames and extend a volume of embedded information after desynchronization attacks. In order to provide invariance to rotation, scaling, and translation attacks, exponential moments on a unit circle were applied. The experiments were conducted with simulation of rotation, salt and pepper noise, Gaussian noise, gamma correction, blurring filter, median filtering, scaling, cropping, and JPEG compression. Also, combination of attacks was simulated. Obtained experimental results show that the proposed algorithm is robust to the most types of attacks but strongly depends of video content.

Chapter 6 describes the multi-threshold analysis of monochromatic images. Typical limitations arise from low signal-to-background ratio in the area of interest, low quality images, excessive quantization, fuzzy boundaries of objects and structures. The original idea is to select and set the optimal threshold value based on the results of the selection of objects in multi-threshold framework to achieve the best selection based on a posteriori information. This approach was originally proposed in [11] for the selection of small-scale objects. Further development of this idea is described in this chapter and includes the evaluation of certain geometric parameters of the object in binary images after multi-threshold processing and the corresponding selection of objects [12, 13]. The optimal threshold value is selected according to the extremum of the selected parameter. This geometric parameters are the area of the object, the ratio of the perimeter square to the area of the entire object, or the ratio of the square of the main axis to the area of the object. The authors develop the idea of reconstructing a three-dimensional hierarchical structure of objects based on the multi-threshold analysis of the raw image. The objects are separated from each other based on the percolation effect. This effect is associated with the elimination of empty pixels that appear below the enhanced threshold from the object content, which ultimately leads to the breakup of the integrity of the object and the emergence of new isolated objects as its fragments. Thus, the objects of interest are represented in the form of 3D structures spanning through a series of binary layers. After 3D reconstruction, one can select the objects of interest using various criteria, such as their percolation properties, geometric characteristics, or texture parameters.

Chapter 7 reports the recognition results of one-handed gestures represented by Russian sign language as a way of communication among deaf and hearing impaired community [14]. The distinguish feature of this research is a combination of hand movements and facial expressions (including lips position). The proposed methodology is applied for the static, dynamic, and both static and dynamic gestures simultaneously. The chapter provides extended review of methods for hand gestures recognition and lip reading in the context of sign language and datasets in this scope. Many techniques for recognition of static and dynamic gestures are analyzed, and deep neural network with long short-term memory cells was chosen for implementation. First, the motion relevant to signs regions (hands and face regions) are detected as the regions of interest. The wrists and mouth regions are localized by certain landmarks. Second, for detection of the hand region and shape of the hand classification, MobileNetV2 as a very effective feature extractor for object detection and segmentation was trained. Finally, deep neural network with long short-term memory cells is applied as the best decision from recurrent neural network modeling time or sequence dependent behavior. The recognition results are obtained on the single-hand part of the collected TheRuSLan database [15] with promising values.

Chapter 8 conducts the investigations in the development of new methods for endoscopic images processing and analysis, which can be used as a base for construction of clinical decision support systems. The important issue for high effective physician analysis is a high quality of images [16]. The propose methods of noise reduction and image enhancement process the endoscopic images with computational cost permitting a real-time realization and high signal/noise ratio. New method of virtual chromoendoscopy consists of two stages. The first stage is a visualization of tissues and surfaces of mucous membranes including vessels structure stressing and the second stage is a tone enhancement [17]. The experimental test of proposed method was conducted on open KVASIR dataset of endoscopic images. For differential diagnostic implementation, methods for polyp and bleeding detection and segmentation in conditional of small database for training were developed. The method of polyp detection is based on combination traditional machine learning technique (random decision forests) and convolutional neural network. The special data augmentation—the sinusoidal image transform is applied in order to solve the problem of insufficiently large endoscopic images dataset. Some original procedures permitted to obtain rather good characteristics of medical images classification under their high variability and, the same time, small dataset for training.

Chapter 9 examines the computational methods for evaluating the indicators of the tissue regeneration process using clinical experiment with mesh nickelide titanium implants. For processing of scanning electron microscopy and classical histological images, a set of algorithms with high accuracy estimates are developed. Algorithms based on the shearlet and wavelet transforms with brightness correction provide better edge information [18, 19]. Algorithms for elastic maps generation with color coding allows to obtain more representative visualization of spatial data. The designed software helps to analyze a sequence of medical images in order to understand a dynamics of reconstructed tissues. The modified fast finite shearlet transform increases the accuracy of selection of linear structures and visual quality of the studied clinical images. Brightness correction using Retinex algorithm [20] allows to obtain a unified average brightness of analyzed images and, in some cases, increase a local contrast. The estimates of morphometric indicators of histological images include a calculation error for the main studied parameters. Evaluation of tissue germination was performed on the basis of scanning electron microscopy images. More objective data were used for images obtained from different angles. As a part of the study, for the evaluation of computer techniques a medical expert specified objects that were defined as tissue, fibers, red blood cells, etc., and the areas with the implant structure were specified separately. For the specified reference samples, parameters were calculated taking into account the indicators of texture characteristics and color code.

Chapter 10 presents the algorithms for histological images segmentation by convolutional neural network with morphological post-filtration [21]. Such algorithms can be used in decision support systems for early diagnosis of breast cancer by pathologists, as well as, a means of training or control for beginners in the field of breast cancer diagnosis. Algorithm 1 based on AlexNet neural network provides the high quality of histological images segmentation. However, this approach cannot be used for direct analysis of medical images in real time due to significant time costs. Thus, Algorithm 1 can be used to create the markup of the training dataset automatically. Algorithm 2 based on U-Net convolutional neural network with subsequent morphological filtering can be successfully used to implement the segmentation of histological images based on automatically obtained markup in real medical practice. Algorithm 2 allows the histological images to be processed 2,700 times faster than Algorithm 1. The segmentation results were evaluated using such segmentation quality assessment metrics as a simple match coefficient, Tversky index, and Sørensen coefficient. Numerical experiments confirmed a necessity to use the morphological filtering as a means of additional processing of histological images binary masks obtained at the output of convolutional neural networks.

1.3 Conclusions

This chapter includes a brief description of the chapters with original mathematical theories, algorithms, and extended experimental results in the image and videos processing as the basic components for creation of intelligent decision making systems, as well as, clinical decision support systems. All investigations included in this book provide the novel ideas, decisions, and applications in computer vision.

References

1. Vasiliev, K., Dementiev, V., Andriyanov, N.: Representation and processing of multispectral satellite images and sequences. Procedia Comput. Sci. **126**, 49–58 (2018)
2. Andriyanov, N.A., Vasiliev, K.K., Dement'ev, V.E.: Analysis of the efficiency of satellite image sequences filtering. J. Phys.: Conf. Ser. **1096**, 012036.1–012036.7 (2018)
3. Andriyanov, N.A., Vasiliev, K.K.: Use autoregressions with multiple roots of the characteristic equations to image representation and filtering. CEUR Work. Proc. **2210**, 273–281 (2018)
4. Krasheninnikov, V.R.: Correlation analysis and synthesis of random field wave models. Pattern Recognit. Image Anal. **25**(1), 41–46 (2015)
5. Krasheninnikov, V.R., Vasil'ev, K.K.: Multidimensional image models and processing. In: Favorskaya, M., Jain, L.C. (eds.) Computer Vision in Control Systems-3, ISRL, vol. 135, pp. 11–64. Springer International Publishing, Switzerland (2018)
6. Blaunstein, N., Yakubov, V.P. (eds.): Electromagnetic and Acoustic Wave Tomography: Direct and Inverse Problems in Practical Applications. CRC, Taylor & Frances Group, Boca Raton, FL (2019)

7. Sergeev A.M., Blaunstein N.S.: Orthogonal matrices with symmetrical structures for image processing. Informatsionno-upravliaiushchie sistemy [Information and Control Systems] **6**, 2–8 (in Russian) (2017)
8. Favorskaya, M., Buryachenko, V.: Fast salient object detection in non-stationary video sequences based on spatial saliency maps. In: De Pietro, G., Gallo, L., Howlett, R.J., Jain, L. C. (eds.) Intelligent Interactive Multimedia Systems and Services, SIST, vol. 55, pp. 121–132. Springer International Publishing, Switzerland (2016)
9. Favorskaya, M., Pyataeva, A., Popov, A.: Texture analysis in watermarking paradigms. Procedia Comput. Sci. **112**, 1460–1469 (2017)
10. Favorskaya, M.N., Jain, L.C. Savchina E.I.: Perceptually tuned watermarking using non-subsampled shearlet transform. In: Favorskaya, M.N., Jain L.C. (eds.) Computer Vision in Control Systems-3, ISRL, vol. 136, pp. 41–69. Springer International Publishing, Switzerland (2018)
11. Volkov, V.: Extraction of extended small-scale objects in digital images. Int. Arch. Photogramm. Remote Sens. Spatial Inf. Sci., XL-5/W6, 87–93 (2015)
12. Krasichkov, A.S., Grigoriev, E.B., Bogachev, M.I, Nifontov, E.M.: Shape anomaly detection under strong measurement noise: An analytical approach to adaptive thresholding. Phys. Rev. E **92**(4), 042927.1–042927.9 (2015)
13. Bogachev, M., Volkov, V., Kolaev, G., Chernova, L., Vishnyakov, I., Kayumov, A.: Selection and quantification of objects in microscopic images: from multi-criteria to multi-threshold analysis. Bionanoscience **9**(1), 59–65 (2019)
14. Ryumin, D., Kagirov, I., Ivanko, D., Axyonov, A., Karpov, A.A.: Automatic detection and recognition of 3D manual gestures for human-machine interaction. Int. Arch. Photogramm. Remote Sens. Spatial Inf. Sci., XLII-2/W12, 179–183 (2019)
15. Ryumin, D., Ivanko, D., Axyonov, A., Kagirov, I., Karpov, A., Zelezny, M.: Human-robot interaction with smart shopping trolley using sign language: data collection. In: IEEE International Conference on Pervasive Computing and Communications Workshops, pp. 1–6 (2019)
16. Obukhova, N., Motyko, A., Alexandr Pozdeev, A.: Review of noise reduction methods and estimation of their effectiveness for medical endoscopic images processing. In: 22nd Conference on FRUCT Association, pp. 204–210 (2018)
17. Obukhova, N., Motyko, A.: Image analysis in clinical decision support system. In: Favorskaya, M.N., Jain, L.C. (eds.) Computer Vision in Control Systems-4, ISRL, vol. 136, pp. 261–298. Springer International Publishing, Switzerland (2018)
18. Cadena, L., Espinosa, N., Cadena, F., Kirillova, S., Barkova, D., Zotin, A.: Processing medical images by new several mathematics shearlet transform. In: International MultiConference on Engineers and Computer Scientists, vol. I, pp. 369–371 (2016)
19. Zotin, A., Simonov, K., Kapsargin, F., Cherepanova, T., Kruglyakov, A., Cadena, L.: Techniques for medical images processing using shearlet transform and color coding. In: Favorskaya, M.N., Jain, L.C. (eds.) Computer Vision in Control Systems-4, ISRL, vol. 136, pp. 223–259. Springer, Cham (2018)
20. Zotin, A.: Fast algorithm of image enhancement based on multi-scale Retinex. Procedia Comput. Sci. **131**, 6–14 (2018)
21. Khryashchev, V., Lebedev, A., Stepanova, O., Srednyakova, A.: Using convolutional neural networks in the problem of cell nuclei segmentation on histological images. In: Dolinina, O., Brovko, A., Pechenkin, V., Lvov, A., Zhmud, V., Kreinovich, V. (eds.) Recent Research in Control Engineering and Decision Making. ICIT 2019. SSDC, vol. 199, pp. 149–161. Springer, Cham (2019)

Part I
Technical Applications

Chapter 2
Image Representation and Processing Using Autoregressive Random Fields with Multiple Roots of Characteristic Equations

Konstantin K. Vasil'ev and Nikita A. Andriyanov

Abstract An analytical review of mathematical models of images was performed, and their main advantages and disadvantages were noted. It is proposed to use Random Fields (RF) generated by AutoRegressive (AR) models with multiple roots of characteristic equations for describing images with a smooth change in brightness. Results of the study of the proposed models probabilistic properties are presented. The results obtained for Random Sequences (RS) are generalized to multidimensional RF. The filtering efficiency of simulated images is investigated. Analytical expressions are obtained for the relative variance of the filtering error of the arbitrary dimension and multiplicities RF against the background of white noise. Algorithm for identifying the parameters and the multiplicity of the model using the Yule–Walker equations is proposed. The possibilities and efficiency of application of the developed algorithms on real images are considered.

Keywords Autoregressive models · Roots of characteristic equations · Random fields · Image analysis · Covariance function · Correlation interval · Optimal filtering · Kalman filtering · Multidimensional wiener filtering · Model parameters identification

K. K. Vasil'ev · N. A. Andriyanov (✉)
Ulyanovsk State Technical University, 32 Severny Venets st., Ulyanovsk 432027,
Russian Federation
e-mail: nikita-and-nov@mail.ru

K. K. Vasil'ev
e-mail: vkk@ulstu.ru

N. A. Andriyanov
Ulyanovsk Civil Aviation Institute, 8/8 Mozhaiskogo st., Ulyanovsk 432071,
Russian Federation

© Springer Nature Switzerland AG 2020   11
M. N. Favorskaya and L. C. Jain (eds.), *Computer Vision in Advanced Control
Systems-5*, Intelligent Systems Reference Library 175,
https://doi.org/10.1007/978-3-030-33795-7_2

2.1 Introduction

Nowadays methods of multidimensional statistical analysis are widely used in various fields of science and technology. One of the most important classes of applied tasks for such an analysis is the representation and processing of multidimensional images. Examples of such images are aerospace images, remote sensing data (Earth remote sensing), medical image sequences, etc. In recent years, sensors have been increasingly used to obtain multispectral (up to 10 spectral ranges) and hyperspectral (up to 300 ranges) images. As a result, multidimensional arrays of information are obtained, which are described by coordinates in the space, time, and range of the spectrum. Thus, there is a rapid increase in the amount of information received, and new methods of presenting and analyzing data as a single multidimensional set are required.

It is obvious that obtaining and processing large amounts of information is a very complex task and requires significant computational cost [1–8]. The most important stage of image preprocessing is filtering stage [9–13]. The effectiveness of the filtering largely determines the results of post-processing. Errors obtained at this stage can have a significant impact when solving subsequent problems, such as image clustering or detecting anomalies. In this regard, it is important to use various methods of noise suppression in the images to be received [14–17].

Another important task is the identification of model parameters [18, 19]. It is clear that the more accurately the model describes a real image, the better its model-based processing will be. However, choice of a model assumes the necessity of its complexity analysis. For example, the development of algorithms for some models can be a simple task from a mathematical point of view. However, the processing efficiency based on such models will be low. On the other hand, increasing the complexity of the model leads to significant computational cost. Thus, it is necessary to describe images using models that combine possibility of analytical study and do not require significant computational cost for image processing tasks.

This chapter is devoted to development and investigation of new mathematical models of multidimensional images, which allow to solve simple recurrent processing algorithms synthesis problems and to analyze the effectiveness of using such algorithms.

The rest of the chapter describes the advantages and disadvantages of known mathematical models of images (Sect. 2.2), the one-dimensional AR with multiple roots model and its processing (Sect. 2.3), the properties and processing of RF generated by multidimensional AR with multiple roots (Sect. 2.4), and the real image processing results (Sect. 2.5). Section 2.6 contains more significant conclusions of the work.

2.2   Mathematical Models of Images

When solving problems of image processing, an important step is the choice of an adequate model for observations. Currently, there is no universal way to form RF with arbitrarily specified characteristics. In addition, there is no sufficiently complete solution to the problem of describing real images. Therefore, the well-known models of RF correspond to real images only by a limited number of parameters, such as the form of Correlation Function (CF), the distribution of amplitudes, etc. There are a large number of methods for simulating RF. In [20], all models of RF are divided into two classes. First class models describe fields with continuous distributions. Gauss and Markov RF [21] models can be categorized into this class. Such models are usually obtained either using spectral transformations or by shaping filter method. Given the discrete nature of real systems of spatial information sensors and additional time sampling when transmitting signals over digital communication channels, it is possible to consider only those models that represent RF on multidimensional space-time grids [22–26].

Let us analyze a number of well-known RF models that can be used for description of images during the synthesis of various image processing procedures, such as, for example, filtering, segmentation, or restoration and prediction. AR stochastic models are usually considered as the most well-known models.

It is possible to describe the images by RF, defined on multidimensional grids. In this case, a general description of RF is achieved using tensor difference stochastic equations [27]. Then the sequence of multidimensional frames is defined as changing in the discrete time RF, specified on the multidimensional grid $J_t = \{\bar{j} = (j_1, j_2, \ldots, j_N), j_1 = 1, M_l, l = 1, 2, \ldots, N\}$, where $j_1, j_2, \ldots, j_N$ are the space coordinates. Figure 2.1 shows an example of such image.

Fig. 2.1 Multizone image frames

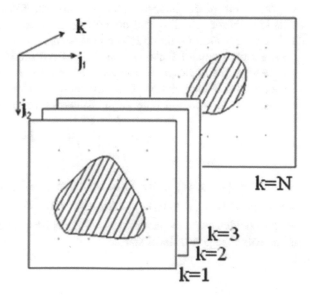

The elements of RF are scalar values, which describe brightness of the image at a given point. Thus, the sequence of changing frames of the analyzed image can be considered as RF on the direct product $J_t \otimes T$ [3], elements of which will be denoted as x (the value of RF at the time moment t at the point $\bar{j}$).

In some cases, the following linear tensor stochastic difference equation [27] can be taken as the mathematical model of RF:

$$x_{\bar{j}}^t = \rho_{\bar{j}\,\bar{l}}^t x_{\bar{j}}^{t-1} + \vartheta_{\bar{j}\,\bar{l}}^t \xi_{\bar{l}}^t \bar{j}, \bar{l} \in J_t, \tag{2.1}$$

where $\{\xi_{\bar{l}}^t, \bar{l} \in J_t\}$ is RF of independent standard Gaussian Random Variables (RV), $\rho_{\bar{j}\bar{l}}^t$, $\vartheta_{\bar{j}\bar{l}}^t$ are the tensors with two group indices. This ratio determines Gaussian Markov RF on the direct product $J_t \otimes T$. Such RF supposes that previous elements ($\Gamma_t^- = \{x_{\bar{j}}^q, \bar{j} \in J_q, q < t\}$) and future elements ($\Gamma_t^+ = \{x_{\bar{j}}^q, \bar{j} \in J_q, q > t\}$) are frame $\Gamma_t = \{x_{\bar{j}}^t, \bar{j} \in J_q\}$ independent. Problems of analysis and synthesis of this model are considered in [27].

However, such a representation of a multidimensional RF leads to considerable computational difficulties. In this regard, it is advisable to use the representation of RF by recurrent procedures both in time and in spatial coordinates [28]:

$$x_{\bar{j}} = \Phi_{\bar{j}}(x_{\bar{l}}, \xi_{\bar{l}}), \bar{l} \in G_{\bar{j}}, \tag{2.2}$$

where $G_{\bar{j}}$ are the areas of elements $\bar{l} \in J$, on which the previous values of RF $\{x_{\bar{j}}\}$ are already determined, i.e. causal window.

In 1956, Levi [29] was the first to introduce Markov RF (MRF) models. Discrete two-dimensional MRF based on the continuous case proposed by Levi have been described by Woods [30]. The discrete MRF model describes each pixel as a weighted sum of neighboring pixels and normal RV. Such RF provides a probabilistic basis for modeling and integrating prior knowledge of images and scenes, and is widely used in digital image processing.

The most studied RF class is the class of AR models [31–33]. One of the main reasons for the widespread use of AR models is the mathematical apparatus developed for RS simulation. The class of AR models of RF is generated by linear stochastic difference equations of the following form [27]:

$$x_{\bar{i}} = \sum_{\bar{j} \in D} \alpha_{\bar{j}} x_{\bar{i}-\bar{j}} + \beta \xi_{\bar{i}}, \ \bar{i} \in \Omega, \tag{2.3}$$

where $X = \{x_{\bar{i}}, \bar{i} \in \Omega\}$ is RF to be simulated, which is determined on N-dimensional grid $\Omega = \{\bar{i} = (i_1, i_2, \ldots, i_N) : \{i_k = \overline{1 \ldots M_k}\}, k = \overline{1 \ldots N}\}$; $\{\alpha_{\bar{j}}, \beta, \bar{j} \in D\}$ are the coefficients of the model; $\{\xi_{\bar{i}}, \bar{i} \in \Omega\}$ is RF of standard Gaussian RV; $D \subset \Omega$ is the causal region of local states.

It is quite convenient and simple to choose a normally distributed RF with independent components for the generating process. In this case, RF X also has a Gaussian distribution. Let us consider the formation of RF $X = \{x_{\bar{i}}, \bar{i} \in \Omega\}$, using AR model supposed by Habibi [33]:

$$x_{i,j} = \rho_x x_{i-1,j} + \rho_y x_{i,j-1} - \rho_x \rho_y x_{i-1,j-1} + \xi_{i,j}, i = \overline{1...M_1}; j = \overline{1...M_2}, \quad (2.4)$$

where ρ_x and ρ_y are the correlation coefficients of neighboring elements column-wise and row-wise, respectively; $\{\xi_{i,j}\}$ is two-dimensional field of independent Gaussian RV with zero mean $M\{\xi_{i,j}\} = 0$ and variance $\sigma_\xi^2 = M\{\xi_{i,j}^2\} = (1 - \rho_x^2)(1 - \rho_y^2)\sigma_x^2$; $\sigma_x^2 = M\{x_{i,j}^2\}$, $M_1 \times M_2$ is the size of simulated image.

RF generated in this way is anisotropic, and its CF due to anisotropy is a generalization of CF of a one-dimensional first-order AR to the two-dimensional case. It can be shown [34] that it is described by the following expression:

$$B(k_1, k_2) = \sigma_x^2 \rho_x^{|k_1|} \rho_y^{|k_2|}, \quad (2.5)$$

where σ_x^2 is the variance of RF X, ρ_x and ρ_y are the model parameters, k_1 and k_2 are the distances between the elements of RF X along the axes x and y, respectively.

The analysis of probabilistic properties of RF is considerably simplified if their spectral density can be factorized. So called separable RF is a convenient object for research. Since these fields have normalized CF $R(\bar{k}) = \prod_{i=1}^{N} R_i(k_i)$ which can also be factorized, then to solve the problem of statistical analysis of CF of a multidimensional RF, it suffices to use the properties of the RS generated by one-dimensional AR with characteristics $R_i(k_i), i = \overline{1...M_i}$, where M_i characterizes the multidimensionality of such RF.

AR models have significant drawbacks associated with the limited size of the local state regions, which do not allow it to be fully used as a model of a multi-zone image. Therefore, for an adequate representation of real images, it is necessary to expand the region of local states, which leads to a significant increase in computational costs when simulating RF.

Models [35–39] based on the possibility of extending to the multidimensional case of AR of the second and higher orders with multiple roots of characteristic equations serve as a certain compromise. For example, for the second order AR process with multiple roots of characteristic equations:

$$x_i = 2\rho x_{i-1} - \rho^2 x_{i-2} + \xi_i \quad (2.6)$$

the corresponding eight-point model of a two-dimensional RF can be obtained:

$$x_{ij} = 2\rho_x x_{i-1,j} + 2\rho_y x_{i,j-1} - 4\rho_x \rho_y x_{i-1,j-1}$$
$$- \rho_x^2 x_{i-2,j} - \rho_y^2 x_{i,j-2} + 2\rho_x^2 \rho_y x_{i-2,j-1} \quad (2.7)$$
$$+ 2\rho_y^2 \rho_x x_{i-1,j-2} - \rho_x^2 \rho_y^2 x_{i-2,j-2} + b\xi_{ij},$$

where b is the normalization coefficient, which allows to simulate RF with a given variance.

The analysis shows that increase of multiplicity of the model makes RF realization form close to isotropic. Obviously, the corresponding changes should influence the form of CF. Thus, we first consider the two-dimensional case with multiplicity $m = 2$. Then CF of one-dimensional sequence takes the following form:

$$R_i(k) = \left(1 + \frac{(1 - \rho_i^2)}{(1 + \rho_i^2)} |k|\right) \rho_i^k, i = 1, 2, \ldots. \quad (2.8)$$

If $1 - \rho_i \ll 1$, the asymptotical expression for CF of two-dimensional RF assumes the following form:

$$R(k_1, k_2) = R(k_1)R(k_2) = 1 - \frac{k_1^2}{a^2} - \frac{k_2^2}{b^2} + k_1^2 k_2^2 (1 - \rho_x)^2 (1 - \rho_y)^2 A(\rho_x) B(\rho_y),$$
$$(2.9)$$

where $a = \sqrt{\frac{1 + \rho_x^2}{(1 - \rho_x^2)(1 - \rho_x)}}$, $b = \sqrt{\frac{1 + \rho_y^2}{(1 - \rho_y^2)(1 - \rho_y)}}$, $A(\rho_x) = \frac{1 + \rho_x}{1 + \rho_x^2}$, $B(\rho_y) = \frac{1 + \rho_y}{1 + \rho_y^2}$.

It is obvious that cross sections of CF $R(k_1, k_2)$ at the level near $R(k_1, k_2) = 1$ can be approximated by ellipsoids.

It can be shown that subject to minor assumptions, CF of two-dimensional RF (Eq. 2.9) generated by AR with multiple roots of characteristic equations can be written as follows [34]:

$$R(k_1, k_2) = R(k_1)R(k_2) = 1 - \frac{k_1^2}{a^2} - \frac{k_2^2}{b^2} + k_1^2 k_2^2 (1 - \rho_x)^{m_1} (1 - \rho_y)^{m_2} A(\rho_x) B(\rho_y),$$
$$(2.10)$$

where m_1, m_2 are the multiplicities of roots of one-dimensional AR.

Along with AR or causal models of RF on flat and spatial rectangular grids, there are a number of non-causal models. Non-Causal AR (NCAR) models represent the values of each pixel as a linear combination of the pixel values of local states and the addition of additive white noise. The difference between the MRF and NCAR models is the spatial correlation of these RV. In [40], an iterative estimation method and an algorithm for synthesizing two-dimensional NCAR models were proposed. This work illustrates the application of NCAR model for representing near-real images with local repetitive properties.

It is possible to get an isotropic model by search for an adequate model in a non-autoregressive class. For example, there is the class of wave models [34]. For wave models, RF is formed as follows:

$$S(x, t) = \sum_{\{i: \tau \leq t\}} f(x, t; u_i, \tau_i; \bar{\varpi}_i),$$
(2.11)

where $x = (x_1, x_2, \ldots, x_n)$, $u = (u_{i_1}, u_{i_2}, \ldots, u_{i_n})$ are the points of n-dimensional space, t and τ_i is the time characteristics, $\{(u_i, \tau_i)\}$ is the discrete Field of Random Points (RPF), $\bar{\varpi}_i$ is the function random parameter vector.

Such RF can be interpreted as the result of the cumulative effect of random perturbations or waves $f(x, t; u_i, \tau_i; \bar{\varpi}_i)$, occurring in random places u_i at random times τ_i and varying according to some law in time and space. Choosing a wave formation method f, parameters of RPF and $\bar{\varpi}_i$ allows one to get a wide range of types of RF. Examples of such RF are Poisson fields, a model of weighted sums, models of random walks.

Simulation of such RF requires significant computational costs. At present the problem of analytic representation of the laws of probability distribution of wave models is not resolved and is a very difficult task.

Obviously, discarding the connection of elements throughout the image and taking only the connections between the nearest elements as a basis, it is impossible to fully describe the properties of real images. The crucial reason for the incompleteness of such a description is the fact that in a real image different sections are usually characterized by different statistical properties. A typical example would be images containing outlines. In [41, 42], a version of the model is proposed that considers the image as the sum of two independent components namely a piecewise-smooth (contour) spatial component defining global brightness changes, and a high-frequency component defining texture, noise, and small details.

Also, non-causal models include Gibbs fields [43, 44], when the value of RF at a point depends on the nearest pixels that do not precede this point in the order of a certain image scanning. The stochastic equations of such RF have the following form [44]:

$$x(p) = f(\{x(u), u \in G_p\}, \xi_p),$$
(2.12)

where $x(p)$ is the value of RF at point p; G_p is the set of nearest neighbors; ξ_p is the random disturbance.

In linear models, Eq. 2.12 can be written as:

$$x(p) = A \cdot (x(u), u \in G_p) + \xi_p,$$
(2.13)

where A is the matrix, $(x(u), u \in G_p)$ is the vector of RF samples by the set of nearest neighbors G_p, ξ_p is the random disturbance. When simulating RF using this model, disturbances ξ_p are first generated and then linear equation system (Eq. 2.13) is solved.

If all the eigenvalues of the matrix A are modulo less than one, then it is possible to implement the model provided by Eq. 2.13 by running through, starting, for example, with the identically zero approximation. Such a case occurs, when Eq. 2.13 is the averaging over the nearest samples with the sum of the modules of coefficients less than one.

In [43, 45], the image synthesis of Gibbs model is considered as an iterative process using Monte Carlo–Markov Chain (MCMC) methods, for example Metropolis Exchange Algorithm (MEA). Given the symmetry of the neighborhood, it is possible to determine Gibbs energy on the grid. The joint probabilities corresponding to the Gibbs energies form Gibbs RF:

$$P(x) = \frac{1}{Z}\exp\left\{-\frac{1}{T}E(x)\right\},\tag{2.14}$$

where Z is the constant; T is RF temperature.

It is considered [43] that the parameter T changes on each iteration step according to the law:

$$T_i = \frac{c}{\lg\left(1 + \frac{i}{i_{cq}}\right)},\tag{2.15}$$

where $i = 0, 1, \dots$ is the current iteration number, i_{cq} is the total iteration number, c is the constant.

By varying these parameters, it is possible to get different fields. The main disadvantage of this approach is the need to perform a large number of iterations in the simulation. In addition, the analytical description of this model is very cumbersome.

In image processing, textures are widely used [46, 47]. This is due to the large range of the resulting images, which can be matched quite close to real. In addition, aerospace observations have rather complex space-time properties, which makes it difficult to analyze data only by spectral features. Unfortunately, there are currently no methods for the synthesis of textures that could provide recognition with a minimum average error.

In [48], the simulation of multispectral spatially inhomogeneous dynamic brightness fields is carried out based on the use of random texture synthesis methods. The phase spectrum method is used as a basic mathematical model for generating realizations of two-dimensional stochastic textures. Such RF are realizations of a two-dimensional spatial spectral model that generalizes the results of processing images of the brightness fields of natural formations obtained in aerospace experiments for different observation conditions.

The power spectral density can be represented as:

$$G(v_x, v_y) = \begin{cases} v^{-p} & \text{if } f(v_x, v_y) \leq 0 \\ Qv^{-q} & \text{if } f(v_x, v_y) > 0 \end{cases}, \qquad (2.16)$$

where $f(v_x, v_y)$ is RF shape function; p and q are the tilt parameters; Q is some coefficient.

This approach enables to simulate multi-zone images that are close in their properties to real. However, the complexity of the described procedure for the synthesis of multi-zone images causes significant difficulties in solving problems of analytic representation of the laws of probability distribution of such a model.

The considered models of RF are intended to describe a spatially uniform real material. However, many types of images and their sequences contain areas with different statistical characteristics. To simulate a multi-zone image, more complex mathematical models with varying multidimensional parameters are needed. One of these classes is the mixed or doubly stochastic models [49–54].

In 1987, Woods et al. [49] proposed a two-dimensional Doubly Stochastic Gaussian (DSG) model. It is also proposed to use combinations of different methods of RF formation for modeling images in one of the first works concerned with mixed models [54]. Example of such combination can be written as follows:

$$B(x) = \eta(x)B_1(x) + (1 - \eta(x))B_2(x), \qquad (2.17)$$

where $B_1(x)$ is the implementation of a single microstructure, for example, an object in the image; $B_2(x)$ is the implementation of a different microstructure, for example, the underlying surface; $\eta(x)$ is the implementation of the field defining interactions of microstructures. Multipliers $\eta(x)$ and $(1 - \eta(x))$ switch model from one implementation $B_1(x)$ to another implementation $B_2(x)$.

Let us consider RF given on a rectangular multidimensional grid Ω so that its values $x_{\bar{i}} = F(x_{\bar{j}}, \alpha_{\bar{i}}, \xi_{\bar{i}})$, where $\bar{i} \in \Omega, \bar{j} \in D$; $F(\cdot)$ is some transformation, $\alpha_{\bar{i}}$ are the model parameters that are independent of $\xi_{\bar{i}}$. Then the probability density function (PDF) of a RV $x_{\bar{i}}$ can be defined as:

$$W(x_{\bar{i}}) = \int_D W(x_{\bar{i}}|\alpha_{\bar{i}})W(\alpha_{\bar{i}})d\alpha_{\bar{i}}. \qquad (2.18)$$

The rejection of Gaussian distributions when modeling hyperspectral data is currently gaining popularity. Such modeling mainly focuses on attempts to describe the distribution of all layers of the image. In work [55], it is proposed to use a mixed model of a hyperspectral image provided that the presence of large materials in the scene is not of random nature. Variability in this model is associated with noise or other factors that exhibit random behavior. The proposed RF is a linear mixed model with a structured background. It was proposed to use two models: the first is based on a multidimensional t-distribution; the second is based on independent components of the exponential distribution. However, a disadvantage of such models is the lack of proximity of the resulting image to the original with an increase in the number of components.

Consider an image simulation process based on the following three steps. At the first step a frame of a uniform RF is generated. It is called basic RF. Next, the values of the RF in the resulting frame $\wp$ are converted into a set of correlation parameters $\{\rho_{\bar{i}}, \bar{i} \in (i_1, i_2, \ldots, i_M)\}$, where M is the dimension of the simulated image. These parameters characterize the magnitude of the connection of the current pixel of the simulated image with adjacent image elements. Then an image is formed as a RF with varying correlation parameters $\rho_{\bar{i}}$.

To simulate the frames of the basic RF, it is possible to use different models. Let us take, for example, the implementation of a two-dimensional RF $X = \{x_{\bar{i}}, \bar{i} \in \Omega\}$; $\bar{i} = \{i, j\}$, which is generated by Habibi AR model (Eq. 2.4). Its use necessitates the preliminary simulation of two basic RF, the brightness values of one of which will be converted into a set of correlation parameters $\{\rho_{1ij}, i = 1, 2, \ldots, M_1, j = 1, 2, \ldots, M_2\}$, and the brightness values of other will be converted into a set of correlation parameters $\{\rho_{2ij}, i = 1, 2, \ldots, M_1, j = 1, 2, \ldots, M_2\}$, respectively:

$$\rho_{1ij} = r_{11}\rho_{1(i-1)j} + r_{12}\rho_{1i(j-1)} - r_{11}r_{12}\rho_{1(i-1)(j-1)} + \xi_{1ij},$$
$$\rho_{2ij} = r_{21}\rho_{2(i-1)j} + r_{22}\rho_{2i(j-1)} - r_{21}r_{22}\rho_{2(i-1)(j-1)} + \xi_{2ij}, \tag{2.19}$$

where $\{\xi_{1ij}\}$ and $\{\xi_{2ij}\}$ are two-dimensional RF of independent Gaussian RV with zero mean and variances $M\{\xi_{1ij}^2\} = (1 - r_{11}^2)(1 - r_{12}^2)\sigma_{\rho_1}^2$ and $M\{\xi_{2ij}^2\} = (1 - r_{21}^2)(1 - r_{22}^2)\sigma_{\rho_2}^2$; $\sigma_{\rho_1}^2 = M\{\rho_{1ij}^2\}$, $\sigma_{\rho_2}^2 = M\{\rho_{2ij}^2\}$.

In the basic RF, the correlation coefficients $\{\rho_{1ij}\}$ and $\{\rho_{2ij}\}$ characterize the size and shape of objects in the simulated image. An increase in these parameters will lead to an increase in the average size of objects in the simulated image.

The choice of the method of converting the brightness values into a set of correlation parameters enables to control CF values, which makes it possible to simulate RF that are close in their correlation properties to real satellite images.

As the RF $\{\xi_{1ij}\}$ and $\{\xi_{2ij}\}$ have zero mean, then mean values for RF $\{\rho_{1ij}\}$ and $\{\rho_{2ij}\}$ are also equal to zero in Eq. 2.19. Therefore, it is necessary to choose such transformation of the values in basic RF, in which the mathematical expectations m_{ρ_1} and m_{ρ_2} at each point will be added to the value of RF $\{\rho_{1ij}\}$ and $\{\rho_{2ij}\}$:

$$\rho_{1ij} = \rho_{1ij(X)} + m_{\rho_1},$$
$$\rho_{2ij} = \rho_{2ij(Y)} + m_{\rho_2}. \tag{2.20}$$

Doubly stochastic models allow the simulation of heterogeneous images using a mathematical apparatus developed for AR models. They, as well as textures, have been still insufficiently studied. However, in the case of describing images containing large objects, within which the brightness changes quite slowly, it is impractical to use such complex models.

The properties of heterogeneity and multidimensionality inherent in real large data remain outside the bounds of most mathematical models that either describe spatially homogeneous signals or turn out to be too complex for the subsequent formation of processing and prediction algorithms. To describe such multidimensional data, it is proposed to use a deep Gaussian processes and their multidimensional generalization. A feature of these processes is the characteristic nested structure, when the parameters of the model are themselves implementations of a lower level process. In [56], authors showed a good approximation quality, which is achieved when describing real processes by deep Gaussian models. In [57], the possibility of describing non-stationary stochastic processes using this approach was shown. However, to date, two important tasks are unsolved. First, most of the work in this area is devoted to the description of one-dimensional processes [58]. At the same time, the majority of real information arrays are multidimensional.

2.3 Representation and Processing of One-Dimensional Autoregressions with Multiple Roots of Characteristic Equations

Autoregressive RS are widely used to describe changes in the state of real physical objects, to simulate signals and interference in various information and communication systems [59, 60]. AR processes of order m can be given by the following stochastic difference equation:

$$x_i = \rho_1 x_{i-1} + \rho_2 x_{l-2} + \cdots + \rho_m x_{i-m} + \xi_i, \ i = 2, 3, \ldots, n, \quad (2.21)$$

where ξ_i, $i = 1, 2, \ldots, n$ are Gaussian independent RV with zero mean and variance $M\{\xi_i^2\} = \sigma_\xi^2$. Coefficients $\rho_1, \rho_2, \ldots, \rho_m$ can be calculated on the basis of experimental material using Yule–Walker equations [59] or obtained on the basis of the known properties of dynamic systems. CF of sequence with different roots z_v, $v = 1, 2, \ldots, m$, of characteristic equation:

$$z^m - \rho_1 z^{m-1} - \rho_2 z^{m-2} - \cdots - \rho_m = 0 \quad (2.22)$$

and meeting the stability condition $|z_v| < 1$, $v = 1, 2, \ldots, m$, is represented by the following sum:

$$B_x(k) = A_1 z_1^{|k|} + A_2 z_2^{|k|} + \cdots + A_m z_m^{|k|}. \quad (2.23)$$

Using the simplest equation of the first order $(m = 1)$ $x_i = \rho x_{i-1} + \beta \xi_i$, $i = 1, 2, \ldots$, leads to a spiky-type realizations of such RS (Fig. 2.2a) and such RS have exponential CF $B_x(k) = \sigma_x^2 \rho^{|k|}$ (Fig. 2.2b).

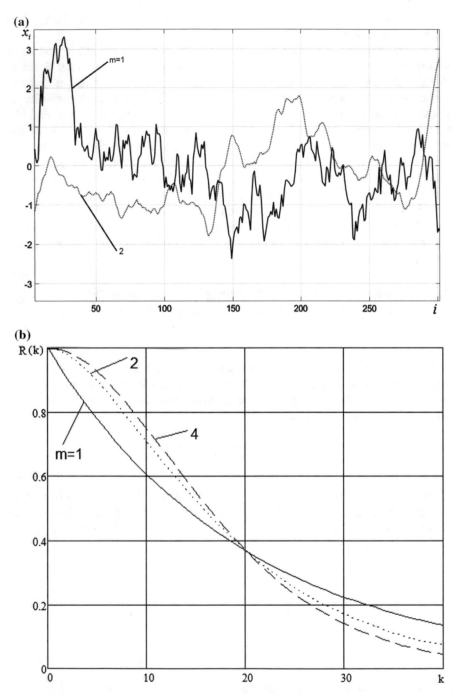

Fig. 2.2 Realizations of the first order and second order autoregressions: **a** with multiple roots of the characteristic equations, **b** their correlation functions

AR processes with multiple roots of the characteristic equation can serve as a compromise between the simplicity of the description and the possibility of modeling close-to-real and smooth-form RS. Consider the properties of such RS [61, 62], the possibilities of optimal recurrent estimation and the parameters identification for AR models with multiple roots.

Let us write the following characteristic equation:

$$z^m - \rho_1 z^{m-1} - \rho_2 z^{m-2} - \cdots - \rho_m = 0. \tag{2.24}$$

AR model of the mth order can be written based on Eq. 2.24 as follows:

$$x_i = \rho_1 x_{i-1} + \rho_2 x_{i-2} + \cdots + \rho_m x_{i-m} + \xi_i, \tag{2.25}$$

where ξ_i is random additive with zero mean and variance σ_ξ^2, which depends on the coefficients $\rho_1, \rho_2, \ldots, \rho_m$.

Characteristic equation (Eq. 2.24) with the root $z = \rho$ of m multiplicity assumes the form $(z - \rho)^m = 0$, and AR equation (Eq. 2.25) assumes the following form:

$$(1 - \rho z^{-1})^m x_i = \xi_i, \tag{2.26}$$

where $z^{-k} x_i = x_{i-k}$.

Using Eq. 2.26, it is possible to explicitly write the expressions for models with roots of the characteristic equation with various multiplicities m:

$$x_i = \rho x_{i-1} + \xi_i, m = 1,$$
$$x_i = 2\rho x_{i-1} - \rho^2 x_{i-2} + \xi_i, m = 2,$$
$$x_i = 3\rho x_{i-1} - 3\rho^2 x_{i-2} + \rho^3 x_{l-3} + \xi_i, m = 3,$$
$$x_i = 4\rho x_{i-1} - 6\rho^2 x_{i-2} + 4\rho^3 x_{i-3} - \rho^4 x_{i-4} + \xi_i, m = 4.$$

Figure 2.3 shows the simulation of such RS for the case of coefficients that provide the same correlation interval.

Analysis of Fig. 2.3 shows that increasing the multiplicity of the roots of the characteristic equation leads to the formation of RS smoother and often closer to real-world processes.

Characteristic equation (Eq. 2.22) with root $z = \rho$ of multiplicity m takes the form $(z - \rho)^m = 0$ and AR process (Eq. 2.21) can be written in operator form as follows:

$$(1 - \rho z^{-1})^m x_i = \beta \xi_i, \tag{2.27}$$

where $z^{-k} x_i = x_{i-k}$.

If the variance $\sigma_x^2 = M\{x_i^2\}$ is given, then $\beta^2(m) = (1 - \rho^2)^{2m-1}$ $\sigma_x^2 / \sum_{l=0}^{m-1} (C_{m-1}^l \rho^l)^2$.

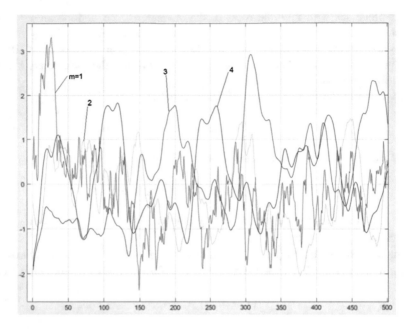

Fig. 2.3 Realizations of AR with multiple roots having the same correlation interval

The considered models of different orders m have the following CF:

$$B_m(k) = M\{x_i x_{i-k}\} = \beta^2(m)\rho^k \sum_{l=0}^{m-1} g(m,l,k) \frac{\rho^{2(m-l-1)}}{(1-\rho^2)^{2m-l-1}}, \qquad (2.28)$$

where $g(m,l,k) = \frac{(m+k-1)!(2m-l-2)!}{l!(m-1)!(m-l-1)!(m+k-l-1)!}$.

If in different models the value of root ρ is the same, then CF decreases slower as the selected order m is increased. A comparative analysis of models of different multiplicities m and having the equal correlation intervals k_0 is of practical interest. The correlation interval is characterized by the following expression:

$$R_m(k_0) = 1/e \qquad (2.29)$$

where $R_m(k) = B_m(k)/\sigma_x^2$ is normalized CF.

Using Eqs. 2.28–2.29, it is possible to find the coefficients ρ, at which the correlation interval is k_0 for a given model order m. Unfortunately, an analytical solution of the corresponding nonlinear equations can only be found, when $m = 1$: $k_0 = -1/\ln\rho \approx 1/(1-\rho)$. Table 2.1 presents the results of numerical calculations for different multiplicities m and correlation intervals k_0.

Analysis of Fig. 2.2a shows that increasing of the order of the AR model results in smoother realizations of the process with the same correlation interval $k_0 = 20$.

Table 2.1 Calculation of model parameters at different correlation intervals

k_0	1	10	20	50	100	500	1000	10,000
$\rho_{m=1}$	0.368	0.9048	0.9512	0.9802	0.99	0.998	0.999	0.99999
$\rho_{m=2}$	0.1908	0.808	0.8985	0.958	0.9789	0.9957	0.99786	0.99979
$\rho_{m=3}$	0.1286	0.7496	0.8651	0.9436	0.9714	0.9942	0.9971	0.99971
$\rho_{m=4}$	0.097	0.706	0.8393	0.9322	0.9655	0.993	0.9965	0.99965

The same fact is confirmed by the graphs of Fig. 2.2b, where successive CF samples have stronger connections at higher orders of the model.

Figure 2.4 presents dependencies of magnitude $\gamma = (1 - \rho)k_0$ on correlation interval k_0 for different orders of AR. Since $k_0 = \gamma/(1 - \rho)$, coefficient γ shows how many times the correlation interval of an m order AR process greater than the correlation interval of the first order AR.

Analysis of the curves in Fig. 2.4 enables to conclude that the value $k_0(1 - \rho)$ at large values of k_0 tends to a constant value. For example, when $m = 1$: $k_0(1 - \rho)$ $\to 1$ if $k_0 \to \infty$ and when $m = 2$: $k_0(1 - \rho) \to 2.15$. However, a significant difference is observed only between models of the second and first orders, and the difference between the properties of AR of the second and higher orders is not so significant.

Consider the possibility of constructing of equations for asymptotic values $\gamma = k_0(1 - \rho)$, when $k_0 \to \infty$, under different orders m of AR models with multiple roots of the characteristic equation.

After substitution of the coefficient $\beta^2(m)$ with $\sigma_x^2 = 1$ in Eq. 2.28 and simple transformations, it is possible to write the following expression for the normalized CF:

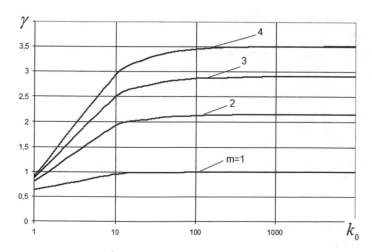

Fig. 2.4 Dependencies of γ versus correlation interval

$$R_m(k) = \frac{\rho^k g(m,0,k)}{\sum\limits_{l=0}^{m-1} (C_{m-1}^l \rho^l)^2} \sum_{l=0}^{m-1} \frac{g(m,l,k)}{g(m,0,k)} \rho^{2(m-l-1)} (1 - \rho^2)^l, \qquad (2.30)$$

where $g(m,0,k) = (2m - 2)!/(m - 1)!(m - 1)!$.

Thus, the correlation interval can be found as a solution to Eq. 2.31:

$$R_m(k_0) = \left(1 - \frac{\gamma}{k_0}\right)^{k_0} \frac{C_{2m-2}^{m-1}}{\sum\limits_{l=0}^{m-1} (C_{m-1}^l \rho^l)^2} \sum_{l=0}^{m-1} \varphi(m,l,k_0) \rho^{2(m-l-1)} \gamma^l = e^{-1}, \qquad (2.31)$$

where

$$\varphi(m,l,k_0) = A(m,l,k_0) \frac{(1+\rho)^l (m-1)(m-2)\ldots(m-l)}{l!(2m-2)(2m-3)\ldots(2m-l-1)},$$

$$A(m,l,k_0) = \frac{(m+k_0-1)(m+k_0-2)\ldots(m+k_0-l)}{k_0^l}.$$

Asymptotically with $k_0 \to \infty$, this equality is greatly simplified given $C_{2m-2}^{m-1} = \sum_{l=0}^{m-1} (C_{m-1}^l)^2$ and $A(m,l,k_0) \to 1$. After taking the logarithm we obtain the basic relation for finding the coefficient γ at different values of the multiplicity of characteristic equation roots m which takes the following form:

$$\ln \sum_{l=0}^{m-1} \varphi(m,l)\gamma^l = \gamma - 1, \qquad (2.32)$$

where $\quad \varphi(m,l) = \frac{(m-1)(m-2)\ldots(m-l)}{l!(m-1)(m-1,5)\ldots(m-0,5(l+1))}; \varphi(m,l) = \frac{m-l}{l(m-0,5(l+1))} \varphi(m,l-1);$ $\varphi(m,l=0) = 1$.

Figure 2.5 shows the dependence of the steady-state coefficient γ versus the multiplicity of the model.

Thus, the obtained asymptotic formula (Eq. 2.32) allows us to find the correlation interval of AR sequence with the roots of the characteristic equation of any multiplicity m. It is necessary to use direct expression for CF $R_m(k)$ with relatively short correlation interval k_0.

Now we will consider the problem of optimal recurrent estimation of RS based on AR with multiple roots if there are observations of the additive mixture

$$z_i = x_i + n_i, \ i = 1, 2, \ldots, \qquad (2.33)$$

with white Gaussian noise n_i, $i = 1, 2, \ldots$, and variance σ_n^2. Since the investigated RS is an m-connected Markov process, let us introduce an extended state vector:

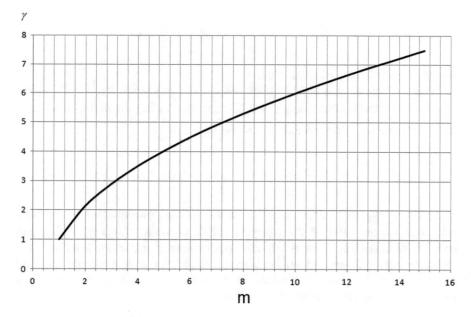

Fig. 2.5 Dependence of coefficient γ versus model multiplicity

$$\bar{x}_i = (x_i x_{i-1} \ldots x_{i-m+1})^T. \tag{2.34}$$

Then the observation model can be written as:

$$z_i = C\bar{x}_i + n_i, \ i = 1, 2, \ldots, \tag{2.35}$$

where $C = (1 \ 0 \ldots 0)$.

Equation of state

$$z^m - \rho_1 z^{m-1} - \rho_2 z^{m-2} - \cdots - \rho_m = 0 \tag{2.36}$$

can also be presented in vector matrix form:

$$\bar{x}_i = \wp\bar{x}_{i-1} + \bar{\xi}_i, \ i = 1, 2, \ldots, \tag{2.37}$$

where
$$\wp = \begin{pmatrix} \rho_{11} & \rho_{12} & \cdots & \rho_{1m} \\ 1 & 0 & \cdots & 0 \\ 0 & 1 & \cdots & 0 \\ 0 & 0 & \cdots & 1 \end{pmatrix}, \qquad \rho_{1j} = (-1)^{j+1} C_m^j \rho^j,$$

$\bar{\xi}_i = (\xi_i 0 \ldots 0)^T$, $V_\xi = M\{\bar{\xi}_i \bar{\xi}_i^T\}$.

After the performed transformations, standard equations of the linear Kalman filter can be used to find the optimal estimates of the information RS [61, 63]:

$$\hat{x}_i = \hat{x}_{\ni i} + P_i C^T \frac{1}{\sigma_n^2}(z_i - C\hat{x}_{\ni i}), P_i = P_{\ni i}(E + \frac{1}{\sigma_n^2}C^T CP_{\ni i})^{-1}, \qquad (2.38)$$

where $P_{\ni i} = \wp P_{i-1}\wp^T + V_\xi; \hat{x}_{\ni i} = \wp\hat{x}_{i-1}$.

Each ith step of estimation implies the optimal prediction $\hat{x}_{\ni i} = \sum_{j-1}^m \rho_{1j}\hat{x}_{i-j}$ based on previous estimates $\hat{x}_{i-j}, j = 1, 2, \ldots, m$ and finding the best estimation in terms of minimum error variance $P_{11i} = M\left\{(x_i - \hat{x}_i)^2\right\}$:

$$\hat{x}_i = \hat{x}_{\ni i} + P_{11i}\frac{1}{\sigma_n^2}(z_i - \hat{x}_{\ni i}), \qquad (2.39)$$

where $P_{11i} = P_{\ni 11i}/(1 + P_{\ni 11i}/\sigma_n^2)$. The remaining components of the vector $\hat{\vec{x}}_i$ are calculated on the basis of the interpolation of previous estimates, taking into account the next observation z_i and cross covariances of estimation errors.

Now consider the case of Wiener filtering, implying a weight summation of all observations with optimal coefficients in the sense of minimum variance of the filtering error. Let there be RS of the form:

$$x_i = \rho_1 x_{i-1} + \rho_2 x_{i-2} + \cdots + \rho_m x_{i-m} + \xi_i, \ i = 2, 3, \ldots, n, \qquad (2.40)$$

where $\xi_i, \ i = 1, 2, \ldots, n$ are Gaussian independent RV with zero means and variances $M\{\xi_i^2\} = \sigma_\xi^2$. Coefficients $\rho_1, \rho_2, \ldots, \rho_m$ can be found based on a single root ρ of model with multiple roots.

Consider the problem of optimal filtering of such RS based on observations of the additive mixture

$$z_i = x_i + n_i, \ i = 1, 2, \ldots, \qquad (2.41)$$

with discrete white Gaussian noise $n_i, \ i = 1, 2, \ldots$, having zero mean and variance σ_n^2.

The variance of the filter error is determined by Eq. 2.42:

$$\sigma_\varepsilon^2 = \sigma_n^2 g_0, \qquad (2.42)$$

where $g_0 = \frac{1}{2\pi i}\oint_C H(z)z^{-1}dz, z = e^{i\omega}$.

The transfer function can be found using Eq. 2.43:

$$H(z) = \frac{F(z)}{F(z) + \sigma_n^2}, \qquad (2.43)$$

where $F(z)$ is spectrum of source signal x_i and it can be found as z-transform of CF: $F(z) = \sum_{j=-\infty}^{\infty} B_x(j)z^{-j}$.

In other words $F(z) = \frac{\beta_m^2}{(1-\rho z^{-1})^m(1-\rho z)^m}$ is modulus square of transfer function $\frac{1}{(1-\rho z^{-1})^m}$, multiplied by the variance β_m^2 of random increment ξ_i.

After some transformation the expression takes the following form:

$$F(z) = \frac{\beta_m^2}{(1 - \rho(z^{-1} + z) + \rho^2)^m}. \tag{2.44}$$

Thus, for a non-realizable Wiener filter the following expression can be written:

$$\frac{\sigma_\varepsilon^2}{\sigma_n^2} = \frac{1}{2\pi i} \oint_C \frac{z^{-1}\beta_m^2/\sigma_n^2}{\beta_m^2/\sigma_n^2 + (1 - \rho(z^{-1} + z) + \rho^2)^m} dz. \tag{2.45}$$

After replacement $z = e^{i\omega}$, $dz = iz d\omega$, $z^{-1} + z = 2\cos\omega$ expression for the relative variance of the filter error will have the following form:

$$\frac{\sigma_\varepsilon^2}{\sigma_n^2} = \frac{1}{2\pi} \int_{-\pi}^{\pi} \frac{\beta_m^2/\sigma_n^2}{\beta_m^2/\sigma_n^2 + (1 - 2\rho\cos\omega + \rho^2)^m} d\omega. \tag{2.46}$$

Thus, Eq. 2.46 depends only on the signal-to-noise ratio and the correlation parameter ρ.

In Wiener filtering, linear weighting of the entire sequence with specified coefficients g_{ik} is assumed:

$$\hat{x}_k = \sum_{i \in D_k} g_{ik} z_i, \tag{2.47}$$

where D_k is area of time points of observations.

Equation 2.47 should be an optimal estimate in the sense of the minimum error variance, therefore $\sigma_{\varepsilon k}^2 = M\{(\hat{x}_k - x_k)^2\}$ should be minimal for all possible coefficients g_{ik}.

After simple transformations, it is possible to write the following expression for the error variance:

$$\sigma_{\varepsilon k}^2 = M\left\{\left(\sum_{i \in D_k} g_{ik} z_i - x_k\right)^2\right\} = \sigma_{xk}^2 + M\left\{\sum_{i \in D_k}\sum_{j \in D_k} g_{ik} g_{jk} z_i z_j - 2x_k \sum_{i \in D_k} g_{i,k} z_i\right\}, \tag{2.48}$$

where $\sigma_{xk}^2 = M\{x_k^2\}$.

To find the optimal weights, it is necessary to differentiate Eq. 2.48 by $\{g_{ik}\}$ and equate the result to zero. So, it leads to the following system of equations:

$$M\left\{\left(\sum_{j\in D_k} g_{jk}z_j - x_k\right)z_i\right\} = 0, i \in D_k \tag{2.49}$$

or

$$\sum_{j\in D_k} g_{jk}B_z(i,j) = B_{zx}(i,k), i \in D_k, \tag{2.50}$$

all coefficients of which, except for the desired weights, are known. Taking into account Eq. 2.50, the minimal variance of the filtering error is easily found as follows:

$$\sigma_{\varepsilon k}^2 = \sigma_{xk}^2 - \sum_{i\in D_k} g_{ik}B_{zx}(i,k). \tag{2.51}$$

To find weights g_{ik} it is possible, for example, to use the well-known relations for CF and the power spectrum. Thus, algorithms of optimal filtering of RS with multiple roots of characteristic equations are synthesized.

We now perform a comparative analysis of the effectiveness of the described algorithms for filtering RS with multiple roots of the characteristic equations ($m = 2$). Figure 2.6 shows the dependencies of the steady-state variance of the filtering error for models of the first order (solid lines) and second order (dashed lines) for different values of the signal-to-noise ratio $q = \sigma_x^2/\sigma_n^2$. At large values of correlation intervals for the second-order model there is a significant gain compared with the filtering of the first-order AR processes.

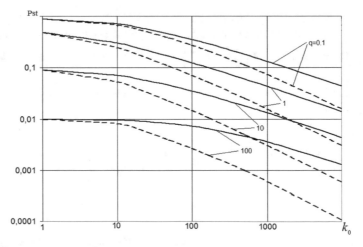

Fig. 2.6 Recurrent filtering effectiveness

Let us also consider the operation of Wiener filtering with weighted summation with the obtained coefficients for processing RF with multiple roots of the characteristic equations of the first and second orders. Figure 2.7 shows simulated RF with noise, as well as their estimates. Figure 2.7a corresponds to the correlation interval $k_0 = 10$, and Fig. 2.7b corresponds to the correlation interval $k_0 = 100$. The dashed line presents the source signal, the dotted line presents the noisy signal, and the solid line presents the result of filtering. The curves of the variance of the filtering error on the signal-to-noise ratio for the first-order model are presented by dashed lines, for the second-order model—by solid line.

Analysis of Fig. 2.7 shows that the actual processing of simulated sequences preserves an advantage when filtering a second-order model.

Finally, let us consider the problem of identifying the parameters of such a model. To solve the problem of identification, it is possible to use AR models of an arbitrary order:

$$x_i = \rho_1 x_{i-1} + \rho_2 x_{i-2} + \cdots + \rho_m x_{i-m} + \xi_i, i = 1, 2, \ldots, M, \tag{2.52}$$

where m is AR order.

The choice of parameters $\rho_1, \rho_2, \ldots, \rho_m$ allows to get Gaussian RS $\{x_i\}$, $i = 1, 2, \ldots, M$ with various correlation properties. In this case, for the values of CF, the following expression can be used:

$$R_x(k) = \rho_1 R_x(k-1) + \rho_2 R_x(k-2) + \cdots + \rho_m R_x(k-m), k > 0. \tag{2.53}$$

The general formula for models of different multiplicities can be written in the form:

$$x_{i,j} = \beta \xi_{i,j} - \sum_{i_1=0}^{N_1} \sum_{j_1=0}^{N_2} \alpha_{i_1,j_1} x_{i-i_1,j-j_1}, \tag{2.54}$$

where N_1 and N_2 characterize the multiplicity of the model; coefficients α_{i_1,j_1} ($\alpha_{0,0} = 0$) are the products of the corresponding coefficients of one-dimensional AR along the axes x and y:

$$\alpha_{i_1,j_1} = \alpha_{xi_1} \alpha_{yj_1}. \tag{2.55}$$

The coefficients of one-dimensional AR (Eq. 2.55) can be obtained using the expressions:

$$\alpha_{xi_1}(\rho_x, N_1) = (-1)^{i_1+1} C_{N_1}^{i_1} \rho_x^{i_1}, \alpha_{yj_1}(\rho_y, N_2) = (-1)^{j_1+1} C_{N_2}^{j_1} \rho_y^{j_1}, \tag{2.56}$$

where $C_n^m = \frac{n!}{m!(n-m)!}$. Coefficient β of two-dimensional model is the normalized product of the corresponding coefficients of one-dimensional AR along the axes x and y:

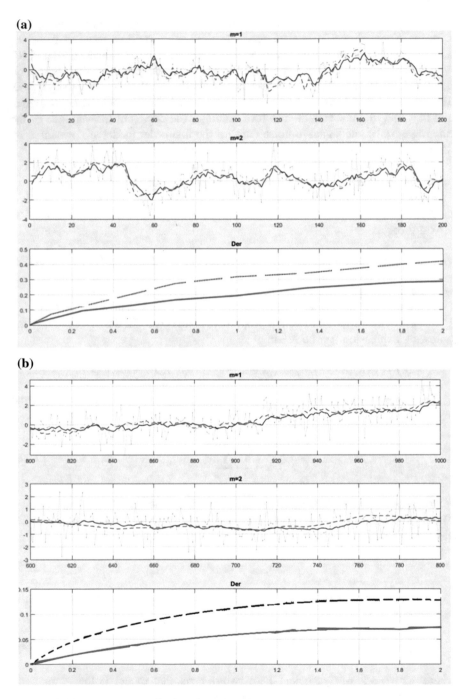

Fig. 2.7 Filtering of autoregression with multiple roots: a $k_0 = 10$, b $k_0 = 100$

$$\beta = \frac{\sigma_x}{\sigma_\xi} \beta_x \beta_y. \tag{2.57}$$

CF for different roots z_v, $v = 1, 2, \ldots, m$, of characteristic equation

$$z^m - \rho_1 z^{m-1} - \rho_2 z^{m-2} - \cdots - \rho_m = 0, \tag{2.58}$$

and meeting the stability condition $|z_v| < 1$, $v = 1, 2, \ldots, m$, can be written as:

$$R_x(k) = A_1 z_1^{|k|} + A_2 z_2^{|k|} + \cdots + A_m z_m^{|k|}. \tag{2.59}$$

Substitution in Eq. 2.53 values $k = 1, 2, \ldots, m$ leads to the well-known Yule–Walker system of equations [64], which for second-order systems assumes the form:

$$\begin{aligned} \rho_1 + \rho_2 R(1) &= R(1), \\ \rho_1 R(1) + \rho_2 &= R(2). \end{aligned} \tag{2.60}$$

The solution of this system enables to find the coefficients $\rho_1, \rho_2, \ldots, \rho_m$ of AR (Eq. 2.52) for pre-set or experimentally estimated CF values $R_x(1)$, $R_x(2)$, ..., $R_x(m)$.

Parameter identification for models with multiple roots of characteristic equations of the first to fourth orders will be performed. In this case, the order can be identified if we take into account only the coefficients that make a significantly nonzero contribution.

Using statistical modeling a procedure was performed to identify the correlation parameters from the values of CF of AR with multiple roots. Analysis of the results shows that the correlation coefficients are estimated the more accurately, the higher the multiplicity is. If the estimated multiplicity exceeds the real, then the additional coefficients are either zero or very close to it. Thus, the process of identifying the order of the model can be performed first for a certain large multiplicity. If there are no zeros in the resulting coefficients, then a calculation should be carried out for a larger multiplicity until zeros appear. If there are zero coefficients, then the order corresponds to the number of the last significant coefficient.

If a multiple root model is used, it is enough to calculate only the first correlation coefficient and then calculate the correlation parameter using the first coefficient and use the correlation parameter to find the remaining coefficients according to Eq. 2.56. For example, if $\rho_1 = 3.6$ and $m = 4$, it is easy to obtain correlation parameter of AR with multiple roots using the following expression:

$$3.6 = (-1)^{1+1} C_4^1 \rho^1, \rightarrow \rho = 0.9. \tag{2.61}$$

After that, it is easy to calculate the second, third, and fourth correlation coefficients:

$$\rho_2 = (-1)^{1+2} C_4^2 \rho^2 = -4.86,$$
$$\rho_3 = (-1)^{1+3} C_4^3 \rho^3 = 2.916, \qquad (2.62)$$
$$\rho_4 = (-1)^{1+4} C_4^4 \rho^4 = -0.6561.$$

Thus, the identification of parameters is not a serious difficulty if the order of the model is known.

2.4 Representation and Processing of Multidimensional Random Fields Generated by Autoregressions with Multiple Roots of Characteristic Equations

Based on the separation property of CF of AR with multiple roots, it is possible to generalize the one-dimensional model to the multidimensional case. Let us consider models that allow forming autoregressive RF with roots of characteristic equations of arbitrary multiplicities and dimensions.

AR Gaussian RF in the general case can be given by the following equations:

$$x_{\bar{i}} = \sum_{\bar{j} \in D} \alpha_{\bar{j}} x_{\bar{i}-\bar{j}} + \sigma_x \beta_0 \xi_{\bar{i}}, \ \bar{i} \in \Omega, \qquad (2.63)$$

where $X = \{x_{\bar{i}}, \bar{i} \in \Omega\}$ is simulated RF, defined on N-dimensional grid $\Omega = \{\bar{i} = (i_1, i_2, \ldots i_N) : i_k = 1 \ldots M_k, k = 1 \ldots N\}$, $\{\beta_0, \alpha_{\bar{j}}, \bar{j} \in D\}$ are the model coefficients, $\{\xi_{\bar{i}}, \ \bar{i} \in \Omega\}$ is the white Gaussian RF with zero mean and unit variance, σ_x^2 is the variance of RF $x_{\bar{i}}$, $D \subset \Omega$ is the causal region of local states.

AR models with multiple roots correspond to a spatial linear filter with following transfer function:

$$H(\bar{z}) = \frac{\sigma_x \beta_0}{1 - \sum_{\bar{j} \in D} \alpha_{\bar{j}} \bar{z}^{-\bar{j}}}, \qquad (2.64)$$

where $\bar{z}^{-\bar{j}} = z_1^{-j_1} z_2^{-j_2} \ldots z_N^{-j_N}$.

At the same time, the power spectrum of RF X can be written as follows:

$$S_x(\bar{z}) = H(\bar{z}) H(\bar{z}^{-1}). \qquad (2.65)$$

The analysis of probabilistic properties of RF is simplified if the transfer function of the multidimensional filter can be factorized: $H(\bar{z}) = \prod_{k=1}^{N} H_k(z_k)$. Then the power spectrum $S_x(\bar{z}) = \prod_{k=1}^{N} S_k(z_k)$ and CF $B(\bar{r}) = \prod_{k=1}^{N} B_k(r_k)$ can also be factorized. Simple and very useful for applications multidimensional separable RF $x_{\bar{i}}$ can be represented using spatial AR

$$\prod_{k=1}^{N} \left(1 - \rho_k z_k^{-1}\right)^{m_k} x_i = \sigma_x \beta_0 \xi_{\bar{i}}, \ \bar{i} \in \Omega \qquad (2.66)$$

with roots ρ_k of characteristic equations of multiplicities m_k, $k = 1, 2, \ldots, N$. In this case

$$H(\bar{z}) = \sigma_x \beta_0 / \prod_{k=1}^{N} \left(1 - \rho_k z_k^{-1}\right)^{m_k}, \qquad (2.67)$$

where $\beta_0 = \prod_{k=1}^{N} \beta_k$; $\beta_k(m_k) = (1 - \rho_k^2)^{2m_k - 1} / \sum_{l=0}^{m_k-1} \left(C_{m_k-1}^l \rho_k^l\right)^2$.

It should be noted that the causal region of the preceding values of RF $x_{\bar{i}}$ consists of $\left(\prod_{k=1}^{N} (m_k + 1)\right) - 1$ elements. For example, if $N = 2$ and $(m_1, m_2) = (2, 2)$ such region will include 8 elements.

In order to find the CFs of multidimensional RF, one can use expressions for one-dimensional CF of AR with multiple roots of characteristic equations. Figure 2.8 presents CF of two-dimensional RF for models of different orders, providing the same correlation radius $k_0 = 10$ at the level $1/e$. Figure 2.8a shows the characteristics of model with multiplicities (1, 1). Figure 2.8b shows the characteristics of model with multiplicities (2, 2). Figures 2.8c and 2.8d show the implementation of model with multiplicities (1, 1) and (2, 2), respectively. Analysis of the figures shows a noticeable flattening of the top of the CF, even with a minimal increase in the multiplicity of the root of the characteristic equation.

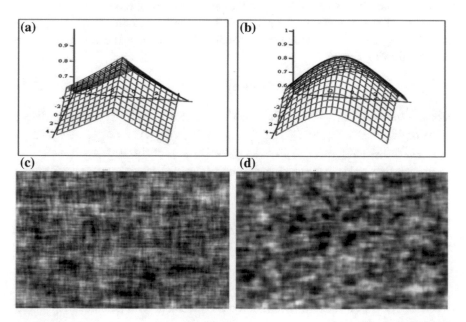

Fig. 2.8 Correlation functions of random fields: **a** characteristics of model with multiplicities (1, 1), **b** characteristics of model with multiplicities (2, 2), **c** implementation of model with multiplicities (1, 1), **d** implementation of model with multiplicities (2, 2)

Table 2.2 Equivalent correlation radii

k_0	1	10	20	50	100	500	1000	10,000
$\rho_{m=1}$	0.606	0.9048	0.9512	0.9672	0.9802	0.99004	0.99502	0.9994998
$\rho_{m=2}$	0.338	0.7657	0.8737	0.9137	0.9472	0.9732	0.98653	0.998644
$\rho_{m=3}$	0.2345	0.685	0.8257	0.8797	0.926	0.96225	0.98093	0.998077
$\rho_{m=4}$	0.1795	0.6275	0.7895	0.854	0.9095	0.9536	0.97653	0.99763

However, RFs with the same correlation radius are most interesting for investigation. Representation of equivalent models of different orders has a significant novelty. For such two-dimensional RF correlation parameters were obtained that ensure the same correlation radius based on the calculations of the CF of the two-dimensional RF. Table 2.2 shows these calculations.

In addition, it is possible that the condition of equivalence is the same correlation interval on separate axes. Such a case is also easily investigated due to the separation of CF, however, the table of parameters will coincide with that obtained earlier for the one-dimensional case. For increasing correlation radii, the asymptotic properties of probabilistic characteristics were investigated.

Thus, the parameter characterizing how many times the radius of correlation of high-order models differs from the first-order model can be found similarly, and this parameter tends to a constant value for large radii of correlation. Recurrent and power expressions are obtained for its calculation in the transition from the discrete case to the continuous one.

To calculate it let us consider again the one-dimensional model and consider the transition of the discrete autoregressive RS to continuous time. Equation $x_i(1 - \rho z^{-1})^m = \beta_m \xi_i$ with $\Delta t \to 0$ and given condition $(1 - \rho) = a\Delta t$ can be transformed to the following form:

$$x(t)(p + a)^m = \xi(t), \qquad (2.68)$$

where $px = dx/dt$, $\xi(t)$ is white noise with spectral density $N_\xi = \beta_m^2/\Delta t^{2m-1} = \sigma_x^2(2a)^{2m-1}/C_{2m-2}^{m-1}$, $\sum_{l=0}^{m-1}(C_{m-1}^l)^2 = C_{2m-2}^{m-1}$, $\beta^2(m) = (1 - \rho^2)^{2m-1}\sigma_x^2/\sum_{l=0}^{m-1}(C_{m-1}^l\rho^l)^2 \to \sigma_x^2 2^{2m-1}(1 - \rho)^{2m-1}/C_{2m-2}^{m-1}$.

For the first order, AR equation the limit transition is performed as follows:

$$\begin{aligned}
x_i &= \rho x_{i-1} + \beta_1 \xi_i \to x_i - x_{i-1} \\
&= -(1 - \rho)x_{i-1} + \beta_1 \xi_i \to (x_i - x_{i-1}) \\
&= -(1 - \rho)x_{i-1} + \sqrt{\sigma_x^2(1 - \rho^2)}\xi_i \to dx(t) \\
&= -ax(t)\Delta t + \Delta\chi(t),
\end{aligned} \qquad (2.69)$$

where $\Delta\chi(t) = \sigma_x\sqrt{2a\Delta t}\xi$ is Wiener process increment with parameter $N_\xi = 2a\sigma_x^2$ during time interval Δt.

Derivative of the Wiener process $\Delta\chi(t)/\Delta t = n(t)$ was called white noise with a power spectrum $N_\xi = 2a\sigma_x^2$. Similar expressions can be written for $m = 2$:

$$
\begin{aligned}
x_i &= 2\rho x_{i-1} - \rho^2 x_{i-2} + \beta_2 \xi_i \rightarrow ((x_i - x_{i-1}) - (x_{i-1} - x_{i-2}))/\Delta t^2 \\
&= (-2a\Delta t(x_{i-1} - x_{i-2}) - a^2\Delta t^2 x_{i-2})/\Delta t^2 + \beta_2 \xi/\Delta t^2 \rightarrow d^2 x(t)/dt^2 \quad (2.70)\\
&= -2a(dx(t)/dt) - a^2 x(t) + d\chi(t)/dt,
\end{aligned}
$$

where $d\chi(t)/dt = n(t)$ is white noise with spectral density $N_\xi = \beta_2^2/\Delta t^3 = 4a^3\sigma_x^2$. In general, such model provides the process $x(t)$ with power spectrum:

$$
G_x(\omega) = \frac{N_\xi}{(\omega^2 + a^2)^m}, \quad (2.71)
$$

where $N_\xi = \beta_m^2/\Delta t^{2m-1} = \sigma_x^2 a^{2m-1} 2^{2m-1}/C_{2m-2}^{m-1}$.

Power spectrum width $\Delta\omega_m$ at the level $1/M$ decreases with growth m. Indeed, using the equation $1/(\omega^2/a^2 + 1)^m = 1/M$ it is possible to find $\Delta\omega_m = a\sqrt{(\sqrt[m]{M} - 1)}$. Parameter γ is described by the equation $\gamma = \Delta\omega_1/\Delta\omega_m = \sqrt{M-1}/\sqrt{(\sqrt[m]{M} - 1)}$ at small values of $1 - \rho$. At this level M, which determines the width of the spectrum, usually can be set within $M = 2 - 3$.

Figure 2.9 shows dependence of γ versus different values of parameter M. The solid line presents the dependence obtained by Eq. 2.32, circles present the calculated values for $M = 10.9$.

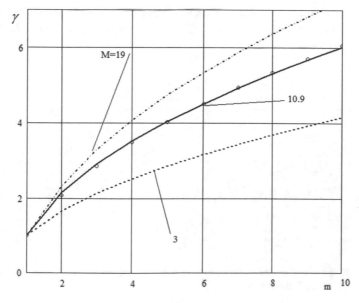

Fig. 2.9 Approximate calculation of γ dependency versus model multiplicity

Analysis of the obtained curves shows that when $M = 10.9$ it is possible to get a fairly accurate solution of the complex Eq. 2.32 using a simple formula. At the same time, sufficient accuracy is ensured at acceptable values of the multiplicity of the model.

Based on processing algorithms for one-dimensional AR with multiple roots quasi-optimal and optimal filtering algorithms for RF generated by AR with multiple roots of characteristic equations can be developed. In particular, on the basis of the Kalman filter, a solution of sequential row-wise and column-wise estimation was obtained, which makes it possible to reduce computational costs in comparison with optimal vector filtering. It uses recurrent filtering procedures discussed earlier for the one-dimensional case and then takes the average estimate of the pixel brightness value based on the results of the row-wise and column-wise filtering. For models of the 1st and 2nd orders, this algorithm loses no more than 10% in terms of the filtering error variance. Also, for the first time, asymptotic dependences of the relative variance of the filtering error for Kalman and Wiener filters in the continuous and discrete cases were obtained. The analysis of the obtained dependencies shows that an increase in the signal-to-noise ratio slightly influences the difference between the variances of filtering errors of the continuous and discrete filters for small signal-to-noise ratios. However, with an increase of the multiplicity of the model and with large signal-to-noise ratios, the error variances calculated for different filters begin to diverge. On the other hand, proximity can be achieved by increasing the correlation interval. So, with the correlation interval $k = 166$, even at a signal-to-noise ratio $q = 10$, the difference between the variances is less than 10% for the 1st order model.

Figure 2.10 shows the dependence of the relative variance of the filtering error on the correlation radius k_0 with different signal-to-noise ratios: $q = 0.1$ (Fig. 2.10a) and $q = 1.0$ (Fig. 2.10b).

Based on filtering algorithms, theoretical values of the variance of filtering errors can be obtained, depending on the correlation order, multiplicity, and signal-to-noise ratio.

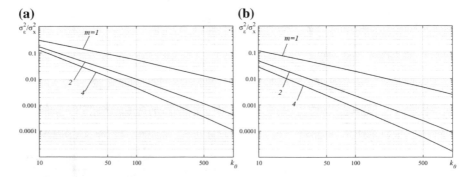

Fig. 2.10 Relative variances of two-dimensional RF filtering errors: **a** with signal-to-noise ratio q = 0.1, **b** with signal-to-noise ratio q = 1.0

Consider the problem of optimal filtering of a discrete RF based on observations of an additive mixture

$$z_i = x_i + n_i, \ i = 1, 2, \ldots, \tag{2.72}$$

with discrete white Gaussian noise n_i, $i = 1, 2, \ldots$, having zero mean and variance σ_n^2.

The variance of the filter error is determined by the expression:

$$\sigma_\varepsilon^2 = \sigma_n^2 g_0, \tag{2.73}$$

where $g_0 = \frac{1}{2\pi i} \oint_C H(z) z^{-1} dz$, $z = e^{i\omega}$.

The transfer function is found using the expression:

$$H(z) = \frac{F(z)}{F(z) + \sigma_n^2}, \tag{2.74}$$

where $F(z)$ is spectrum of source signal x_i and it can be found as z-transformation of CF:

$$F(z) = \sum_{j=-\infty}^{\infty} B_x(j) z^{-j}. \tag{2.75}$$

In other words, $F(z) = \frac{\beta_m^2}{(1-\rho z^{-1})^m (1-\rho z)^m}$ is a square of modulus of transfer function $\frac{1}{(1-\rho z^{-1})^m}$ multiplied by the variance of the random additive ξ_i, i.e. β_m^2.

After a simple conversion, the transfer function is written as:

$$F(z) = \frac{\beta_m^2}{(1 - \rho(z^{-1} + z) + \rho^2)^m}. \tag{2.76}$$

Thus, a non-realizable Wiener filter provides the following relation:

$$\frac{\sigma_\varepsilon^2}{\sigma_n^2} = \frac{1}{2\pi i} \oint_C \frac{z^{-1} \beta_m^2 / \sigma_n^2}{\beta_m^2 / \sigma_n^2 + (1 - \rho(z^{-1} + z) + \rho^2)^m} dz. \tag{2.77}$$

After replacement $z = e^{i\omega}$, $dz = iz d\omega$, $z^{-1} + z = 2\cos\omega$, relation for variance filtering error assumes form:

$$\frac{\sigma_\varepsilon^2}{\sigma_n^2} = \frac{1}{2\pi} \int_{-\pi}^{\pi} \frac{\beta_m^2 / \sigma_n^2}{\beta_m^2 / \sigma_n^2 + (1 - 2\rho\cos\omega + \rho^2)^m} d\omega. \tag{2.78}$$

It should be noted that Eq. 2.78 depends only on the signal-to-noise ratio and the correlation parameter ρ and it can be calculated numerically.

Let us perform a limit transition to continuous time. Equation $x_i(1 - \rho z^{-1})^m = \beta_m \xi_i$ with $\Delta t \to 0$ and given $(1 - \rho) = a\Delta t$ takes the following form:

$$x(t)(p + a)^m = \xi(t), \tag{2.79}$$

where $px = dx/dt$; $\xi(t)$ is white noise with spectral density $N_\xi = \beta_m^2/\Delta t^{2m-1}$.

Power spectrum of the process $x(t)$ is determined by Eq. 2.80:

$$G_x(\omega) = \frac{N_\xi}{(\omega^2 + a^2)^m}, \tag{2.80}$$

where $N_\xi = \frac{\sigma_x^2(2a)^{2m-1}}{C_{2m-2}^{m-1}}$, and CF can be written as:

$$B_x(\tau) = \frac{1}{2\pi} \int_{-\infty}^{\infty} G_x(\omega)e^{j\omega\tau}d\tau. \tag{2.81}$$

Then it is possible to write the equivalent of Wiener filter in continuous time:

$$\hat{x}_0 = \int_{-\infty}^{\infty} h(\tau)z(\tau)d\tau, \tag{2.82}$$

where $z(t) = x(t) + n(t)$; $n(t)$ is the white noise with CF $N_0\delta(\tau)$ or limit $\Delta\chi(t)/\Delta t$, wherein $\chi(t)$ is Wiener process with parameter $N_0 = \sigma^2\Delta t$.

Let us transform an approximately continuous version (Eq. 2.82) into a discrete version:

$$\hat{x}_0 \cong \sum_{i=-\infty}^{\infty} h(\tau_i) \int_{\tau_i}^{\tau_i + \Delta\tau} z(\tau)d\tau \cong \sum_{i=-\infty}^{\infty} h(\tau_i)(x(\tau_i)\Delta\tau + \Delta\chi(\tau_i)) = \sum_{i=-\infty}^{\infty} g_i z_i. \tag{2.83}$$

The optimal filter in continuous time is described by the transfer function. $H(\omega) = \frac{G_x(\omega)}{N_0 + G_x(\omega)}$, therefore, the variance of the filter error can be found as:

$$\sigma_\varepsilon^2 = N_0 h(0) = \frac{N_0}{2\pi} \int_{-\infty}^{\infty} H(\omega)d\omega = \frac{N_0}{2\pi} \int_{-\infty}^{\infty} \frac{G_x(\omega)}{N_0 + G_x(\omega)} d\omega. \tag{2.84}$$

After transformation Eq. 2.84 can be written as:

$$\frac{\sigma_\varepsilon^2}{\sigma_x^2} = \frac{1}{2\pi} \int_{-\infty}^{\infty} \frac{K_m dz}{\left((z^2+1)^m + K_m \frac{\sigma_x^2}{\sigma^2(1-\rho)}\right)}, \tag{2.85}$$

where $K_m = 2^{2m-1}/C_{2m-2}^{m-1}$.

The obtained Eqs. 2.78 and 2.85 can be used to study the ratio of the variance of filtering error for the discrete filter to the variance of filtering error for the continuous filter. Figure 2.11 shows the dependencies of relative error variances for models of various multiplicities with correlation interval $k_0 = 100$ (Fig. 2.11a) and $k_0 = 20$ (Fig. 2.11b).

Figure 2.12 shows similar graphs for models of different multiplicities with a signal-to-noise ratio $q = 0.1$ (Fig. 2.12a) and $q = 2$ (Fig. 2.12b) versus the correlation parameter.

The analysis of the obtained curves enables to conclude that the ratio of the variances of filter errors of the discrete and continuous filters is close to unity. In this case, high-order models behave similarly for different q, and as the correlation parameter increases, the filtering results of the discrete and continuous filters approach each other. However, with an increase in the signal-to-noise ratio for models of the second and higher orders, there is a slight divergence between the discrete filter and the continuous one.

The problem of identifying model parameters can also be generalized to the multidimensional case. Thus, a ratio similar to Eq. 2.53, but for a two-dimensional model (Eq. 2.4), is written as:

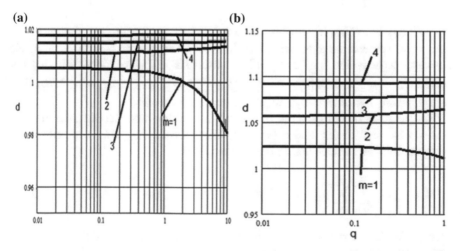

Fig. 2.11 Comparison of relative variances of discrete and continuous filter errors versus signal-to-noise ratio: **a** $k_0 = 100$, **b** $k_0 = 20$

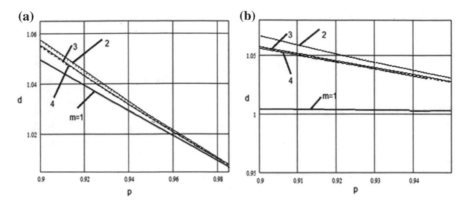

Fig. 2.12 Comparison of relative variances of discrete and continuous filter errors depending on correlation parameter: **a** $q = 0.1$, **b** $q = 2$

$$R(k_1, k_2) = \rho_{10}R(k_1 - 1, k_2) + \rho_{01}R(k_1, k_2 - 1) + \rho_{11}R(k_1 - 1, k_2 - 1), k_1 > 0, k_2 > 0. \tag{2.86}$$

It is easy to verify that solving a two-dimensional system of Yule–Walker type equations for CF (Eq. 2.86) gives coefficients identical to the coefficients of a two-dimensional first-order AR model. To increase the order of AR, as in the one-dimensional case, it is necessary to increase the number of correlation coeffi- cients. In this case, the RF model can be written as follows:

$$x_{i,j} = \sum_{l=0}^{m_j} \sum_{k=0}^{m_i} \rho_{kl} x_{i-k,j-l} - \rho_{00} x_{i,j} + \xi_{i,j}, i = \overline{1, M_1}, j = \overline{1, M_2}, \tag{2.87}$$

where $\{x_{i,j}\}$ is RF implementation or simulated image; ρ_{kl} are correlation coeffi- cients for elements lagging behind each other along the axes i and j on k and l pixels, respectively; $\{\xi_{i,j}\}$ is two-dimensional RF of independent Gaussian RV with $M\{\xi_{i,j}\} = 0$ and $M\{\xi_{i,j}^2\} = \sigma_\xi^2 = [1 - \sum_{l=0}^{m_j} \sum_{k=0}^{m_i} \rho_{kl} R(k, l) + \rho_{00} R(0, 0)]\sigma_x^2$, m_i and m_j are AR orders; M_1 and M_2 are the image sizes.

The number of components of the model, taking into account the random addition is equal to $(m_i + 1) \times (m_j + 1)$. Using Eqs. 2.86–2.87, it is possible to write the relation for calculating the values of CF:

$$R(k_1, k_2) = \sum_{l=0}^{m_j} \sum_{k=0}^{m_i} \rho_{kl} R(k_1 - k, k_2 - l) - \rho_{00} R(k_1, k_2), k_1 > 0, k_2 > 0. \tag{2.88}$$

Equation 2.88 can also be used for the case of non-separable CF if the param- eters of an arbitrary AR field of any order are identified. In the case of a model with multiple roots, the procedure will be similar to the one-dimensional case. The orders of the models are calculated along separate axes, and then, based on the CF, the value of the correlation parameter that forms all the coefficients of the model is found.

Studies have shown that AR models with multiple roots of characteristic equations are a convenient tool for the synthesis of various processing algorithms for multidimensional RF.

Based on the obtained dependences of the filtering error variance an expression can be obtained for RF of arbitrary dimension. Indeed, using the multidimensional Z-transform leads to an expression for the variance of the relative error [36]:

$$\frac{\sigma_\varepsilon^2}{\sigma_x^2} = \frac{1}{(2\pi)^N} \int\limits_{-\pi}^{\pi} \cdots \int\limits_{-\pi}^{\pi} \frac{\beta_0^2}{\prod\limits_{k=1}^{N} \left(1 + \rho_k^2 - 2\rho_k \cos \lambda_k\right)^{m_k} + q\beta_0^2} d\bar{\lambda}, \qquad (2.89)$$

where $q = \sigma_x^2/\sigma^2$, N is the dimension of filtered RF, m_k is the multiplicity of the model in kth dimension, ρ_k is the correlation parameter in kth dimension.

Figure 2.13a, where k_0 is quite small, and Fig. 2.13b, where k_0 is quite large, show dependence of the filtering error relative variance versus the correlation interval k_0 for models of different dimensions and orders, when $q = 0.1$.

The analysis of the obtained dependencies shows that at sufficiently small correlation intervals ($k_0 < 10$) the variances of filtering errors of AR fields of the 1st and 2nd orders are quite close. An increase in the dimensions and a further increase in the correlation interval lead to divergence of the curves. The smallest values of the relative variances of filtering errors were obtained for the cases $N = 3$, $m = (2, 2, 2)$, $N = 4$, $m = (2, 2, 2, 2)$. This is explained by the fact that, for $m = 1$, AR processes along the axes are quite spiky, and their filtering leads to large errors.

(a)   (b)

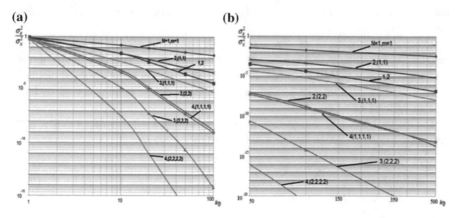

Fig. 2.13 Relative variances of multidimensional random fields filtering errors, where: **a** k_0 is quite small, **b** k_0 is quite large

2.5 Real Image Processing

Despite the previously obtained dependences, confirming the advantage of AR with multiple roots in filtering, the case of processing real images seems to be the most interesting for comparison. For statistical modeling and checking the adequacy of the proposed algorithms, images with smooth changes in brightness and large objects were selected.

There are a lot of noise models in literature [65]. The examples of noise models are Gaussian noise, Brownian noise, Salt and Pepper (Impulse Valued) noise, Periodic noise, Quantization noise, Speckle noise, Photon (Poisson) noise, Poisson–Gaussian noise, Structured noise, Gamma noise, Rayleigh noise, etc. [65]. However in this chapter, we limited ourselves to the study of additive white Gaussian noise reducing. One of the tasks of investigation is to compare filtering efficiency provided by AR models and models of AR with multiple roots. The known researches show the good possibilities of AR models in reducing additive white Gaussian noise. Thus, the comparison analysis was performed for such noise and respectively observation models. Using such models is appropriate in situations when image may be corrupted by significant number of different noises so that its distribution may be given as Gaussian. As for AR with multiple roots, such models may represent slow changing in brightness on real satellite images providing simple mathematical description.

The filtering algorithm based on a multiple-root model was tested on a multi-dimensional satellite image. Figure 2.14a shows source image, Fig. 2.14b shows noisy image, Fig. 2.14c shows filtering results based on first order AR, and Fig. 2.14d shows filtering results based on AR with multiple roots.

Analysis of the presented images shows that the model with multiple roots provides the best results in the sense of the filtering error variance. The relative variance of the filter error for Fig. 2.14c equals 0.782, and the relative variance of the filtering error for Fig. 2.14d equals 0.358. The signal-to-noise ratio is 0.5 for both filters.

Consider the process of the model parameters identification based on the proximity of the model CF and the source data. Moreover, in the first case, the AR with separable CF will be considered and separate calculation of the coefficients for the row and the column will be performed. In the second case, the equations based on Eq. 2.88 will be used.

Let there be a real image represented as $I(i, j), i \in 1, \ldots, M_1, j \in 1, \ldots, M_2$. Expression for estimation its CF can be written as follows:

$$\hat{R}_I(k_1, k_2) = \frac{1}{\hat{\sigma}_I^2} \sum_{i=k_1+1}^{M_1} \sum_{j=k_2+1}^{M_2} (I(i, j) - \hat{m}_I)(I(i - k_1, j - k_2) - \hat{m}_I), \qquad (2.90)$$

where $\hat{m}_I$ is the average brightness over the entire image, $\hat{\sigma}_I^2$ is the brightness variance estimate.

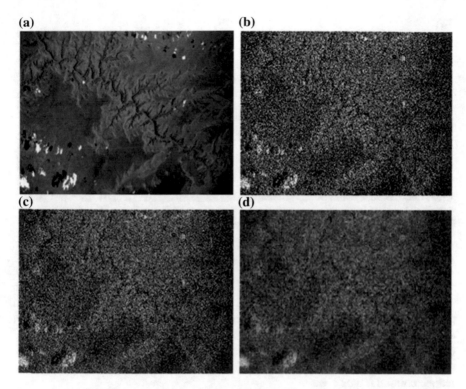

Fig. 2.14 Satellite image filtering: **a** source image, **b** noisy image, **c** filtering results based on first order AR, **d** filtering results based on AR with multiple roots

Figures 2.15a, b show the investigated image and its CF, respectively. Identification is performed for the 4th order AR model:

- results for RF with separable CF are the following:

$$\rho_{10} = 1.098, \rho_{20} = -0.39; \rho_{30} = 0.364, \rho_{40} = -0.082,$$
$$\rho_{01} = 0.828; \rho_{02} = 0.0047, \rho_{03} = 0.111, \rho_{04} = 0.038, \varepsilon = (\hat{R} - R)^2 = 0.387.$$

- results for RF with non-separable CF are the following:

$$\rho_{10} = 0.998, \rho_{20} = -0.514, \rho_{30} = 0.43, \rho_{40} = -0.133,$$
$$\rho_{01} = 0.185; \rho_{11} = 0.152; \rho_{21} = -0.194, \rho_{31} = 0.142, \rho_{41} = -0.092,$$
$$\rho_{02} = -0.174, \rho_{12} = 0.309, \rho_{22} = -0.265, \rho_{32} = 0.217, \rho_{42} = -0.097,$$
$$\rho_{03} = -0.048, \rho_{13} = 0.186, \rho_{23} = -0.205, \rho_{33} = 0.153, \rho_{43} = -0.079;$$
$$\rho_{04} = -0.122; \rho_{14} = 0.267, \rho_{24} = -0.242, \rho_{34} = 0.22, \rho_{44} = -0.102,$$
$$\varepsilon = (\hat{R} - R)^2 = 0.014.$$

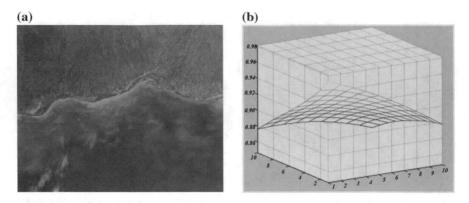

Fig. 2.15 Example of processing: **a** real image, **b** its CF

Analysis of the obtained values of the error variances shows that the use of the model with a non-separable CF ensures a greater closeness between CF of real and that of simulated images. This is explained by the fact that for such a model 24 correlation parameters were calculated, while for the model with a separable CF only 8 parameters. At the same time, a sufficient proximity of CF is ensured, especially in the vicinity of zero. Therefore, it is advisable to use such models to reduce computational cost.

Figure 2.16 presents CF cross sections for the source image (solid line), as well as, for RF with separable CF (dash-dotted line) and RF with non-separable CF (dashed line).

An analysis of the curves presented shows that as k increases, the discrepancy between real and simulated CF increases too. It is possible to achieve greater proximity by increasing AR order, but this leads to an increase in computational costs.

Similar studies were conducted on a sample of 100 images. Analysis of the results shows that the use of non-separable models in 85% of cases provides a much greater CF proximity than the use of separable models. However, in 15% of cases, models with split CF were successfully used to describe the real image, which made it possible to significantly reduce computational costs during the processing.

Thus, for images that contain objects with slowly varying brightness, it is possible to apply models with multiple roots of characteristic equations. Their use provides smaller error variance when filtering against the background of white Gaussian noise, as well as, closer CF when describing such an image using an autoregressive RF. In addition, the developed algorithms can be applied to the processing of one-dimensional signals, in particular, models with multiple roots and doubly stochastic models are used in analyzing the data of the taxi service in [66–68].

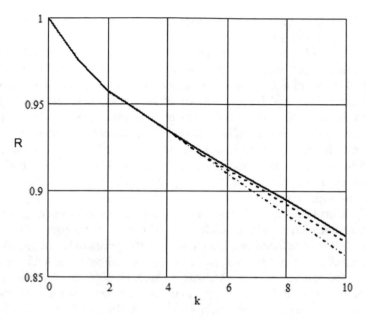

Fig. 2.16 CF cross sections of the source image and models to be identified

2.6   Conclusions

The main results of the work are as follows:

1. An analytical review of existing methods for describing images is presented. During the survey, special attention was paid to an autoregressive RF with multiple roots of the characteristic equations, which are based on a convenient mathematical apparatus and provide sufficient proximity of CF when models describe real images.
2. Based on the procedures of vector Kalman filtering, optimal recurrent estimates of AR sequences with multiple roots of characteristic equations are constructed. For the first time, an approach to the analysis of AR with multiple roots of characteristic equations of different orders, having equal correlation intervals of the processes simulated, was investigated. In addition to Kalman algorithm, Wiener filter was synthesized and investigated when processing RS generated by AR with multiple roots of characteristic equations.
3. Compact analytical relations have been obtained for analyzing the effectiveness of RF filtering algorithms with multiple roots of characteristic equations. Obtained relations are based on results of Wiener filter, generalized for the multidimensional case. The analysis shows that when describing real processes using the proposed models, a significant decrease in the variance of the optimal estimation error is observed. In particular, for large correlation intervals

($k > 1000$) for the model with multiple roots, the filtering error variance is 5–10 times less than that for the ordinary AR model.

4. Using statistical modeling of the proposed algorithms, representative catalogs of the filtering error variances have been obtained. The behavior of filters was studied at various correlation intervals and in the processing of RF with the roots of characteristic equations of various multiplicities. It is established that the larger is the multiplicity of the model, the more effective is the filtering. However, increasing the multiplicity leads to an increase of computational costs. At the same time, with an increase in the multiplicity, the difference between models of neighboring multiplicities decreases. Therefore, the use of very high-order models is in most cases impractical due to a significant increase of computational cost.

5. In the course of the conducted research, approaches to identifying the parameters of AR models with multiple roots of characteristic equations were considered. It is established that the solution of this problem for high-order AR is associated with computational difficulties. An algorithm for identification of the parameters and orders of the AR with multiple roots of characteristic equations based on solving Yule–Walker system of equations is proposed. A characteristic feature of AR with multiple roots is the possibility of obtaining all the correlation coefficients by a single parameter, found, for example, from the recurrent relation for CF values.

6. Using the property of separability of CF of AR with multiple roots, in this chapter a generalization of the one-dimensional model to the multidimensional case was performed. Models are proposed that enable to generate autoregressive RF with multiple roots of characteristic equations of arbitrary multiplicities and dimensions. For two-dimensional RF, correlation parameters have been obtained that provide the same correlation radius based on the calculations of the CF of two-dimensional RF. The asymptotic properties of probabilistic characteristics with increasing correlation radii were investigated. It is shown that there is a parameter characterizing the correlation radius of high-order models, tending to a constant value for large correlation radii. Recurrent and power expressions are obtained for its calculation in the transition from the discrete case to the continuous one.

7. Quasi-optimal and optimal filtering algorithms for RF based on AR with multiple roots of characteristic equations have been developed and investigated. In particular, on the basis of Kalman filter, a solution of sequential row-by-row and column-by-column estimation was obtained, which makes it possible to reduce computational costs in comparison with optimal vector filtering. For models of the 1st and 2nd orders, the proposed algorithm loses no more than 10% in terms of the variance of the filtering error. The asymptotic dependencies of the relative filtering error variance for the Kalman and Wiener filters are obtained for the first time when proceeding to a continuous multidimensional space.

8. Theoretical relations for the filtering error variances have been obtained depending on the correlation parameter, multiplicity, and signal-to-noise ratio. Graphs of dependences of the filtering error variances versus the correlation interval and the signal-to-noise ratio are plotted using these relations.

9. The obtained algorithms are tested on real images. The analysis shows that for the studied images, the use of AR models with multiple roots provides a significantly lower filtering error variance than traditional approaches based on first-order models.

References

1. Vasiliev, K.K., Sluzhivny, M.N.: Mathematical modeling of communication systems: tutorial, 2nd edn. UlSTU, Ulyanovsk (2010)
2. Schowengerdt, R.: Remote Sensing: Models and Methods for Image Processing. Elsevier Inc., Amsterdam (2007)
3. Dudgeon, D.E., Mersereau, R.M.: Multidimensional Digital Signal Processing. Prentice-Hall, Inc., Englewood Cliffs (1984)
4. Cuzzocrea, A., Song, I.-Y., Davis, K.C.: Analytics over large-scale multidimensional data: the big data revolution! In: 14th International Workshop on Data, pp. 101–104 (2011)
5. Bondarev, A.E.: The procedures of visual analysis for multidimensional data volumes. In: The International Archives of the Photogrammetry, Remote Sensing and Spatial Inform. Sciences, vol. XLII-2/W12, pp. 17–21 (2019)
6. Sembiring, R.W., Zain, J.M.: The design of pre-processing multidimensional data based on component analysis. Comput. Inf. Sci. 4(3), 106–115 (2011)
7. Banalagay, R., Covington, K.J., Wilkes, D.M., Landman, B.A.: Resource estimation in high performance medical image computing. Neuroinformatics 12(4), 563–573 (2014)
8. Andriyanov, N.A.: Software complex for representation and processing of images with complex structure. In: CEUR Workshop Proceedings, vol. 2274, pp. 10–22 (2018)
9. Tehrani, M., Bagheri, M., Ahmadi, M., Norouzi, A., Karimi, N., Samavi, S.: Artistic instance-aware image filtering by convolutional neural networks (2018). arXiv:1809.08448
10. Vigneshwari, K., Kalaiselvi, K.: Adaptive median filter based noise removal algorithm for big image data. Int. J. Adv. Stud. Sci. Res. 3(10), 154–159 (2019)
11. Piao, W., Yuan, Y., Lin, H.: A digital image denoising algorithm based on Gaussian filtering and bilateral filtering. In: 4th Annual International Conference on Wireless Communication and Sensor Network, vol. 17, pp. 01006.1–01006.8 (2018)
12. Bhattacharyya, S., Pal, S.K., Pan, I., Das, A. (eds): Recent trends in signal and image processing. In: Proceedings of ISSIP 2018, AISC. Springer Nature Singapore Pte LtD., Singapore (2019)
13. Jiang, G., Luo, M., Chen, S., Bai, K.: An improved filtering algorithm for the CT image processing. In: International Conference on Applied System Innovation, pp. 679–682 (2017)
14. Milanfar, P.: A tour of modern image filtering: New insights and methods, both practical and theoretical. IEEE Signal Process. Mag. 30(10), 106–128 (2013)
15. Andriyanov, N.A., Dementiev, V.E.: Developing and studying the algorithm for segmentation of simple images using detectors based on doubly stochastic random fields. Pattern Recognition and Image Analysis 29(1), 1–9 (2019)
16. Andriyanov, N.A., Vasiliev, K.K., Dement'ev, V.E.: Analysis of the efficiency of satellite image sequences filtering. J. Phys. Conf. Ser. 1096, 012036.1–012036.7 (2018)

17. Vasiliev, K., Dementiev, V., Andriyanov, N.: Representation and processing of multispectral satellite images and sequences. Proc. Comput. Sci. **126**, 49–58 (2018)
18. Yemez, Y., Anarim, E., Istefanopulos, Y.: Causal and semicausal AR image model identification using the EM algorithm. IEEE Trans. Image Process. **2**(4), 523–528 (1993)
19. Li, Y., Di, X.: Image mixed blur classification and parameter identification based on cepstrum peak detection. In: Proceedings of 35th Chinese Control Conference, pp. 4809–4814 (2016)
20. Jain, A.K.: Advances in mathematical models for image processing. Proc. IEEE **69**(5), 502–528 (1981)
21. Lafferty, J., McCallum, A., Pereira, F.: Conditional random fields: probabilistic models for segmenting and labeling sequence data. In: 18th International Conference on Machine Learning, pp. 282–289 (2001)
22. Vasil'ev, K.K., Dement'ev, V.E., Andriyanov, N.A.: Doubly stochastic models of images. Pattern Recognit. Image Anal. **25**(1), 105–110 (2015)
23. de Vries, A., Westerveld, T.: A comparison of continuous vs. discrete image models for probabilistic image and video retrieval. In: 2004 International Conference on Image Processing, pp. 2387–2390 (2004)
24. Bouman, C.A.: Model Based Imaging. Purdue University (2013)
25. Vasiliev, K.K., Andriyanov, N.A.: Synthesis and analysis of doubly stochastic models of images. In: CEUR Workshop Proceedings 2005, pp. 145–154 (2017)
26. Andriyanov, N.A., Vasil'ev, K.K., Dement'ev, V.E.: Investigation of the filtering and objects detection algorithms for a multizone image sequence. In: International Archives of the Photogrammetry, Remote Sensing and Spatial Information Sciences—ISPRS Archives, vol. XLII-2/W12, pp. 7–10 (2019)
27. Vasilyev, K.K., Dragan, Ya.P., Kazakov, V.A.: Applied theory of random processes and fields. Ulyanovsk UlSTU (in Russian) (1995)
28. Vasyukov, V.N., Gruzman, I.S., Raifeld, M.A., Spektor, A.A.: New approaches to solving problems of processing and image recognition. High-Tech Technol. **3**, 44–51 (2002)
29. Levi, P.: A special problem of Brownian motion, and a general theory of Gaussian random functions. In: Third Berkeley Symposium Mathematical Statistics and Probability, vol. 2. California Press, Berkeley California (1956)
30. Woods, J.W.: Two-dimensional Kalman filtering. In: Huang, T.S. (ed.) Two-Dimensional Digital Signal Processing I: Linear Filters, TAP, vol. 42, pp. 155–205. Springer, Berlin (1981)
31. Shalygin, A.S., Palagin, YuI: Applied Methods of Statistical Modeling. Mashinostroenie, Leningrad (1985). (in Russian)
32. Vaishali, D., Ramesh, R., Christaline, J.A.: 2D autoregressive model for texture analysis and synthesis. In: International Conference on Communication and Signal Processing, pp. 1135–1139 (2014)
33. Habibi, A.: Two-dimensional Bayesian estimation of images. Proc. IEEE **60**(7), 878–883 (1972)
34. Krasheninnikov, V., Vasil'ev, K.: Multidimensional image models and processing. In: Favorskaya, M.N., Jain, L.C. (eds.) Computer Vision in Control Systems 3: Aerial and Satellite Image Processing, ISRL, vol. 135, pp. 11–64. Springer International Publishing, Cham (2018)
35. Andriyanov, N.A., Gavrilina, Yu.N.: Image models and segmentation algorithms based on discrete doubly stochastic autoregressions with multiple roots of characteristic equations. In: CEUR Workshop Proceedings, vol. 2076, pp. 19–29 (2018)
36. Andriyanov, N.A., Vasiliev, K.K.: Use autoregressions with multiple roots of the characteristic equations to image representation and filtering. In: CEUR Workshop Proceedings, vol. 2210, pp. 273–281 (2018)
37. Boswijka, H.P., Franses, P.H., Niels, H.: Multiple unit roots in periodic autoregression. J. Econom. **80**(1), 167–193 (1997)
38. Andriyanov, N.A., Vasilyev, K.K.: Properties of autoregression with multiple roots of characteristic equations. Bull. Ulyanovsk State Tech. Univ. **1**, 36–40 (2019). (in Russian)

39. Andriyanov, N.A., Vasilyev, K.K.: Optimal filtering of multidimensional random fields generated by autoregression with multiple roots of characteristic equations. In: V International Conference on Youth School Information Technologies and Nanotechnologies (ITNT-2019), vol. 2, pp. 349–354 (2019)
40. Sharma, G., Chellappa, R.: Two-dimensional spectrum estimation using noncausal autoregressive models. IEEE Trans. Inf. Theory **32**(2), 268–275 (1986)
41. Alaghi, A., Li, C., Hayes, J.: Stochastic circuits for real-time image-processing applications. In: 50th Annual Design Automation Conference, pp. 136.1–136.6 (2013)
42. Hajjar, A., Chen, T.: A VLSI architecture for real-time edge linking. IEEE Trans. Pattern Anal. Mach. Intell. **21**(1), 89–94 (1999)
43. Picard, R.W., Pentland, A.P.: Temperature and Gibbs image modeling. M.I.T. Media Laboratory Perceptual Computing Section Technical report no 254 (1995)
44. Krasheninnikov, V.R.: Patterns of random fields on surfaces. Proc. Samara Sci. Center Russ. Acad. Sci. **4**(3), 812–816 (2012). (in Russian)
45. Methropolis, N., Rosenbluth, A., Rosenbluth, M., Teller, A., Teller, E.: Equation of state calculations by fast computing machines. J. Chem. Phys. **21**(6), 1087–1092 (1953)
46. Armi, L., Fekri-Ershad, S.: Texture image analysis and texture classification methods—a review. Int. Online J. Image Process. Pattern Recognit. **2**(1), 1–29 (2019)
47. Musgrave, F.K., Peachey, D., Perlin, K., Ebert, D.S., Worley, S.: Texturing and Modeling. A Procedural Approach. Morgan Kaufmann, San Francisco (2002)
48. Bondur, V.G., Arzhenenko, N.I., Linnik, V.N., Titova, I.L.: Simulation of multispectral aerospace images of dynamic brightness fields. Study Earth Sp. **2**, 3–17 (2003)
49. Woods, J.W., Dravida, S., Mediavilla, R.: Image estimation using doubly stochastic Gaussian random field models. Pattern Anal. Mach. Intell. **2**(9), 245–253 (1987)
50. Vasil'ev, K.K., Dement'ev, V.E., Andriyanov, N.A.: Application of mixed models for solving the problem on restoring and estimating image parameters. Pattern Recognit. Image Anal. **26**(1), 240–247 (2016)
51. Vasiliev, K.K., Dementiev, V.E., Andriyanov, N.A.: Filtering and restoration of satellite images using doubly stochastic random fields. In: CEUR Workshop Proceedings, vol. 1814, pp. 10–20 (2017)
52. Andriyanov, N.A., Dement'ev, V.E.: Application of mixed models of random fields for the segmentation of satellite images. In: CEUR Workshop Proceedings, vol. 2210, pp. 219–226 (2018)
53. Andriyanov, N.A., Dementiev, V.E., Vasiliev, K.K.: Developing a filtering algorithm for doubly stochastic images based on models with multiple roots of characteristic equations. Pattern Recognit. Image Anal. **29**(1), 10–20 (2019)
54. Palagin, Y.I., Fedotov, S.V., Shalygin, A.S.: Parametric models for statistical modeling of vector inhomogeneous random fields. Avtomat. Telemekh. **6**, 79–89 (1990). (in Russian)
55. Bajorski, P.: Non-Gaussian linear mixing models for hyperspectral images. J. Electr. Comput. Eng. **2012**, 818175.1– 818175.8 (2012)
56. Viroli, C., McLachlan, G.J.: Deep Gaussian Mixture models (2017). arXiv:1711.06929v1 [stat.ML]
57. Tran, G.-L., Bonilla, E.V., Cunningham, J.P., Michiardi, P., Filippone, M.: Calibrating deep convolutional Gaussian processes (2018). arXiv:1805.10522 [stat.ML]
58. Knoblauch, J.: Robust Deep Gaussian Processes (2019). arXiv:1904.02303 [stat.ML]
59. Box, G., Jenkins, G.: Time Series Analysis: Forecasting and Control. Holden-Day, San Francisco (1970)
60. Vasiliev, K.K.: Autoregression with multiple roots of characteristic equations. Radiotekhnika **11**, 74–76 (2014). (in Russian)
61. Vasiliev, K.K., Andriyanov, N.A.: Analysis of autoregression with multiple roots of characteristic equations. Radiotekhnika **6**, 13–17 (2017). (in Russian)
62. Vasiliev, K.K., Andriyanov, N.A.: Investigation of the asymptotic efficiency of filtering autoregression processes with multiple roots of characteristic equations. REDS: Telecommun. Dev. Syst. **8**(3), 12–15 (2018). (in Russian)

63. Vasiliev, K.K.: Optimal signal processing in discrete time. Radio Engineering, Moscow (2016). (in Russian)
64. Vasiliev, K.K.: Signal processing techniques: tutorial (2001). (in Russian)
65. Boyat, A.K., Joshi, B.K.: A review paper: noise models in digital image processing. Signal Image Process. Int. J. (SIPIJ) **6**(2), 63–75 (2015)
66. Azanov, P., Danilov, A., Andriyanov, N.: Development of software system for analysis and optimization of taxi services efficiency by statistical modeling methods. In: CEUR Workshop Proceedings, vol. 1904, pp. 232–238 (2017)
67. Andriyanov, N.A., Sonin, V.A.: Using mathematical modeling of time series for forecasting taxi service orders amount. In: CEUR Workshop Proceedings 2258, pp. 462–472 (2018)
68. Danilov, A.N., Andriyanov, N.A., Azanov, P.T.: Ensuring the effectiveness of the taxi order service by mathematical modeling and machine learning. J. Phys. Conf. Ser. **1096**, 012188.1–012188.8 (2018)

Chapter 3
Representation and Processing of Spatially Heterogeneous Images and Image Sequences

Vitaly E. Dement'ev, Victor R. Krasheninnikov and Konstantin K. Vasil'ev

Abstract In the present chapter, the problems of multidimensional images and image sequences representation and processing within the framework of the Earth remote sensing problems solving are considered. The issues of correlated data multidimensional arrays description and optimal and suboptimal processing thereof based on formalized mathematical models are investigated. It is proposed to use doubly stochastic autoregressive models as the basis for such models. It is shown that similar models enable to describe two essential properties of the actual satellite material. Firstly, the construction of doubly stochastic models allows for description of multidimensional random fields and time sequences thereof. This is of fundamental importance for real-time sequences of multispectral images each being considered as brightness values 3D array composed of separate two-dimensional frames, which correspond to the Earth surface registration results in a separate spectral band. A time sequence of such images is equivalent to four-dimensional random field with one-dimension corresponding to discrete time. Secondly, doubly stochastic models enable to carry out estimation of spatially heterogeneous images, i.e. images having probabilistic properties, which vary with spatial coordinates. Such variations are typical for actual satellite snapshots containing objects of various nature: rivers, forests, fields, etc. The chapter contains investigation of probabilistic and correlation properties of doubly stochastic models. Algorithms for these models parameters estimation based on actual satellite signals observations are proposed. It is shown that basing on the proposed models it is possible to synthesize multidimensional images filtering algorithms allowing for spatial heterogeneity of the images. Several classes of such algorithms enabling also to

V. E. Dement'ev · V. R. Krasheninnikov (✉) · K. K. Vasil'ev
Ulyanovsk State Technical University, 32 Severny Venets St.,
Ulyanovsk 432027, Russian Federation
e-mail: kvrulstu@mail.ru

V. E. Dement'ev
e-mail: 986514@mail.ru

K. K. Vasil'ev
e-mail: vkk@ulstu.ru

© Springer Nature Switzerland AG 2020
M. N. Favorskaya and L. C. Jain (eds.), *Computer Vision in Advanced Control Systems-5*, Intelligent Systems Reference Library 175,
https://doi.org/10.1007/978-3-030-33795-7_3

carry out processing of real-time sequences of multispectral satellite images are proposed. It has been established that the found algorithms possess higher effectiveness in comparison with known analogues. In this study, the problem of extended objects detection against the background of multispectral images time sequences is considered. A family of detection algorithms based on preliminary nonlinear filtering of these images is synthesized. Application of the developed algorithms, when solving problems of satellite images classification and natural or man-made objects monitoring, is briefly considered.

Keywords Multidimensional image model · Autoregressive model · Tensor model · Double stochastic model · Image processing · Potential efficiency · Prediction · Filtration · Anomaly detection · Recognition · Adaptive algorithm · Pseudo-gradient algorithm

3.1 Introduction

In recent years, a lot of papers concerning processing of various kinds of images and image sequences have been published. It is caused by a broad range of applications for image processing methods. Solution for the following problems is directly connected with processing of various types of images: the Earth remote sensing, video data processing, medical images analysis etc. Despite the achieved progress in solving these and other problems at present time there is a number of yet unsolved significant problems, impeding further development of computer vision. Among these problems we should single out the issues concerning description of correlated data multidimensional arrays and the corresponding optimal and suboptimal processing based on formalized mathematical models. These issues were considered in a fairly limited number of papers. Substantially, specialists in image processing tend to use neural networks and big data [1–3], which is a prospective area for research but it does not enable to solve problems of multidimensional images understanding and image processing optimal procedures forming.

Among the methods for multidimensional images description, the most promising ones include models of images specified on sequences of multidimensional grids [4–9]. Such representation of images enables to quite simply do operations with available image models both element-wise and as a whole. However, constructing such models by specifying probability distributions in general form is rather cumbersome and therefore various simplifications are used. The most important of the latter are Markov property and homogeneity. Autoregressive models are the most suitable for representation [6, 7, 10–12] and processing [13–20] of images with such properties. However, models of this type transformed for representation of random fields on multidimensional grids do not enable to appropriately describe many real images with heterogeneity.

Analysis of the models different from autoregressive ones shows that most of them cause significant problems of characteristics identification though they enable to form heterogeneous images. Among these models, we should single out the wave model [21], Gibbs random fields [22], wavelet and shearlet decomposition [23]. Formation of such fields often requires a fairly large number of iterations to guarantee the specified properties. However, their more significant drawback consists in complexity of analytical representation and as a consequence the problem of design and practical implementation of large-size image sequences processing algorithms.

Search of new probabilistic models for multispectral image sequences can be done among composite models. For example, recently morphological image analysis have been extensively used in image processing [24, 25]. Application of the latter is explained to a large extent by a wide range of the acquired images, which can have properties very close to real images. Besides, aerospace observations have rather complex properties. This makes it difficult to analyze the data basing only on spectral features. Although, an integral representation of a real snapshot using textures (excluding their assemblage) seems to be hardly probable. In this case, the description of a model is partitioned into several sections. Forming of random fields by means of nonlinear transformations enables to generate non-Gaussian distributions. Mixed Gaussian models [4, 26] and models based on Gaussian copula function and wavelet decomposition produce fairly good results, when applying them to known databases though they lack for a developed mathematical apparatus to investigate them.

Thus, we conclude that full solution for the image description problem has not been yet obtained. At the same time, the issues concerned with description of spatially heterogeneous and non-stationary in time real material remain beyond the framework of the existing mathematical models. In this connection, the construction and investigation of mixed multidimensional images is an urgent problem: on the one hand, the images must have properties very similar to properties of real images including the Earth surface snapshots and, on the other hand, the images must enable to synthesize multidimensional images processing algorithms having practically acceptable computational complexity. Along with that when developing new models it is necessary to investigate their characteristics and the effectiveness of the processing algorithms for the images generated by means of these models.

The chapter is organized as follows. In Sect. 3.2, the issues concerning synthesis and analysis of doubly stochastic models based on spatial autoregressive models are considered. Section 3.3 is devoted to the issues concerned with development and effectiveness evaluation of multidimensional images filtering and estimation algorithms based on the proposed Doubly Stochastic (DS) models [14, 26–30]. The problem of detection of extended objects with unknown parameters against the background of spatially heterogeneous images time sequences is considered in Sect. 3.4. Section 3.5 discusses how the developed algorithms are concerned to the problems of satellite images and image sequences processing. Models of heterogeneous images with radial-circular structure defined on the cylinder and circle [31, 32] are given in Sect. 3.6. Section 3.7 concludes the chapter.

3.2   Doubly Stochastic Autoregressive Models of Multidimensional Random Fields

In this section, the variants of synthesis and analysis of double stochastic models based on autoregressive models are considered, as well as, these models parameters estimation methods for real signals are proposed in Sects. 3.2.1–3.2.3, respectively.

3.2.1   Doubly Stochastic Models of Random Fields

Let Eq. 3.1 be the mathematical model of the Random Field (RF) assigned on rectangular N-dimensional grid $\Omega = \{\bar{i} = (i_1, (i_2, \ldots, i_N) : (i_k = 1, 2, \ldots, M_k), k = 1, 2, \ldots, N\}$, where $\Xi = \{\xi_{\bar{i}}, \bar{i} \in \Omega\}$ is the RF to be modeled, $\{\alpha_{\bar{i}\bar{j}}, \beta_{\bar{i}} : \bar{i} \in \Omega, \bar{j} \in D_{\bar{i}}\}$ are the model coefficients, $\Xi = \{\xi_{\bar{i}}, \bar{i} \in \Omega\}$ is the generating white RF, $D_{\bar{i}}$ is the causal domain of local states for the point $\bar{i}$.

$$x_{\bar{i}} = \sum_{\bar{j} \in D_{\bar{i}}} \alpha_{\bar{i}\bar{j}} x_{\bar{i}-\bar{j}} + \beta_{\bar{i}} \xi_{\bar{i}}, \ \bar{i} \in \Omega \tag{3.1}$$

Suppose that the coefficients $\alpha_{\bar{i}\bar{j}}$ of the given model are random variables determined by Eq. 3.2, where $\{r_{\bar{l}}, \gamma_{\bar{i}} : \bar{i} \in \Omega, \bar{l} \in D_{\bar{i}}\}$ are the constant coefficients, $\Sigma = \{\zeta_{\bar{i}}, \bar{i} \in \Omega\}$ is the auxiliary white RF.

$$\alpha_{\bar{i}\bar{j}} = \sum_{\bar{l} \in D_{\bar{i}}} r_{\bar{l}} \alpha_{\bar{i}-\bar{l}\bar{j}} + \gamma_{\bar{i}} \zeta_{\bar{i}}, \ \bar{i} \in \Omega \tag{3.2}$$

The random nature of the model coefficients in Eq. 3.1 enables to use this model also for description of spatially heterogeneous and time non-stationary multidimensional signals, i.e. the signals with various probabilistic and correlation characteristics at different time points and for different spatial coordinates. An important special case for this model is DS model based on AutoregRessive with Multiple Roots (ARMR) models of characteristic equation. The ARMR model can be defined as $F_{ARMR(\bar{K})}(\bar{\rho}, b\xi_{\bar{i}})$ using Eq. 3.3, where K_i are the coefficients defining multiplicity of the model, β is the normalizing coefficient, z_k^{-1} is the shift operator $z_k^{-1} x_{\bar{i}} = x_{i_1, \ldots, i_k-l, \ldots, N}$, $\alpha_{\bar{i}} = F_{ARMR(\bar{K}_a)}(\bar{r}_\alpha, \gamma_\alpha \xi_{\alpha \bar{i}})$, $\rho_{\bar{i}} = \{F_{ARMR(\bar{K}_{\rho 1})}(\bar{r}_{\rho 1}, \gamma_1 \xi_{1 \bar{i}j}),$ $F_{ARMR(\bar{K}_{\rho N})}(\bar{r}_{\rho N}, \gamma_N \xi_{N \bar{i}j})\}$, $b_{\bar{i}} = F_{ARMR(\bar{K}_b)}(\bar{r}_b, \gamma_b \xi_{b \bar{i}})$ are random variables defined by the ARMR model, $\xi_{\bar{i}}, \xi_{\alpha \bar{i}}, \xi_{1 \bar{i}}, \ldots, \xi_{N \bar{i}}, \xi_{\beta \bar{i}}$ are Gaussian white RFs, $\bar{i} = \{i_1 = 1, \ldots, M_1, \ldots, i_N = 1, \ldots, M_N\}$.

$$\prod_{i=1}^{N} \left(1 - \rho_i z_i^{-1}\right)^{K_i} x_{\bar{i}} = b\beta \, \xi_{\bar{i}} \tag{3.3}$$

The DS model can be written as Eq. 3.4, where A, B are certain numbers determining average values of mathematical expectation and variance of the image, $\bar{m}_\rho$ is the vector of the parameters defining averaged correlation properties of the image:

$$x_{\bar{i}} = A + \alpha_{\bar{i}} + F_{ARMR(\bar{K})}(\bar{m}_\rho + \bar{\rho}_{\bar{i}}, (B + b_{\bar{i}})\beta_{\bar{i}}\xi_{\bar{i}}) \tag{3.4}$$

The present model, as well as, a more general model in Eq. 3.1, allows variation of mathematical expectation, variance, and correlation properties of the simulated signals depending on the current spatial coordinates. An important property of the model in Eq. 3.2 is in quasi-isotropy of Correlation Functions (CF) of RF to be generated. In Fig. 3.1, certain realizations of DS model for the two-dimensional case are presented. Here, Fig. 3.1b depicts an image corresponding to the auxiliary RF $\{a_{i,j}\}$, which determines dynamics of mathematical expectation variation of the simulated image. Figure 3.1c shows an image corresponding to the auxiliary RF $\{b_{i,j}\}$, which determines dynamics of variance variation of the simulated image. Figure 3.1d depicts an image corresponding to the auxiliary RF $\{\rho_{i,j}\}$, which determines dynamics of correlation properties variation of the simulated image. In Fig. 3.1a, DS model realizations can be seen.

3.2.2   Analysis of Double Stochastic Models

Unfortunately, CF of random fields and processes generated by DS model is described in general case by fairly cumbersome expressions specified by the

(a)   **(b)**   **(c)**   **(d)**

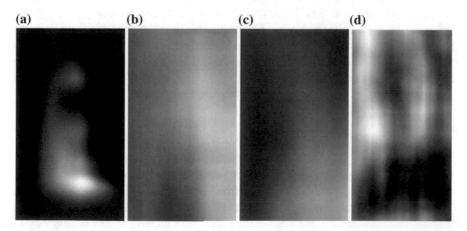

Fig. 3.1 Realization of DS model and auxiliary RF: **a** DS model realizations, **b** image corresponding to the auxiliary RF $\{a_{i,j}\}$, **c** image corresponding to the auxiliary RF $\{b_{i,j}\}$, **d** image corresponding to the auxiliary RF $\{\rho_{i,j}\}$

presence of double correlation dependencies of the process itself and its basic parameters. Although, CF can be found for an important special case of slow (compared to the correlation interval of RF to be formed) variation of basic RF. Then, for CF of one-dimensional first-order DS model we can obtain the following Eq. 3.5, where $\left[\frac{k}{2}\right]$ is the rounding to a smaller value, $C_k^j = \frac{k!}{j!(k-j)!}$, $(2j-1)!!$ is determined as the product of all natural odd numbers on the interval $[1,(2j-1)]$.

$$B(k) = \sigma_x^2 \sum_{j=0}^{\left[\frac{k}{2}\right]} C_k^{2j} \sigma_\rho^{2j} m_\rho^{k-2j} (2j-1)!! \tag{3.5}$$

For CF of the two-dimensional DS model based on ARMR models with multiplicity (2,2), we obtain:

$$B(k_1, k_2) = \sigma_x^2 \left(1 - |k_1| + \frac{2|k_1|}{2 + m_{\rho_1} + m_{\rho_1}^2}\right) \sum_{j_1=0}^{\left[\frac{k_1}{2}\right]} C_{k_1}^{2j_1} \sigma_\rho^{2j_1} m_{\rho_1}^{k_1-2j_1} (2j_1-1)!!$$

$$\times \left(1 - |k_2| + \frac{2|k_2|}{2 + m_{\rho_2} + m_{\rho_2}^2}\right) \times \sum_{j_2=0}^{\left[\frac{k_2}{2}\right]} C_{k_2}^{2j_2} \sigma_v^{2j_2} m_{\rho_2}^{k_2-2j_2} (2j_2-1)!! \tag{3.6}$$

Figure 3.2 depicts the graphs of CFs and exact values for CF of DS model obtained by means of numerical methods. The results are considered for the section $k_2 = 0$. The model parameters are the following: $m_{\rho_1} = m_{\rho_2} = 0.7$, $\sigma_{\rho_1}^2 = \sigma_{\rho_2}^2 = 0.014$.

Having analyzed the obtained graphs we can conclude that increase of the distance k_1 leads to a slower decrease of CF of the doubly stochastic RF than CF of the RF based on ARMR model. Consequently, the dependencies between the elements of such a model are stronger. In case when the elements of the parametric vectors $\bar{r}_{\rho i}$, $i = 1,\ldots,N$ tend to zero the CF of DS model converges to the corresponding CF of the ARMR model.

3.2.3 Estimation of Parameters of Double Stochastic Models

When describing real signals it is also necessary to solve the problem of identification of DS model parameters. For two-dimensional case, it means that RF formed on the basis of DS model must have characteristics of a real image $\{z_{i,j}\}$. Here, we propose an identification technique based on a combination of DS model varying parameters estimation by means of a sliding window and by means of pseudo-gradient procedures [6, 33]. The parameters $\bar{\theta} = (\sigma_a^2, \rho_{1a}, M_{1a}, \ldots, \rho_{Na}, M_{Na})$,

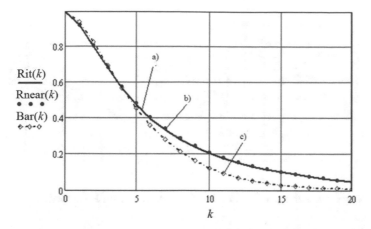

Fig. 3.2 Comparative characteristic for CF of doubly stochastic RF on the basis of combined ARMR model, where (a) is the graph of CF determined from Eq. 3.2, (b) are the exact values for CF of DS model obtained by means of numerical methods, (c) graph of CF for RF based on ARMR with multiplicity (2,2) with constant correlation coefficients m_{ρ_1} and m_{ρ_2}

defining $\{a_i\}$ can be determined by means of pseudogradient relaxation from the maximum of the expressions Eq. 3.7, where B_{z_1} is the diagonal matrix $M_j \times M_j$ composed of the estimates of CF of RF, $\{z_i\}$ are the corresponding to the rows of the $\bar{i}$ th measurement, B_{a_i} is the CF of one-dimensional ARMR model.

$$F_i(\sigma^2_{a,ia}, M_{ia}) = \sum_{j=1}^{M_i} \left(\sum_{j_1=1}^{M_1} .. \sum_{j_N=1}^{M_N} z_{j_1,\dots,j_N} \right.$$
$$\left. \times \left(B_{z_i}^{-1} - B_{a_i}(\sigma^2_{a,ia}, M_{ia})^{-1} \right) z_{j_1,\dots,j_i+j,\dots,j_N} \right) \tag{3.7}$$

Estimation of the parameters determining the behavior of the auxiliary random fields $\{\bar{\rho}_i\}$ is concerned with search of the roots of the functional Eq. 3.8 at each point $\bar{i} \in \Omega$.

$$\varphi(\bar{\rho}_i) = F_{ARMR(\bar{K})}(\bar{m}_\rho + \bar{\rho}_i, 0) - z_i \tag{3.8}$$

Here, the vectors $\bar{K}$ and $\bar{m}_\rho$ can be found on the basis of minimization of the difference between observations CF and CF of the supposed ARMR model $B_z - B_{ARKK(\bar{K}, \bar{m}_\rho)}$ according to the norm L_2. Taking into account the separability of CF of ARMR model such minimization along with determination of the roots (Eq. 3.4) can be carried out by means of Pseudo-Gradient (PG) algorithms. For the aggregate of the estimates $\{\bar{\rho}_i\}$ and $\{\hat{b}_i = (z_i - \hat{A})^2 - \hat{B}\}$, the parameters $\{\bar{r}_{\rho 1}, \dots, \bar{r}_{\rho N}, \bar{r}_b, \gamma_1, \dots, \gamma_N, \gamma_b\}$ can be determined in accordance with the same scheme. In Fig. 3.3 the dependence of the identification average error for specific

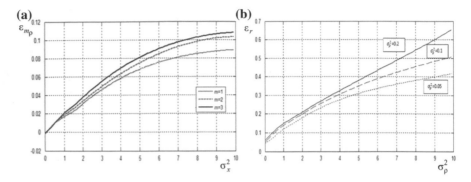

Fig. 3.3 Dependence of parameters identification accuracy versus RF properties

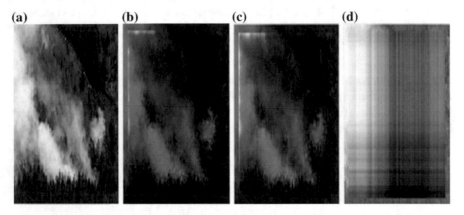

Fig. 3.4 Forming of images with varying properties: **a** real image of size 250×250, **b** RF realization generated by DS model based on ARMR model of multiplicity (2,2), **c** RF realization generated by DS model based on ARMR model of multiplicity (1,1), **d** RF realization generated by first-order AR model

correlation parameters $\varepsilon_{m_\rho} = \sqrt{M\left(\hat{m}_{\rho i} - m_{\rho i}\right)^2}$ and $\varepsilon_r = \sqrt{M\left(\hat{m}_{rij} - m_{rij}\right)^2}$ versus variance σ_x^2 of the basic RF and variance σ_ρ^2 of an auxiliary RF is shown.

In Fig. 3.4, an example of application of the proposed technique for identification and simulation of real images is shown. Fitting error variances found as the average squared value of brightness divergence for real image and RF realizations are equaled to 0.49, 0.37, and 0.96 for Fig. 3.4b, Fig. 3.4.c, and Fig. 3.4d, respectively.

3.3 Doubly Stochastic Filtering

Extraction of useful image components based on noisy observations is an important problem, since it gives an opportunity to improve an image against noise background. However, there exist many different approaches to solve the problem of determining the potential accuracy of estimation and construction of multidimensional RF filtering algorithms. On the basis of doubly stochastic models, it is possible to synthesize special types of filters that can perform processing of spatially heterogeneous signals. This synthesis is discussed in Sect. 3.3.1. Section 3.3.2 shows the possibility of using doubly stochastic filter for processing two-dimensional images. Section 3.3.3 is devoted to the search for algorithms that provide fast image processing based on doubly stochastic filter. Sections 3.4–3.5 deal with the issues of filtering and recovery for satellite multispectral images and time sequences of images.

3.3.1 Synthesis of Double Stochastic Filter

To synthesize filtering algorithms for doubly stochastic RF, the model in Eq. 3.1 can be represented in the form of vector stochastic Eq. 3.9, where $\bar{\xi}_i$ is the vector composed of normal random variables with CF $V_{\xi i}$.

$$\bar{x}_i = \varphi_i(\bar{x}_{i-1}) + \bar{\xi}_i \tag{3.9}$$

On the basis of the central limit theorem for martingales, it has been shown that for DS model the distribution of the random variables vector $\bar{x}_i$ can be approximated by the normal distribution as the index i increases. Now assuming that at each i th step we have an observation $\bar{z}_i$ of the vector $\bar{x}_i$ distorted by the additive noise $\bar{n}_i$ with covariance matrix V_{ni}, i.e. $\bar{z}_i = C_i\bar{x}_i + \bar{n}_i$, we can write the following expression for the joint Probability Density Function (PDF) $w(X_k, Z)$:

$$w(X_k)w(Z|X_k) \cong \frac{1}{\sqrt{(2\pi)^{mk}\det V_x}} \frac{1}{\sqrt{(2\pi)^n\det V_{ni}}} \exp(-W_k(X_k)), \tag{3.10}$$

where

$$W_k(X_k) = \frac{1}{2}\sum_{i=1}^{k}(\bar{x}_i - \varphi_{i-1}(\bar{x}_{i-1}))^T V_{\xi i}^{-1}(\bar{x}_i - \varphi_{i-1}(\bar{x}_{i-1}))$$
$$+ \frac{1}{2}\sum_{i=1}^{k}(\bar{z}_i - C_i\bar{x}_i)^T V_{ni}(\bar{z}_i - C_i\bar{x}_i).$$

The search of the best estimate $\bar{\hat{x}}_{Wi}$ according to Bayesian performance criterion can be represented as the problem of search of conditional extremum $W_k(X_k, \xi_k) = \sum_{i=1}^{k}(\frac{\bar{\xi}_i^T V_{\xi i}^{-1} \bar{\xi}_i}{2} - f_i(\bar{x}_i, \bar{z}_i))$ with restriction $\bar{x}_i = \varphi_{i-1}(\bar{x}_{i-1}) + \bar{\xi}_i$. Its solution is representable in the form of a system of the following $2k$ equations of Eq. 3.11, where $\bar{\lambda}_i$ are Lagrange multipliers, $i = 1, 2, \ldots, k$, $\bar{\lambda}_k = 0$, $\bar{\hat{x}}_0 = 0$.

$$\bar{\lambda}_{i-1} - \varphi_i'(\bar{\hat{x}}_{Wi})^T \bar{\lambda}_i = f_{i'}(\bar{\hat{x}}_{Wi}, \bar{z}_i), \quad V_{\xi i} \bar{\lambda}_{i-1} = \bar{\hat{x}}_{Wi} - \varphi_{i-1}(\bar{\hat{x}}_{Wi-1}) \tag{3.11}$$

It is important that for DS model $\varphi_i'(\bar{\hat{x}}_{Wi})$ and $f_i'(\bar{\hat{x}}_{Wi}, \bar{z}_i) = C_i^T V_{ni}(\bar{z}_i - C_i \bar{\hat{x}}_{Wi})$ are linearized matrix transforms.

To obtain recurrent estimates $\bar{\hat{x}}_i$ from $\bar{\hat{x}}_{Wi}$, let us utilize the invariant embedding method. Then after expansion of the vector function $\varphi_i(\bar{x}_i)$ for DS model in view of sparseness of the tensor $\varphi_i''(\bar{a})\bar{x}_i$, we can obtain the following expression Eq. 3.12, where $P_k = P_{Pk}(E + C_k^T V_{nk} C_k P_{Pk})^{-1}$.

$$\bar{\hat{x}}_k = \varphi_k(\bar{\hat{x}}_{k-1}) + P_k C_k^T V_{nk}(\bar{z}_k - C_k \varphi_k(\bar{\hat{x}}_{k-1})) \tag{3.12}$$

On condition that the filtering errors matrices P_i and prediction error matrices P_{pi} calculated at forward pass of the filter are known the given relations Eq. 3.5 enable to refine the found estimates on the information in observations following the value to be estimated in compliance with the formula Eq. 3.13, where $A_i = P_{i-1}\varphi_{i-1}'(\bar{\hat{x}}_{Wi-1})^T P_{Pi}^{-1}$, $i = k \ldots 1$, $\bar{\hat{x}}_{Wk} = \bar{\hat{x}}_k$.

$$\bar{\hat{x}}_{Wi-1} = \bar{\hat{x}}_{i-1} + A_i(\bar{\hat{x}}_{Wi} - \bar{\hat{x}}_{Pi}) \tag{3.13}$$

3.3.2 Analysis of Double Stochastic Filter

Let the two-dimensional image $\{x_{i,j}\}$ be described by the model (Eq. 3.3) at $N = 2$, $\bar{K} = \bar{K}_{\rho 1} = \bar{K}_{\rho 2} = (2, 2)^T$. Here, to reduce the necessary memory capacity we suppose that the parameters $a_{i,j} = 0$ and $b_{i,j} = \sigma_x^2$, $i = 1 \ldots M_1, j = 1 \ldots M_2$. Let the observations of the mixture $z_{i,j} = x_{i,j} + n_{i,j}$ of information RF and additive white Gaussian RF $\{n_{i,j}\}$ with variance $\sigma_n^2 = M(n_{i,j}^2)$ are obtained. Let us solve the problem of restoration of the samples $\{x_{i,j}\}$ on the observations $\{z_{i,j}\}$.

For this purpose, let us compose the following vector of elements $\bar{x}_{i,j} = (\bar{x}_{xi,j}, \bar{P}_{xi,j}, \bar{P}_{yi,j})$ of the length $6M_1 + 6$, where

$$\bar{x}_{xi,j} = \left(x_{i,j}x_{i,j-1} \ldots x_{i,1}x_{i-1,M_1} \ldots x_{i-1,1}x_{i-2,M_1} \ldots x_{i-2,j-2}\right)^T,$$

$$\bar{\rho}_{1i,j} = \left(\rho_{1i,j}\,\rho_{1i,j-1} \cdots \rho_{1i,1}\,\rho_{1i-1,M_1} \ldots \rho_{1i-1,1}\,\rho_{1i-2,M_1} \cdots \rho_{1i-2,j-2}\right)^T,$$

$$\bar{\rho}_{2i,j} = \left(\rho_{2i,j}\,\rho_{2i,j-1} \cdots \rho_{2i,1}\,\rho_{2i-1,M_1} \ldots \rho_{2i-1,1}\,\rho_{2i-2,M_1} \cdots \rho_{2i-2,j-2}\right)^T.$$

Then to carry out sequential filtering let us use a window of the size $2M_1 + 2$ elements, which can be moved on the image and carries out sequential processing by means of line-by-line scanning.

Then, the whole model (Eq. 3.3) can be written in a compact vector form $\bar{x}_{i,j} = \varphi_{i,j}(\bar{x}_{ij}) + \bar{\xi}_{i,j}$, $\bar{\xi}_{ij} = (\xi_{i,j}, \xi_{\rho 1i,j}, \xi_{\rho 2i,j})$. Using these relations we can construct the following two-dimensional DS filter (Eq. 3.14):

$$\hat{\bar{x}}_{i,j} = \hat{\bar{x}}_{Pi,j} + B_{i,j}(z_{i,j} - \hat{x}_{Pi,j}) \tag{3.14}$$

where $\hat{x}_{Pi,j}$ is the first element of the vector $\hat{\bar{x}}_{pi,j}$, $B_{ij} = P_{Pi,j}C^T D_{i,j}^{-1}$, $C = (1, 0, \ldots, 0)$, $D_{i,j} = CP_{Pi,j}C^T + \sigma_n^2$, $P_{Pi,j} = \varphi_{i,j-1}{}'(\hat{\bar{x}}_{i,j-1})P_{i,j-1}\varphi_{i,j-1}{}'(\hat{\bar{x}}_{i,j-1})^T + V_{\xi i,j}$.

The obtained algorithm possesses important features. Firstly, despite external cumbersomeness it does not require to invert large-size matrices as it is the case, when performing row-wise Kalman estimation [6, 8, 14, 25]. Secondly, when forming an estimate at the point (i, j), all the elements to the left and above this point are used and the elements included in the vector $\bar{x}_{i,j}$ preceding $\hat{x}_{i,j}$ are re-estimated. Thirdly, the result of the filtering includes not only the aggregate of the estimates $\hat{x}_{i,j}$ but also the estimates of the correlation dependencies $\hat{\rho}_{1i,j}$ and $\hat{\rho}_{2i,j}$. Such feature enables to use DS filter not only for noise compensation but also as an element of the texture-correlation analysis algorithms, for example, at image segmentation.

The proposed filtering algorithm is close in its structure to the vector Kalman filter. In Fig. 3.5, the filtering error variances for a row of a spatially homogeneous image in case of vector Kalman filter and nonlinear DS filtering are shown.

It is obvious that DS filter is to a little degree inferior to Kalman filter at the beginning of a row during the stabilization process. When the number of elements increases, we can observe convergence to Kalman filter. The obtained result enables

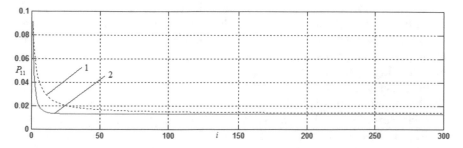

Fig. 3.5 Error variances for DS filter (1) and for linear Kalman filtering (2)

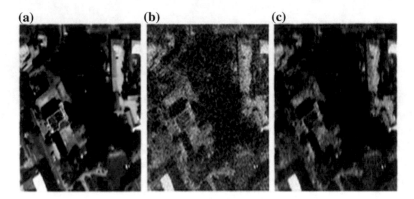

Fig. 3.6 Filtering of satellite images: **a** original image, **b** distorted image, **c** filtering result

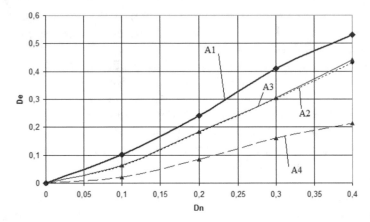

Fig. 3.7 Effectiveness evaluation of satellite image filtering using vector Kalman filter without interpolation (A1), discrete Wiener Filter (A2), Kalman filter with interpolation (A3), and DS filter (A4)

to assert that the procedures have similar effect, when filtering homogeneous images with known correlation parameters. In case of processing of spatially heterogeneous signals, the proposed DS filter provides processing gain in direct proportion to the image heterogeneity degree. In Fig. 3.6, an example of artificially noised satellite image filtering is presented. Figure 3.7 shows the dependencies of image filtering error variance D_e versus noise variance D_n.

Simple analysis of the curves in Fig. 3.7 proves effectiveness evaluation of the proposed filtering method, whose gain in comparison with other filters comes to 120% in terms of estimation error variance. Moreover, there is one stronger result. DS filter has on average an advantage of up to 15–20% in terms of estimation error variance over the filters A1–A3 even if the latter are applied to an image, which has already been segmented. It is explained by the possibility of DS filter parameters adaptive adjustment to gradually varying probabilistic and correlation image parameters.

3.3.3 Block Double Stochastic Filter

The presented solution can be generalized for the case of filtering of the field $\{\bar{x}_i\}$ in windows sliding on the grid Ω. Here it can be shown that, for example, for two-dimensional case any DS model (Eq. 3.1) can be written as the following Eq. 3.15, where $\bar{X}_{i,j}$ is the aggregate of samples of the basic and the auxiliary fields placed in the window centered at the point (i,j), $\bar{\xi}_{i,j}$ is the aggregate of the generating random variables in the same window, $P_{i,j}$ and $Q_{i,j}$ are the tensors depending also on the samples included in $\bar{X}_{i,j-1}$ and $\bar{X}_{i-1,j}$. Figure 3.8 depicts a graphical illustration of relation between the elements $\bar{X}_{i,j}$ and $\bar{X}_{i+1,j+1}$.

$$\bar{X}_{i,j} = P_{i,j}\bar{X}_{i,j-1} + Q_{i,j}\bar{X}_{i-1,j} + \vartheta_{ij}\bar{\xi}_{ij} \tag{3.15}$$

If the correlation properties at each point are assumed to be equal row-wise and column-wise and their rate of change is so slow that within the window $\bar{X}_{i,j}$ the correlation can be described by only one parameter $\rho_{i,j}$ then the filtering process can be significantly simplified by using zigzag processing or line-by-line filtering in parallel flows with a conditional cycle delay. For example, for the model (Eq. 3.3) at $N = 2$ and the window size 3×3, we can write as the following Eq. 3.16:

$$\bar{X}_{i,j} = \bar{X}_{Pi,j} + P_{Dij}V_n^{-1}(z_{i+1,j+1} - Y\bar{X}_{Pij}), \tag{3.16}$$

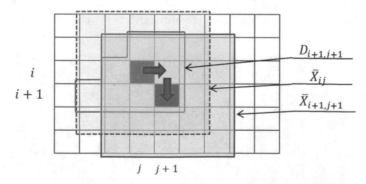

Fig. 3.8 Graphical illustration of relation between the elements $\bar{X}_{i,j}$ and $\bar{X}_{i+1,j+1}$

where

$$\bar{X}_{Pi,j} = P_{i,j}\bar{X}_{i-1,j} + Q_{i,j}\bar{X}_{i,j-1}, \; P_{Di,j} = P_{Pi,j}\left(E + V_n^{-1}P_{Pi,j}\right)^{-1},$$

$$P_{Pi,j} = P_{P1i,j} + P_{P2i,j} + V_{\xi i,j}, \; P_{P1i,j} = P_{i,j}P_{Di,j-1}P_{i,j}^T, \; P_{P2i,j} = Q_{i,j}P_{Di-1,j}Q_{i,j}^T.$$

Here, $P_{Pij}, P_{P1ij}, P_{P2ij}, P_{Di,j-1}$ are the matrices 10×10, $V_n^{-1} = \frac{1}{\sigma_n^2}$, Y is the matrix

$$10\times3, \; Y = \begin{pmatrix} 0 & 0 & 1 & 0 & 0 & 0 & 0 & 0 & 0 & 0 \\ 0 & 0 & 0 & 0 & 0 & 1 & 0 & 0 & 0 & 0 \\ 0 & 0 & 0 & 0 & 0 & 0 & 0 & 0 & 1 & 0 \end{pmatrix}.$$

The obtained procedures can be easily generalized for an arbitrary filtering window size. Thus, for example, for the window size 5×5 the vector $\bar{X}_{i,j}$ length will increase from 10 elements to 26 elements and correspondingly the size of the matrices $P_{i,j}, Q_{i,j}, P_{Pij}, P_{Di,j}$ will amount to 26×26, the matrix Y will have the size of 5×26. However, the structure of all these matrices will not change.

We should also note that all above-mentioned filters at each filtering step also enable to obtain a local estimate of image correlation characteristics. For example, for the last filter such characteristic is represented by the element $\bar{X}_{i,j}(10)$ for each point (i,j). It enables to adjust the filtering window size to the estimated local correlation characteristics during filtering process. Such adaptation can be also interpreted as determination for each point (i,j) a non-causal domain determining the field value at this point.

To analyze effectiveness the developed algorithms, fragments of satellite images presented in Fig. 3.9 were mixed with white noise of various intensities so that the signal-to-noise ratio $q = \frac{\sigma_x^2}{\sigma_n^2}$ was varied in the range from 2 to 10 and sequential filtering of the images was performed.

In Table 3.1, a concise comparative analysis of the above-described procedures effectiveness in comparison with Local Polynomial Approximation with Intersection Of Confidence Intervals (LPA/ICI) algorithm [14], wavelet-filter with Stein's Unbiased Risk Estimate (SURE) threshold [15], separable bilateral filter

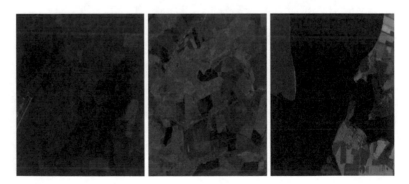

Fig. 3.9 Examples of image fragments used for effectiveness evaluation

Table 3.1 Filtering efficiency of satellite images

Filtering algorithm	Image	Processing time	q = 2	q = 5	q = 10
LPA/ICI algorithm	№1	23 s	0.076	0.042	0.031
	№2	18 s	0.067	0.036	0.028
	№3	28 s	0.083	0.045	0.032
Wavelet filtering with SURE threshold	№1	14 s	0.083	0.065	0.053
	№2	13.5 s	0.081	0.061	0.049
	№3	14 s	0.087	0.064	0.051
Bilateral filter (based on parallel decomposition	№1	0.8 s	0.112	0.091	0.071
	№2	0.7 s	0.097	0.083	0.063
	№3	0.8 s	0.107	0.089	0.076
Doubly stochastic filter	№1	17 min 56 s	0.089	0.048	0.036
	№2	17 min 39 s	0.083	0.042	0.033
	№3	18 min 19 s	0.092	0.051	0.041
Doubly stochastic filter with interpolation	№1	6 h 3 min 9 s	0.071	0.038	0.027
	№2	6 h 1 min 11 s	0.063	0.033	0.024
	№3	6 h 4 min 53 s	0.079	0.043	0.030
Doubly stochastic filtering in parallel sliding windows*	№1	4.4 s	0.076	0.04	0.03
	№2	4.3 s	0.064	0.035	0.026
	№3	4.4 s	0.083	0.045	0.031

*Implementation has been carried out on MYCUDA NVIDIA TESLA M4 without tensor optimization

[16] and a conventional doubly stochastic filter in terms of relative filtering error variance $\left(\text{MSE}/\sigma_x^2\right)$ is given.

An analysis of the presented results indicates superiority of the proposed filtering techniques at various signal-to-noise ratios q. It is also connected with presence of extended regions on the image to be filtered, within which doubly stochastic filters are able to adapt to image local features. As a result, the processing performance using concurrently sliding and interconnected windows turns out to be even better than that of formally optimal doubly stochastic filters. The performed research shows that this is concerned with presence of distinct boundaries on original images leading to local mismatch of the optimal doubly stochastic filter. Its analogue carrying out processing in non-causal windows has significantly more compact operating domain and adapts faster to abrupt changes of image properties. In addition, gain (by a factor of 102–103) in the number of computational resources required per one realization is obvious. As a consequence the algorithms implemented on CUDA architecture turn out to be significantly faster than the algorithms used in graphic editors. In addition to this, the obtained gain can be substantially increased in case of using of special techniques of tensor operations optimization in view of sparseness of tensors to be used.

3.3.4   Double Stochastic Filter for Multispectral Image Processing

The presented results can be generalized for the case of processing of a Separate Multispectral Image (MSI). For this, let us assume that all N two-dimensional frames of this MSI are defined by the RF $\{x_{i,j}^k\}$, $k = 1\ldots n$, placed on the same grid Ω. Let the matrix R of interframe correlations between separate frames be determined as a result of long-term observations. We also assume that closeness of interframe correlation characteristics on all frames results in characteristic properties of object monitoring in these frames. Then, the first frame of such MSI can be described by Eq. 3.2 and to describe the second and the subsequent frames we can use the following Eq. 3.17, in which $v_{i1,j1}$ are the elements of the triangular matrix V such that $VV^T = B$, where B is the covariance matrix of a separate, for example, the first frame, $\xi_{i1,j1}^k$ are the white noise samples with variance $\sigma_{\xi k}^2 = \prod_{p=1}^{k} \sqrt{1 - R(p,k)^2}$. The matrix V can be obtained by Cholesky decomposition.

$$x_{i,j}^k = \sum_{l=1}^{k} R(l,k)x_{i,j}^l + \sum_{i1=1}^{M_1}\sum_{j1=1}^{M_2} v_{i1,j1}\xi_{i1,j1}^k \qquad (3.17)$$

The relations (Eq. 3.17) enable to modify the filter (Eq. 3.14) for the case of MSI. To achieve this, it is necessary to fill in advance the previously considered vector $\bar{x}_{xi,j}$ with scalar samples $x_{i,j}^2, \ldots, x_{i,j}^N$ and to write the nonlinear equality $\bar{x}_{i,j} = \varphi_{i,j}(\bar{x}_{i,j}) + \bar{\xi}_{i,j}$ again in view of the appropriate changes in the vector $\bar{\xi}_{i,j}$. Then the proposed algorithms can be applied for filtering taking into account the fact that each point with the coordinates (i,j) contains not only one scalar observation $z_{i,j}$ but an observation vector, whose elements correspond to the values of a multispectral image at the point (i,j) on the corresponding frame.

On graphs in Fig. 3.10, the dependencies of filtering effectiveness versus number of MSI frames and interframe correlation coefficient are presented. More particularly, in Fig. 3.10a the filtering errors arising at a separate frame processing (dashed line) using several frames (up to 10) are shown. In Fig. 3.10b, dependence of filtering error versus interframe correlation coefficient R is presented. In the latter case, it was assumed that R remains the same for any pair of different two-dimensional frames included in MSI. Here, the curve (1) corresponds to the

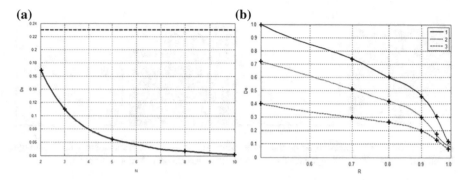

Fig. 3.10 Multispectral image filtering effectiveness: **a** dependence of the filtration dispersion on the number of frames, **b** dependence of the dispersion filtering of interframe correlation

Table 3.2 Filtering efficiency of multispectral images

Number of bands used for filtering	Image 1			Image 2			Image 3		
	$q = 2$	$q = 5$	$q = 10$	$q = 2$	$q = 5$	$q = 10$	$q = 2$	$q = 5$	$q = 10$
1	0.083	0.042	0.033	0.068	0.036	0.028	0.081	0.039	0.034
2	0.067	0.037	0.029	0.059	0.032	0.026	0.067	0.036	0.028
3	0.064	0.036	0.029	0.056	0.029	0.024	0.063	0.035	0.027
4	0.058	0.033	0.027	0.051	0.028	0.022	0.056	0.033	0.026
5	0.053	0.031	0.026	0.048	0.027	0.021	0.054	0.032	0.025
6	0.052	0.031	0.024	0.046	0.026	0.021	0.054	0.031	0.025
7	0.049	0.028	0.023	0.043	0.024	0.020	0.051	0.029	0.024
8	0.049	0.027	0.023	0.042	0.021	0.019	0.050	0.028	0.022

case of simultaneous processing of two frames, the curve (2) – four frames, and the curve (3) – 10 frames.

Table 3.2 lists the ratio between filtering error variance and image variance (MSE/ σ_x^2) for processing of multispectral images presented in Fig. 3.9.

Analysis of the data given in Table 3.2 shows that increase in the number of spectral bands used for processing significantly improves effectiveness of this processing. Thus, exploitation of all available eight spectral bands enables to reduce the filtering error variance by 40% on average in comparison with the results of a unit frame processing. Analysis of the given graphs substantiates the gain achieved by increase in the number of frames and by their correlation degree enhancement.

We should note that use of DS filter as a means for obtaining both brightness and correlation image characteristics also enables to solve the problem of determination of boundaries between objects on an image. For this purpose, in many cases it is sufficient to carry out image filtering in forward and reverse direction

(a) (b) (c) (d)

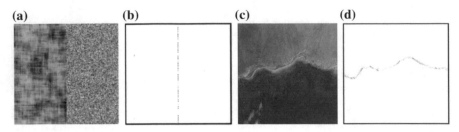

Fig. 3.11 Boundary detection on an image by means of two-pass DS filter: **a** artificial DS image, **b** calculated statistic L, **c** fragment of a real satellite snapshot, **d** calculated statistic L

obtaining correspondingly the estimates $\{\hat{\bar{x}}_i = (\hat{x}_i, \hat{\alpha}_i, \hat{\rho}_{1\bar{i}}, \ldots, \hat{\rho}_{N\bar{i}}, \hat{\beta}_i)\}$ and $\{\bar{\bar{x}}_i = (\bar{x}_i, \bar{\alpha}_i, \bar{\rho}_{1\bar{i}}, \ldots, \bar{\rho}_{N\bar{i}}, \bar{\beta}_i)\}$, and then to determine for the adjacent points $(\bar{i})$ and $(\bar{j})$ the statistic: $L = \sum_{\bar{l} \in D_i} \bar{K}_1(\bar{i})(\hat{\bar{x}}_{\bar{l}} - \hat{x})^T + \sum_{\bar{l} \in D_j} \bar{K}_2(\bar{l})(\bar{\bar{x}}_{\bar{l}} - \bar{x})^T$, where $K_1(\bar{l})$ and $K_2(\bar{l})$ are statistic tensor coefficients. In case when L is greater than the threshold value L_0, a decision of presence of a boundary between the points $(\bar{i})$ and $(\bar{j})$ is made. The meaning of the detector is concerned with the fact that when crossing the explicit boundary between two objects each being described by its own DS model realization the doubly stochastic filter (Eq. 3.6) demonstrates a short-term increase of estimation error variance which is the larger, the more distinct the difference between these objects becomes. This abrupt change of variance can be detected by comparing the estimates from the forward and the reverse filters. Figure 3.11 depicts an illustration of boundary detection on an image by means of two-pass DS filter.

The found solutions can be generalized for the case of MSI time sequence processing. For this purpose, tensor modification of the extended Kalman filter combined with the results of a separate multispectral image filtering can be used. Here, we assume that the spatial grids $J_0, J_1, \ldots, J_N$, on which time sequence of multispectral images is specified are identical and each vector-element of a subsequent multispectral image can be described by the following tensor stochastic equation:

$$\bar{x}_{i,j}^t = \varphi_{i,j}^t(\bar{x}_{k,l}^{t-1}) + \upsilon_{i,j}^t(\bar{x}_{k,l}^{t-1})\bar{\xi}_{i,j}^t, \; k, l \in J_{t-1},$$

where $\varphi_{i,j}^t(\bar{x}_{k,l}^{t-1})$ and $\upsilon_{i,j}^t(\bar{x}_{k,l}^{t-1})$ are certain functionals operating on all the samples $\bar{x}_{k,l}^{t-1}$ at the preceding frame, $\bar{\xi}_{i,j}^t$ are vectors of independent normal random variables with zero mean and unit variance. Let us suppose that for the vector $\varphi_{i,j}^t$ there exists an appropriate matrix $\varphi_{i,j}^{t\prime}(\bar{x}_{k,l}^{t-1})$ of derivatives and the samples $\bar{x}_{i,j}^t$ are observed in additive noise $\bar{n}_{i,j}^t$ so that $\bar{z}_{i,j}^t = C\bar{x}_{i,j}^t + \bar{n}_{i,j}^t$. Then we can develop the tensor filter Eq. 3.18, where $\hat{\bar{x}}_{Pi,j}^t = \varphi_{i,j}^t(\{\hat{\bar{x}}_{i,1j1}^{t-1} : i1, j1 \in J_{t-1}\})$ is the optimal prediction for the tth MSI at the point (i, j) for all points of the $(t - 1)$ th MSI,

$P^t_{Pi,j,k,l} = \varphi''_{i,j}(\hat{\bar{x}}^{t-1}_{i1,j1})P^{t-1}_{i,j,k,l}(\varphi''_{i,j}(\hat{\bar{x}}^{t-1}_{i1,j1}))^T + v^t_{k,l}(\hat{\bar{x}}^{t-1}_{i1,j1})v^t_{k,l}(\hat{\bar{x}}^{t-1}_{i1,j1})^T$ is the prediction error covariance matrix, $\hat{\bar{x}}^1_{Pi,j} = 0$, $P^1_{3i,j,k,l} = M\{\bar{x}^1_{i,j}\bar{x}^1_{k,l}\}$, $P^t_{i,j,k,l} = P^t_{Pi,j,k,l}(E + V^{t-1}_\theta P^t_{Pi,j,k,l})$.

$$\bar{x}^t_{i,j} = \hat{\bar{x}}^t_{Pi,j} + P^t_{i,j,k,l}V^{t-1}_\theta(C^T z^t_{k,l} - \hat{\bar{x}}^t_{Pk,l}) \tag{3.18}$$

Equation 3.18 enables to unite the estimate produced at the preceding $(t-1)$ th frame and observations of the tth frame. If we suppose that MSI are shifted absolutely precisely in time domain and the time interval between recordings of such MSI is small so that the correlation between adjacent MSIs is considerable, then it can be assumed that the spatial domain $G^{t-1}_{ij} = \{k,l : \hat{\bar{x}}^{t-1}_{ij} = \varphi^{t-1}_{ij\_local}(\bar{x}^{t-1}_{i1j1})\}$ on the $(t-1)$ th MSI containing samples used for prediction at the point (i,j) by using the functional $\varphi^{t-1}_{ij\_local}(\cdot)$ is akin to the similar domain G^t_{ij} on t th MSI. Then considering the joint CF of the whole sequence to be separable in time and space so that within G^t_{ij} the equality $M\{\bar{x}^t_{i,j}\bar{x}^s_{k,l}\} = \rho^{|t-s|}R^t_{x1x2}(|i-j|,|k-l|)R_{x3}$ holds true Eq. 3.18 can be rewritten in the following form of Eq. 3.19 with the initial conditions $P^1_{Pi,j} = R^t_{x1x2}(|i-j|,|k-l|)R_{x3}V^{1-1}_{\theta\,ij}$ and $P^t_{i,j,k,l} = P^t_{Pi,j,k,l}(E + V^{t-1}_\theta P^t_{Pi,j,k,l})$, $P^t_{Pi,j,k,l} = \rho^2 P^{t-1}_{i,j,k,l} + (1-\rho^2)R^t_{x1x2}(|i-j|,|k-l|)R_{x3}V^{t-1}_\theta$.

$$\begin{aligned}\bar{x}^t_{i,j} = {} & \rho\,\varphi^{t-1}_{ij_{local}}(\{\bar{x}^{t-1}_{i1,j1} : i_1,j_1 \in G^{t-1}_{i,j}\}) \\ & + P^t_{i,j,k,l}V^{t-1}_\theta(C^T z^t_{k,l} - \rho\,\varphi^{t-1}_{ij\_local}(\{\bar{x}^{t-1}_{i,1j,1} : i_1,j_1 \in G^{t-1}_{i,j}\})),\ k,l \in G^{t-1}_{ij}\end{aligned} \tag{3.19}$$

In Fig. 3.12, the process of estimates sequential integration for the preceding image and observations on the current MSI is presented as a graphical illustration.

The processing algorithms presented in the paper can be unified within the scope of the following flowchart for MSIs time sequence processing Fig. 3.13.

We should note that MSI processing according to the flowchart in Fig. 3.12 can also include additional procedures simplifying the processing. Such procedures can be exemplified by algorithms of image-type related processing of satellite MSI enabling to carry out preliminary segmentation of these images.

Table 3.3 contains the effectiveness analysis results for the developed procedures of satellite images processing in Fig. 3.8 for the cases of processing of one frame, the whole MSI, additional non-causal processing by means of sliding window of alterable size, tensor filtering of MSIs time sequence. For comparison the results of application of anisotropic MultiResolution (MR) LPA denoising algorithm are given. The signal-to noise ratio amounted to $q = 10$ for all cases.

Analysis of the presented data shows that DS filter modifications do enable to perform joint processing for all MSI frames and carry out MSIs time sequence processing. Using all the above-mentioned techniques, the gain in normalized MSE can achieve 50% in comparison with LPA algorithm.

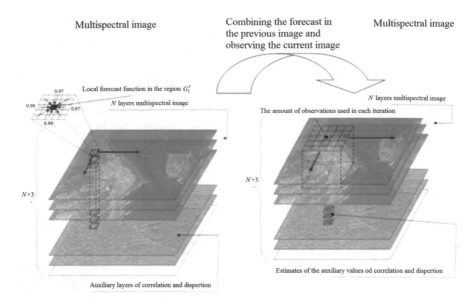

Fig. 3.12 Processing of a sequence of MSIs

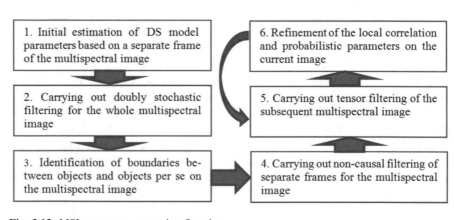

Fig. 3.13 MSIs sequence processing flowchart

3.3.5 Restoration of Multispectral Images Based on Double Stochastic Filtering

The above-described algorithms can be applied to solve an important problem of satellite image restoration in cases, when fragments of these images are damaged or turned out to be inauthentic. For this purpose, the predicted estimates obtained on available observations at preceding and current times can be combined. Such integrated estimate will have the following form Eq. 3.20, in which G_1^T is the

Table 3.3 Filtering efficiency of time sequences of multispectral images

Algorithm	Image 1				Image 2			
	$t = 1$	$t = 2$	$t = 4$	$t = 8$	$t = 1$	$t = 2$	$t = 4$	$t = 8$
Anisotropic MR LPA denoising algorithm. (BLUE band)	0.028	0.026	0.029	0.028	0.02	0.021	0.021	0.022
Algorithm (BLUE band)	0.033	0.035	0.033	0.34	0.028	0.026	0.032	0.031
Algorithm for a separate multispectral image	0.027	0.028	0.028	0.026	0.021	0.023	0.022	0.02
Using non-causal prediction	0.023	0.024	0.023	0.023	0.018	0.019	0.018	0.017
Algorithm based on tensor filtering	0.023	0.020	0.018	0.017	0.018	0.015	0.014	0.014

domain within which restoration is performed, $\hat{x}^T_{P1i,j} = \hat{\tau}_{T-1} \varphi^{T-1}_{i,j_{local}}(\bar{x}^{T-1}_{i1,j1})$ is the estimate at the point (i, j) based on the observations at preceding times with prediction error covariance matrix P_{P1}, $\hat{x}^T_{P2i,j}$ is the predicted estimate based on available observations at current instant of time with prediction error covariance matrix P_{P1}, $\hat{\tau}_{T-1}$ is correlation estimate between multispectral images recorded at times T and $T - 1$, $P^T = P^T_{P1}(E + P^{T-1}_{P2} P^T_{P1})$.

$$\hat{\hat{x}}^T_{i,j} = \hat{x}^T_{P1i,j} + P^T(\hat{x}^T_{P2i,j} - \hat{x}^T_{P1i,j}), \quad \bar{i} \in G^T_1 \qquad (3.20)$$

For illustration in Fig. 3.14 fragments of the original image are presented: local cloud cover (Fig. 3.14a), the restored image (Fig. 3.14b); graph characterizing filtering/restoration error variance as a result of using two-stage restoration procedure (Fig. 3.14c).

A simple visual analysis shows high quality of restoration using the given method. Actually untrained observer cannot distinguish with certainty between real and restored images and determine the domain which was covered by clouds.

3.4 Detection of Objects Against the Background of Multispectral Satellite Images

In many applications, it is often desirable to detect anomalies that may appear on a signal, separate image or on recurrent image of frame sequence [17, 25]. For example, these anomalies may be forest fires, pathological changes in medical images, new objects in a security area, etc. The synthesis of algorithms for detecting objects against the background of spatially heterogeneous images is discussed in Sect. 3.4.1. The research of these algorithms is performed in Sect. 3.4.2.

(a)   **(b)**

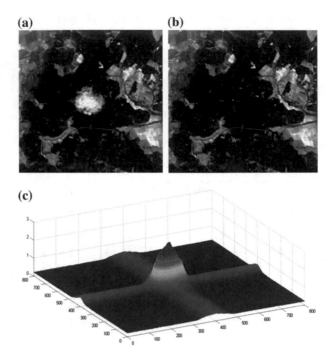

Fig. 3.14 MSI restoration: **a** original image, **b** image restored by using DS filter, **c** restoration error

3.4.1 Synthesis of Object Detection Algorithms

Let us consider the problem of certain object detection against the background of MSI. In doing so, we suppose that the object is known to be absent on $(T - 1)$ th MSI in time sequence but it might be present on T th image. We will also suppose that the form and structure of the object to be detected is known and it is defined by the samples $f_{i,j}^k$, where (i, j) are the spatial coordinates and k is the number of spectral band but its intensity value in each of the spectral bands s^k, $k = 1, \ldots, N$, actual object rotation angle φ, its scale coefficient μ and geometric centre displacement vector $\overline{\Delta} = (\Delta_x, \Delta_y)^T$ are assumed to be unknown. For this case, the observation model with presence of the desired signal (hypothesis H_1) can be written as Eq. 3.21, where $f_{i,j}^k$ are the samples defining the form and structure of the object to be detected in kth spectral zone, $\tilde{G}_0^{k,T} = F(G_0^{k,T}, \Delta_x, \Delta_y, \varphi, \mu)$ is the domain occupied by the reference object when it is displaced by Δ_x and Δ_y in spatial coordinates, rotated by the angle φ or its scale is changed by a factor of μ, $F(\cdot)$ is the affine transform of coordinates.

$$z_{i,j}^{k,T} = x_{i,j}^{k,T} + F(s^k f_{i,j}^k, \Delta_x, \Delta_y, \phi, \mu) + \theta_{i,j}^{k,T}, \ k = 1, 2, \ldots, N, \ (i,j) \in G_{F0}^{k,T}$$

$$z_{i,j}^{k,T} = x_{i,j}^{k,T} + \theta_{i,j}^{k,T}, \ k = 1, 2, \ldots, N, \ (i,j) \notin G_0^{k,T}, \tag{3.21}$$

$$z_{i,j}^{k,t} = x_{i,j}^{k,t} + \theta_{i,j}^{k,t}, \ k = 1, 2, \ldots, N, \ t = 1, \ldots, T-1$$

In the case when the desired signal is absent (hypothesis H_0), the observation model assumes the form:

$$z_{i,j}^{k,t} = x_{i,j}^{k,t} + \theta_{i,j}^{k,t}, \ k = 1, 2, \ldots, N, \ t = 1, \ldots, T.$$

Then using the modified likelihood ratio method, we can write the following decision rule:

$$L = \frac{\max\limits_{\Delta_x, \Delta_y, \alpha, \mu, s^k} \omega(\{z_{i,j}^k\}, (i,j,k) \in F(G_0^{k,T}, \Delta_x, \Delta_y, \alpha, \mu)|H_1)}{\max\limits_{\Delta_x, \Delta_y, \alpha, \mu, s^k} \omega(\{z_{i,j}^k\}, (i,j,k) \in F(G_0^{k,T}, \Delta_x, \Delta_y, \alpha, \mu)|H_0)} \begin{cases} > L_0 - signal, \\ < L_0 - no\ signal. \end{cases}$$

Let $M_1 = M\{z_{i,j}^{k,t}/Z_0, H_1, s^k\} = S + \hat{X}_P$ and $M_0 = \hat{X}_P = M\{X/Z_0\}$ be an optimal predictions of random field at time T, providing the minimum of error variance. The predictions are made on the basis of all observations Z_0, in which there is definitely not a useful signal when hypotheses H_0 and H_1 are right. It is called a prediction in area $F(G_0^{k,T}, \Delta_x, \Delta_y, \phi, \mu)$. And let $V = M\{(Z - M_{0,1})(Z - M_{0,1})^T\} = P_P + V_\Theta$, where $P_P = M\{(X - \hat{X}_P)(X - \hat{X}_P)^T\}$, V_Θ is thw noise covariance matrix, be an error covariance matrix for optimal prediction. Then the conditional probability density distributions $\omega(\{z_{i,j}^k\}, (i,j,k) \in F(G_0^{k,T}, \Delta_x, \Delta_y, \alpha, \mu)|H_{0,1})$ can be approximated by Gaussian distributions:

$$\omega(\{z_{i,j}^k\}, (i,j,k) \in F(G_0^{k,T}, \Delta_x, \Delta_y, \alpha, \mu)|H_{0,1})$$

$$\cong \frac{1}{(2\pi)^{M/2}\sqrt{\det V}} \exp\left\{-\frac{1}{2}(Z - M_{0,1})V^{-1}(Z - M_{0,1})^T\right\}.$$

After elementary transformations, taking into account the fact that $P(E + V_\theta^{-1}P_P) = P_P$, we obtain the detection algorithm given by Eq. 3.22, where $\tilde{f}_{i,j}^k = F(f_{i,j}^k, \Delta_x, \Delta_y, \alpha, \mu)$, $\Delta_{i,j}^k = \sum\limits_{k=1}^{N}\sum\limits_{i,j \in \tilde{G}_k} V_{\theta i,j}^{k-1}(z_{i,j}^k - x_{i,j}^k)$ is the filtering error at the point (i,j) for the k th spectral band which is normalized on the noise $\theta_{i,j}^k$ variance.

$$L = \max\limits_{\Delta_x, \Delta_y, \alpha, \mu}\left(\sum\limits_{k=1}^{N}\sum\limits_{i,j \in \tilde{G}_k} f_{i,j}^k s^k \tilde{f}_{i,j}^k \Delta_{i,j}^k\right)\begin{cases} > L_0 - signal, \\ < L_0 - no\ signal \end{cases} \tag{3.22}$$

At the same time the levels $\hat{s}^k$ can be determined from the system of linear equations:

$$\sum_{k=1}^{N} \sum_{i,j \in \bar{G}_k} \tilde{f}_{i,j}^t \sum_{l,v \in \bar{G}_k} V_{\theta i,j,l,v}^{-1} \tilde{f}_{l,v}^k = \sum_{k=1}^{N} \sum_{i,j \in \bar{G}_k} \tilde{f}_{i,j}^t \sum_{l,v \in \bar{G}_k} V_{\theta i,j,l,v}^{-1} (z_{ij}^k - \hat{x}_{ij}^k), t = 1, 2, \ldots, N.$$

An important feature of the decision rule (Eq. 3.22) is in the fact that the object detection process when used can be represented in two stages. The first stage is concerned with performing of MSI sequence filtering and obtaining the estimates $\Delta_{i,j}^k$. At the second stage, a derivation of the statistic (Eq. 3.22) and determination of the parameters $\bar{\alpha} = \{\Delta_x, \Delta_y, \alpha, \mu\}$ values, at which this statistic assumes the largest values, is carried out.

3.4.2   Analysis of Object Detection Algorithms

In view of spatial heterogeneity of real satellite images at the first stage, it is reasonable to use the above-described DS filtering procedures. In Fig. 3.15, the examples of noisy images containing point objects, results of DS filtering and correct detection probabilities obtained using an algorithm based on DS model and an algorithm based on AR models provided that all the object parameters are known a priori are shown.

An analysis of correct detection probabilities shows that the filtering algorithm based on DS models in most cases provides better detection quality than the detector based on AR models. This is due to the fact that DS filter enables to estimate the model parameters change process. Thus, for the estimated value of the interior correlation parameter $r = 0.8$ for an image fragment we obtain an average gain equal to 24% using DS model, at $r = 0.99$ the gain reduces to 13%. It is interesting to note that even at the values of r very close to unit the signal detection effectiveness is higher, when the algorithms based on DS models are used.

Within the framework of the second stage the detection problem can be interpreted as the problem of object image identification by a template, which in its turn can be reduced to search of a spatial transform which minimizes the distance between the sought image and the template in a specified metric space. One method, which implements such approach is the PseudoGradient Identification Method (PGIM) wherein the identification parameters $\bar{\alpha}$, are sought in recurrent manner at constant template position.

For illustration and analysis of the presented algorithm we provide a satellite image fragment Fig. 3.16 of Volga river basin obtained by means of spacecraft Landsat 8 in visible spectral band (channel 2) and the difference between this fragment and the result of joint DS filtering of these observations and two preceding

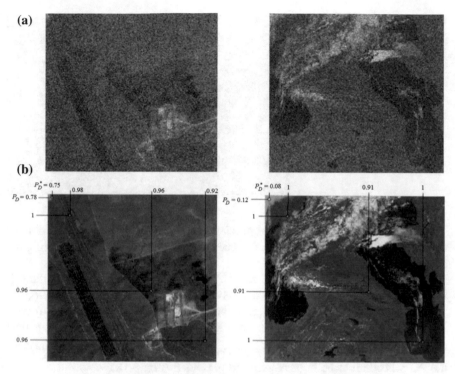

Fig. 3.15 Examples of noisy and filtered images: **a** noisy images, **b** filtered images using an algorithm based on DS model (left) and an algorithm based on AR models (right)

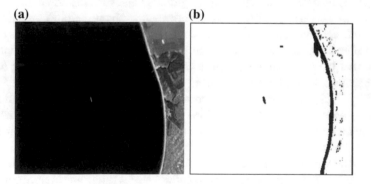

Fig. 3.16 An example of detecting: **a** satellite image fragment of Volga river basin, **b** corresponding artificial image

in time multispectral images Fig. 3.16b. For display convenience, the image in Fig. 3.16b was put through the procedure of white noise suppression and histogram stretching.

Fig. 3.17 The domains
separated against the
background of the original
snapshot

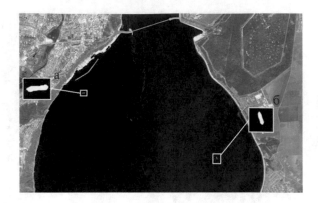

In Fig. 3.17, a separate MSI artificial frame obtained by superposition of domains separated at the stage of spectral mismatch analysis on the original snapshot is presented.

Identification of the objects "a" and "b" by PGIM method resulted in the following values: the object "a" is a "barge-type" floating means (correlation 95,9%) with azimuth –17°, the object "b" is a "bulk carrier-type" floating means (correlation 91,4%) with azimuth 74°. In Fig. 3.3, dependencies of mathematical expectation estimates $m_{\Delta z} = \frac{1}{mes \tilde{G}_{F0}^{k,T}} \sum_{i,j,k \in \tilde{G}_0^{k,T}} \left(\tilde{z}_{i,j}^k - \hat{s} \tilde{f}_{i,j}^k \right)$ versus iteration number of the PG procedure to be implemented are presented (Fig. 3.18).

To estimate the quantitative characteristics of the proposed algorithm effectiveness let us consider certain situations, which might occur when the object "a" is detected. In the first situation, we suppose that all parameters $\bar{\alpha} = \{\Delta_x, \Delta_y, \varphi, \mu\}$ of the object to be detected except for its brightness values s^k are known. In the second situation, we suppose that the available information about the object rotation angle φ is not correct. For definiteness, we assume that the true value of φ and its estimate in use differ from each other by 90°. In the third situation, among other things we assume that the positioning data are incorrect, namely, the estimates Δ_x, Δ_y differ from their true values by 3 pixels each.

In Fig. 3.19, dependencies of correct detection probability versus object brightness average coefficient for all MSI frames for the above-mentioned situations using the algorithm (Eq. 3.22) are presented. In all cases the false alarm probability is $P_F = 0.0001$.

The obtained results indicate similarity of characteristics of the synthesized detector in Eq. 3.22, operating in the conditions of a priori uncertainty concerning parameters vector $\bar{\alpha} = \{\Delta_x, \Delta_y, \varphi, \mu\}$ and the optimal detector in situation when the information on the parameters $\bar{\alpha}$ is a priori unknown. In the case when a part of this information appears to be unknown or incorrect the algorithm [6] is preferable. Thus, at correct detection probability $P_d = 0.5$ in the case of incorrect information on rotation angle and object center the detector [6] produces gain in desired signal value of about 73%.

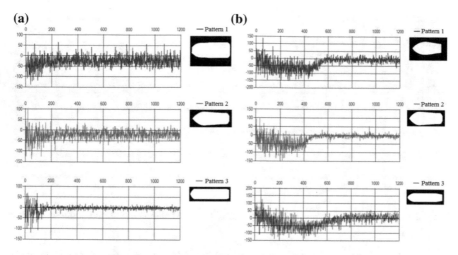

Fig. 3.18 The detected domains against the background of the original snapshot (identification of the corresponding objects): **a** "barge", **b** "bulk carrier"

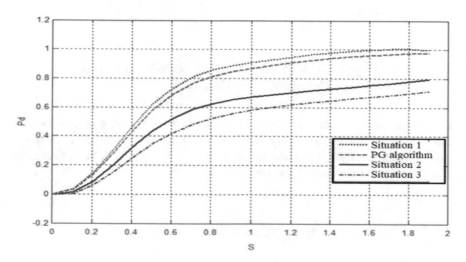

Fig. 3.19 Comparison of detection algorithms effectiveness for various situations

By means of statistical modeling numerical calculations of effectiveness of extended anomalies detection algorithm against the background of MSI sequence have been carried out. In Fig. 3.20, dependencies of correct detection probabilities P_d versus average correlation interval $\tau = 1/(1 - r)$ in the neighborhood of an anomaly and number of preceding images are presented.

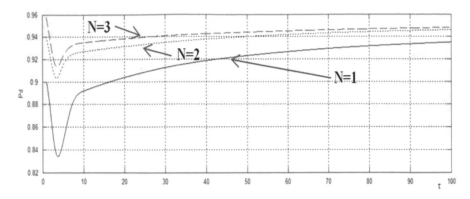

Fig. 3.20 Dependence of correct detection probability on correlation radius and number of frames

It can be concluded that increase in number of frames in time sequence leads to increase of correct detection probability. So, it seems to be possible to decide upon the necessary number of the observed scene frames ensuring detection effectiveness specified characteristics.

3.5 Image-Type Related Mapping of Satellite Images and Objects Condition Prediction

The above algorithms can be directly applied to the processing of real satellite material. In Sect. 3.5.1, a doubly stochastic filtering for classification of multi-spectral satellite images is considered. Section 3.5.2 is devoted to the construction of assessments of the state of dynamic natural objects.

3.5.1 Thematic Mapping of Satellite Images

Implementation of the satellite surveying systems requires carrying out preliminary statistical analysis of the incoming satellite data. During the course of this analysis estimation of satellite image quality is carried out, its reference to geographical coordinates is performed, MSI segmentation and identification of objects found on it are accomplished. For this purpose, we can use a modification of convolutional neural network UNET with Fully-Connected Layers (FCN) [2, 3], in which the network input layer (consisting by default of spectral layers of a multispectral image) is extended by several auxiliary grayscale images obtained from the original image by means of the NDVI, EVI and SAVI transforms, the estimates obtained through DS filtering, and two two-dimensional arrays constituting results of segmentation of the given territory at the preceding time instant one year ago.

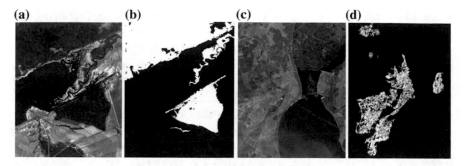

Fig. 3.21 The results of satellite multispectral images automatic classification: **a, c** original images, **b, d** processing results

In Fig. 3.21, the results of automatic classification of multispectral images obtained at various periods of 2017 are presented. Thus, in Figs. 3.21a–3.21b, the forest tract in Melekess district of the Ulyanovsk region to be outlined is presented. Figures 3.21c–3.21d depict the city of Ulyanovsk with suburbs. The network learning itself had been carried out over a period of 50 epochs. In the process of learning each epoch was trained on 500 batches. Each batch in its turn was formed from 128 random fragments picked from a random part of the original snapshot of size 128×128.

Quality analysis for such an algorithm has shown accuracy of processing comparable with the results of a qualified operator performance.

3.5.2   Monitoring of Natural Objects

We should note that availability of the results of satellite material image-type related processing enables to answer questions concerned with Earth surface state features at preceding and current time instants. However, the instruments enabling to carry out state prediction for one or the other objects and produce recommendations for responsible persons are a necessary element of Earth Remote Sensing (ERS) processing system. As an example, let us consider the problem concerned with forming of prediction regarding possibility of hazardous approach of various natural and man-made objects. Such approach can be exemplified by a gradual forest invasion of a territory along roads or electric power transmission lines or avalanche processes bringing about damage to diverse infrastructure.

Let us formulate the given problem as follows. Let there be an aggregate of segments describing a certain extended object T_o (a road, an electric power transmission line etc.). The given aggregate is usually described by a vector object having geographical reference in absolute coordinates. Let us separate on this extended object a set of points $T_{oi} = (x_{0i}, y_{oi})$ having the distance between them equal to Δ_o. Let us assume that at a certain time instant t next to the object there lies a certain extended domain G_R^t defined by a set of points each corresponding to a pixel in the original halftone image. For each point T_{oi} let us construct a perpendicular to the segment $[T_{oi}, T_{oi+1}]$. Let us find the point $T_{Ei}^t = (z_{Exi}^t, z_{Eyi}^t)$ of intersection of this perpendicular and the domain G_R^t. Obviously the set of points $\{T_{Ei}^t, i = 1, \ldots, N_o\}$ describe a conditional boundary of the domain G_R^t as viewed from the extended object. It enables to use estimates of the points coordinates $\{T_{Ei}^t, i = 1, \ldots, N_o\}$ obtained at different times $t = 1, \ldots, T$ as a source of information on the domain G_R^t dynamics for the purpose of forming a prediction about its boundaries at future times $t > T$ as well. In view of errors arising at recording and processing of satellite images, we get Eq. 3.23, where n_{Exi}^t and n_{Eyi}^t are the white noise samples with zero mean and variance σ_n^2.

$$z_{Exi}^t = x_{Ei}^t + n_{Exi}^t, z_{Eyi}^t = y_{Ei}^t + n_{Eyi}^t, i = 1, \ldots, N_o, t = 1, \ldots, T \quad (3.23)$$

Direct measurements based on the results of satellite material image-type related mapping show that MSE $\sqrt{\sigma_n^2}$ is usually approximately equal to $1.5 D_{xy}$, where D_{xy} is resolution of the original images.

Let us suppose that the boundary of the domain G_R^t can move non-uniformly. Thus, for example, the area of a precipice or a ravine can increase by tens of centimeters per each year and at a certain moment it might acquire a dynamics of hundreds of times more than it was before. Then we will use DS model to describe unknown coordinates in the following form Eq. 3.24. In this case, r_{ax}, r_{ay} are the scalar parameters determining change potential for the accelerations a_{Exi}^t and a_{Eyi}^t, ξ_{axi}^t, ξ_{ayi}^t are the independent normal random variables with zero mean and variance σ_ξ^2:

$$\begin{aligned}
x_{Ei}^t &= 2x_{Ei}^{t-1} - x_{Ei}^{t-2} + a_{Exi}^t(x_{Ei}^{t-1} - x_{Ei}^{t-2}), \\
y_{Ei}^t &= 2y_{Ei}^{t-1} - y_{Ei}^{t-2} + a_{Eyi}^t(y_{Ei}^{t-1} - y_{Ei}^{t-2}), \\
a_{Exi}^t &= r_{ax}a_{Exi}^{t-1} + \xi_{axi}^t, \ a_{Eyi}^t = r_{ay}a_{Eyi}^{t-1} + \xi_{ayi}^t.
\end{aligned} \quad (3.24)$$

This model can be rewritten in the form of Eq. 3.25:

$$\bar{X}_{Ei}^t = \varphi_{Exi}^t(\bar{X}_{Ei}^{t-1}) + \bar{\xi}_{xi}^t, \ \bar{Y}_{Ei}^t = \varphi_{Eyi}^t(\bar{X}_{Ei}^{t-1}) + \bar{\xi}_{yi}^t, \quad (3.25)$$

where

$$\varphi_{Exi}^t(\bar{X}_{Ei}^{t-1}) = \begin{pmatrix} x_{Ei}^{t-1} & v_{Exi}^t & 0 \\ 0 & v_{Exi}^{t-1}(1 + a_{Exi}^t) & 0 \\ 0 & 0 & a_{Exi}^{t-1} r_{ax} \end{pmatrix},$$

$$\varphi^t_{Eyi}(\bar{Y}^{t-1}_{Ei}) = \begin{pmatrix} y^{t-1}_{Ei} & v^t_{Eyi} & 0 \\ 0 & v^{t-1}_{Eyi}(1+a^t_{Eyi}) & 0 \\ 0 & 0 & a^{t-1}_{Eyi}r_{ay} \end{pmatrix}, \bar{\xi}^t_{xi} = \begin{pmatrix} 0 \\ 0 \\ \xi^t_{axi} \end{pmatrix}, \bar{\xi}^t_{xi} = \begin{pmatrix} 0 \\ 0 \\ \xi^t_{ayi} \end{pmatrix}.$$

The assigned notations enable to apply DS nonlinear filtering to observations and for construction of predictions of behaviour of the region R_0 in reference to the object T_o. In doing so, let us introduce $\bar{X}^t_{\ni Ei} = \varphi^t_{Exi}(\bar{X}^{t-1}_{Ei})$ and $\bar{Y}^t_{Ei} = \varphi^t_{Eyi}(\bar{X}^{t-1}_{Ei})$ are the extrapolated predictions for the point T_{Ei} coordinates at the time point t based on the preceding observations z^{t-1}_{Exi} and z^{t-1}_{Eyi}. Denote by P^{t-1}_{xi}, P^{t-1}_{yi} are the filtering error covariance matrices at the time point $(t-1)$, $V^t_{x_\xi i} = M\{\bar{\xi}^t_{xi}\bar{\xi}^{tT}_{xi}\}$, $V^t_{y_\xi i} = M\{\bar{\xi}^t_{yi}\bar{\xi}^{tT}_{yi}\}$ are the diagonal covariance matrices for random increments $\bar{\xi}^t_{xi}$. Then error covariance matrices for such extrapolation have the following form:

$$P^t_{\ni xi} = M\{(\bar{X}^t_{\ni Ei} - \bar{X}^t_{\ni Ei})(\bar{X}^t_{\ni Ei} - \bar{X}^t_{\ni Ei})^T\} = \varphi^t_{Exi}{}'(\bar{X}^{t-1}_{Ei})P^{t-1}_x\varphi^t_{Exi}{}'(\bar{X}^{t-1}_{Ei})^T + V^t_{x_\xi i},$$

$$P^t_{\ni yi} = M\{(\bar{Y}^t_{\ni Ei} - \bar{Y}^t_{\ni Ei})(\bar{Y}^t_{\ni Ei} - \bar{Y}^t_{\ni Ei})^T\} = \varphi^t_{Eyi}{}'(\bar{Y}^{t-1}_{Ei})P^{t-1}_{yi}\varphi^t_{Eyi}{}'(\bar{Y}^{t-1}_{Ei})^T + V^t_{y_\xi i}.$$

If we denote by $\hat{x}^t_{PEi}$, $\hat{y}^t_{PEi}$ are the first elements of the vectors $\bar{X}^t_{PEi}$ and $\bar{Y}^t_{PEi}$, $B^t_{xi} = P^t_{Pxi}C^T_xD^{-1}_{xi}$, $B^t_{yi} = P^t_{Pyi}C^T_yD^{-1}_{yi}$, $D^t_{xi} = C_xP^t_{Pxi}C^T_x + \sigma^2_n$, $D^t_{yi} = C_yP^t_{Pyi}C^T_y + \sigma^2_n$, $C_x = C_y = (1 \quad 0 \quad 0)$, then we can write the Eq. 3.26 for DS coordinate filters.

$$\bar{X}^t_{Ei} = \bar{X}^t_{PEi} + B^t_{xi}(z^t_{Exi} - \hat{x}^t_{PEi}), \bar{Y}^t_{Ei} = \bar{Y}^t_{PEi} + B^t_{yi}(z^t_{Eyi} - \hat{y}^t_{PEi}) \tag{3.26}$$

The filtering error variance at each step is determined by the matrices:
$P^t_{xi} = (E - B^t_{xi})P^t_{Pxi}$, $P^t_{yi} = (E - B^t_{yi})P^t_{Pyi}$.

Special attention must be given to the fact that due to specificity of the problem stated, namely, necessity for the distance from the extended object T_o to the domain R_0 to be surveyed, it is possible to simplify the boundaries coordinates filtering process R_0 by processing only one coordinate, namely, the distance x^t_{ri} from the point T_{oi} belonging to T_o up to the boundary R_0 at the time point t. An aggregate of similar observations $z^t_{Ex,i}$ can be processed by a technique identical to the above-described one.

As an illustration of such a technique in Figs. 3.22–3.23, series satellite images fragments for the forest tract in Cherdakly district of the Ulyanovsk region for the period 2001–2017 (Fig. 3.22) and Milanovsky opencast colliery on riverbank of the Volga in the northern part of the city of Ulyanovsk for the period 2013–2017 (Fig. 3.23) are presented. Here for convenience of color image perception and its recovery overlapping of visible spectral bands and superposition of the segmented image fragment and normals to the object to be monitored is carried out. In the first case the number of multispectral images to be processed amounted to 42 snapshots,

(a)   **(b)**   **(c)**

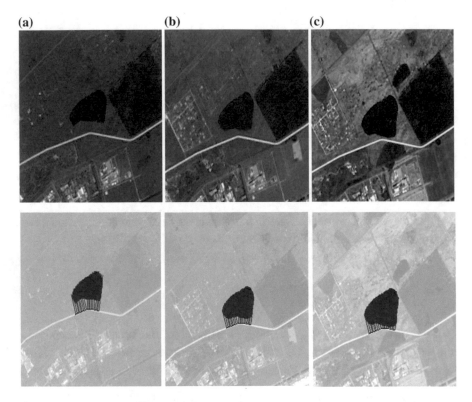

Fig. 3.22 Forest tract satellite images fragments and the results of these images processing: **a** July 2003, **b** June 2011, **c** July 2017

in the second case – 32 snapshots. The minimal time interval for satellite information production amounts to 14 days.

The above-mentioned groups of multispectral images were separated into the training and operating samples. The training samples were used to specify the filtering parameters, in particular, to estimate the parameters r_{ax} and σ_{ξ}^2. The operating part of the sample was processed by three algorithms enabling to carry out the distance prediction x_{ri}^{t+1} basing on the preceding observations. The first algorithm (I) involves constructing a simple prediction $\hat{x}_{ri}^{t+1} = 2z_{ri}^t - z_{ri}^t$ wherein only the variable $\hat{x}_{ri}$ change speed is taken into account for the time interval $(t-1, t)$. The second algorithm (II) assumes linear Kalman filtering of the observations z_{ri}^t and construction of the extrapolated prediction $\hat{x}_{Pri}^{t+1}$ based on the results of the processing. The third algorithm (III) is in the above-described doubly stochastic filtering of an aggregate of the observations z_{ri}^t and construction of the vector $\hat{\hat{x}}_{PEi}^{t+1}$. In Table 3.4, the values of prediction average errors depending on the object type are presented.

(a)   (b)   (c)

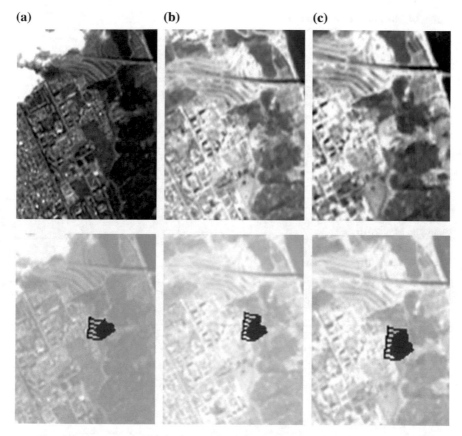

Fig. 3.23 Milanovsky opencast colliery satellite images fragments and the results of these images processing: **a** August 2015, **b** April 2016, **c** May 2016

On average DS filter provides prediction accuracy 6% higher than in case of using Kalman filter and 58% higher than in case of simple linear predictions. It enables to estimate coordinates and rate change dynamics for boundaries of the domain to be monitored by using DS filter. It is essential that the DS filter enables to quicker respond to abrupt rate change of the processes determining the object behaviour. As an illustration, we provide the estimates behaviour for the distance from opencast colliery to one of the points to be monitored (Fig. 3.24a) and the parameter estimate (Fig. 3.24b).

Direct analysis of the given results in comparison with the data of objective monitoring (solid line) indicates superiority of DS filter over conventional linear Kalman filter in filtering accuracy. As it takes place, this superiority makes itself evident in the most distinct manner in case of abrupt change of the rock collapsing process (and the corresponding reduction of the distance between the opencast colliery and the point to be monitored). This change corresponds to a significant

Table 3.4 Boundary estimation errors

	Average error for the algorithm I (m)	Average error for the algorithm II (m)	Average error for the algorithm III (m)
The forest tract snapshot. October 2014	6.7	2.7	2.6
The forest tract snapshot. May 2015	10.7	3.9	3.7
The forest tract snapshot. June 2016	6.2	3.6	3.3
Milanovsky opencast colliery snapshot. May 2014	7.1	3.8	3.6
Milanovsky opencast colliery snapshot. May 2015	7.3	3.9	3.8
Milanovsky opencast colliery snapshot. May 2016	6.9	3.9	3.7
Milanovsky opencast colliery snapshot. April 2016. Beginning of the avalanche processes	12.4	8.9	7.8
Milanovsky opencast colliery snapshot. April 2016. Continuation of the avalanche processes	30.7	32.8	12.6
Milanovsky opencast colliery snapshot. May 2016. Cessation of the avalanche processes	20.3	18.1	17.3

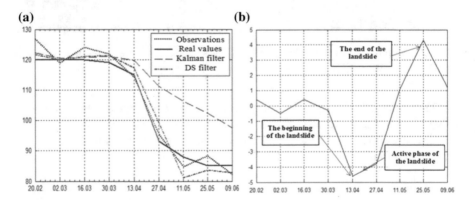

Fig. 3.24 Dependence of the moving point parameters filtering results versus survey time: **a** estimation of coordinates, **b** estimation of acceleration

change of the parameter a_i^t estimate, which enables to register considerable changes in the opencast colliery domain state basing only on this estimate dependence nature versus survey time.

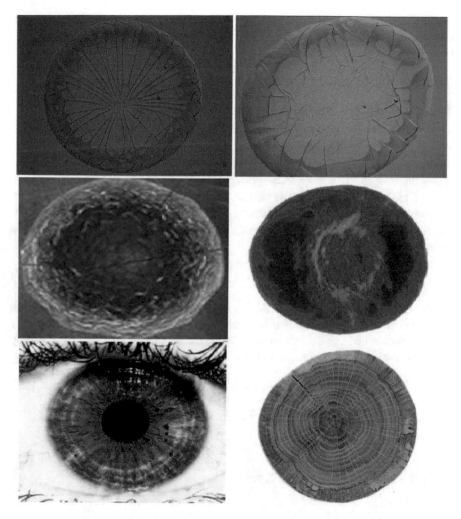

Fig. 3.25 Examples of images of radial-ring structure

3.6 Autoregressive Models of Images on a Circle

At present, the vast majority of known image models are varieties of random fields defined on rectangular flat grids or grids of higher dimension. Such models are discussed earlier in this chapter. In some practical situations, the processed images have the radial, circular or radial-circular structures. For example, images of a biological fluid, eye, biological cell, flower, slice of a tree trunk are shown in Fig. 3.25. These circumstances require their consideration in random field models. Thus, this section proposes autoregressive models of random fields defined on the cylinder and circle (Sects. 3.6.1–3.6.2). Section 3.3.6.3 presents doubly stochastic models of random fields on a circle. Filtration and identification are considered in Sect. 3.6.4.

3.6.1   Models of Random Fields on a Cylinder

We first consider the autoregressive model of a random field on a cylinder. It will be the basis for circular fields. In [6, 7], the model given by Eq. 3.27 was used to represent images on a cylindrical grid, where k is the spiral turn number, $l = 0, \ldots, T$ is the node number in the turn, $x_{k,l} = x_{k+1, l-T}$ when $l \geq T$, T is the period, i.e. the number of points in one turn, $\xi_{k,l}$ are the independent standard random variables. This model is similar to the Habibi model [33] of a flat field with a rectangular grid.

$$x_{k, l} = a\, x_{k, l-1} + b\, x_{k-1, l} - a\, b\, x_{k-1, l-1} + c\, \xi_{k,l} \tag{3.27}$$

For convenience of analyzing this model, we will assume that the pixels are numbered and located on a cylindrical spiral (Fig. 3.25a). Then model in Eq. 3.27 can be represented in an equivalent form as the model of a random process given by Eq. 3.28, which is a scan of the image along the spiral, where $n = kT + l$.

$$x_n = a\, x_{n-1} + b\, x_{n-T} - a\, b\, x_{n-T-1} + c\, \xi_n \tag{3.28}$$

The characteristic equations of model in Eq. 3.28 are the follows:

$$z^{T+1} - a\, z^T - b\, z + a\, b = 0 \text{ or } (z^T - b)\,(z - a) = 0.$$

Applying the z-transform, we have:

$$(z^T - b)\,(z - a)x_n = c\xi_n \quad \text{or} \quad x_n = \frac{c}{(z^T - b)\,(z - a)}\xi_n.$$

Therefore, CF $V(n) = M[x_i x_{i\pm n}]$ of random process $\{x_n, n = 0, 1, \ldots\}$ is expressed as follows:

$$V(n) = \frac{c^2}{2\pi i} \oint\limits_{|z|=1} \frac{z^{n-1} dz}{(z^T - b)(z - a)(z^{-t} - b)(z^{-1} - a)},$$

where integration is carried out along a unit complex circle. Using residues, we obtain Eq. 3.29, where $z_k = \sqrt[T]{b}\exp(i2\pi k/T)$ and $s = a^T$.

$$V(n) = c^2 \left(\frac{1}{(1-b^2)T} \sum_{k=0}^{T-1} \frac{z_k}{(1-az_k)(z_k-a)} z_k^n + \frac{s}{(1-a^2)(1-bs)(s-b)} \rho^n \right)$$

$$(3.29)$$

In particular, when $n = 0$ we obtain the variance of RF:

$$\sigma^2 = \frac{c^2(1+bs)}{(1-a^2)(1-b^2)(1-bs)}.$$

To reduce the calculations, it is possible to calculate only $V(0), V(1), \ldots, V(T)$ by Eq. 3.29, and for the rest of values, use recurrent formula:

$$V(n) = a\,V(n-1) + b\,V(n-T) - ab\,V(n-T-1).$$

This CF decreases with increasing distance n, but at distances divisible by period T, it is high (Fig. 3.26b). Figure 3.26c shows an example of a cylindrical image cut lengthwise and unfolded, simulated using model of Eq. 3.27.

Note that the described model is only the simplest case. By introducing additional terms into Eq. 3.27, one can obtain random fields with more complex CF.

3.6.2   Models of Random Fields on a Circle

The polar coordinate system (r, φ) is convenient for circular images representation. To do this, we will consider the turns of the cylindrical spiral in model Eq. 3.27 as turns of a circle grid showed in Fig. 3.27a. In other words, index k is converted into a polar radius, and index l into a polar angle. Thus, the value $x_{k,l}$ in the pixel (k, l) of the cylindrical image is converted to the same value in the pixel with coordinates $(k\Delta r, l\Delta\phi)$ of the circular image. When using model (Eq. 3.28), it is also convenient to use a spiral grid (Fig. 3.27b), similar to the cylindrical spiral in Fig. 3.26a. Note that this representation in the form of a spiral is made conditionally to

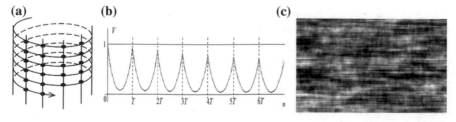

Fig. 3.26 Cylindrical image: **a** cylindrical grid, **b** view of the correlation function, **c** example of imitation

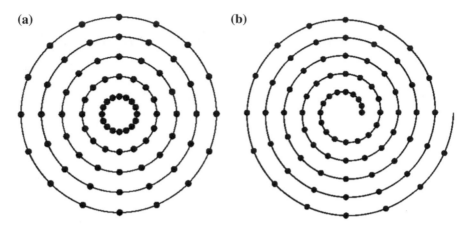

Fig. 3.27 Grids on a circle: **a** circular, **b** spiral

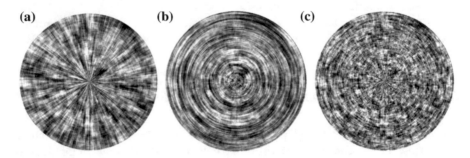

Fig. 3.28 Simulated images on a circle for various values of parameters in the model Eq. 3.27:
a $b = 0.99$, $a = 0.95$; **b** $a = 0.99$, $b = 0.95$; **c** $a = b = 0.95$

facilitate analysis. In the image, a sequence of turns of a conditional spiral is a sequence of expanding circles.

The parameters a and b of model represented by Eq. 3.27 set the degree of correlation in the radial and circular directions. When $a < b$, the image will have a higher correlation in the radial directions. In Fig. 3.28a, the simulated image is shown at $a = 0.95$ and $b = 0.99$. When $a > b$, the image will have a higher correlation in the circular direction. Figure 3.28b shows the simulated image with $a = 0.99$ and $b = 0.95$. In the case $a \approx b$, the image is approximately equally correlated in both directions. Figure 3.28c shows such a simulated image with $a = b = 0.95$.

The CF of the circular images is given by Eq. 3.29, since they are determined by model in Eq. 3.27. But this CF is for linear pixel numbering. Therefore, the resulting image is homogeneous only for this numbering. If we consider the

(a)   **(b)**

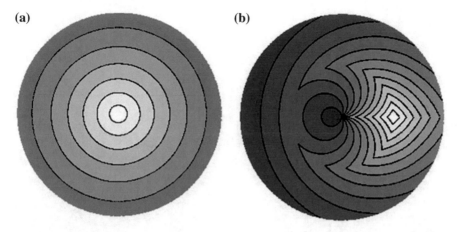

Fig. 3.29 Normalized correlation function of a circular image: **a** relative to the center, **b** relative to the middle of a radius

Euclidean distance, the image circular correlation weakens with distance from the center. And this is quite consistent with the specifics of the radial-ring images.

Let us consider the CF view of circular image. The circular image in this model is actually a geometric transformation of a cylindrical one, so its CF can be obtained from the CF of a cylindrical image. Figure 3.29a shows the values of CF relative to the central pixel of this figure, i.e. $V(m, n) = M[x_{0,0}x_{0+m,0+n}]$. The lighter areas correspond to larger CF values. The isocovariation lines $V(m, n) = Const$ are shown in black. It is natural that these lines are circles. Figure 3.29b shows the values of CF and its isolines relative to the pixel $(R/2,0)$ in the middle of a radius of the image, i.e. $V(m, n) = M[x_{R/2,0}x_{R/2+m,0+n}]$. Isolines at short distances are close to rhombuses, which is typical property of the Habibi model [33].

Note that by applying more complex types of autoregression, it is possible to obtain circular images with very different properties (that is, with covariance functions other than of Eq. 3.29 types).

3.6.3 Doubly Stochastic Models of Random Fields on a Circle

In this chapter, doubly stochastic models were used to represent heterogeneous images with random heterogeneities. In these models, some random field sets the parameters of the resulting random field.

Let $Y = \{y_n\}$ be the realization of a circular image obtained using the model in Eq. 3.27. We take Y as the control image, which forms the variable parameters of the controlled image $X = \{x_n\}$ in accordance with the model in Eq. 3.27, in which

(a)   **(b)**

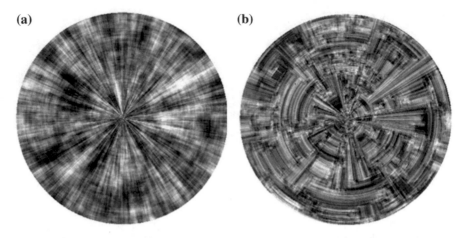

Fig. 3.30 Simulated images with control: **a** control image, **b** controlled image

instead of the constant a and b are used values obtained by autoregressions in Eq. 3.30.

$$a_{1,n+1} = r_1 a_{1,n} + \gamma_1 y_n, \quad b_{2,n+1} = r_2 b_{2,n} + \gamma_2 y_n \tag{3.30}$$

Figure 3.30 shows an example of such a simulation with $r_1 = r_2 = 0.999$, $\gamma_1 = \gamma_2 = 0.005$ and the control image of a predominantly radial structure.

In the described model, two images are unequal: one controls the parameters of the other. In [31], a model of autoregressive images defined on a cylinder was proposed, jointly controlling the parameters of each other. Let us apply this approach to circular images. Let image Y determines the parameters for the next pixel of image X, as in Eq. 3.30. At the same time, image X in the same way sets the parameters for the next element of the image Y. As a result, these two images together control the parameters of each other. Figure 3.31 shows an example of such images. Significant correlation of images is noticeable, which is a consequence of the mutual influence on the autoregressive parameters of their models.

Some of images in Fig. 3.23 have only approximately a radial-circular structure. To represent such images, random fluctuations of the polar radius and angle can be introduced. Let the image $X = \{x_{k,l}\}$ be simulated by the model in Eq. 3.27. Now, the value $x_{k,l}$ in the pixel (k, l) of X is converted to the same value in the pixel with coordinates $(K_k \Delta r, L_l \Delta \varphi)$ of the resulting image Y, where K_k and L_l are some random processes. Figure 3.32a shows an image with a random angle fluctuation K_k and usual regular radius $L_l = l$. In this case, the overall shape of the image is circular. In Fig. 3.32b, the radial coordinate is also random. As a result, the image is generally different from a circle.

(a)   **(b)**

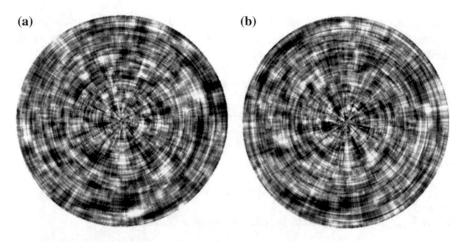

Fig. 3.31 Examples of simulated images with mutual influence

(a)   **(b)**

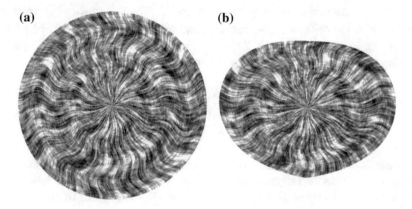

Fig. 3.32 Simulation of deformed images: **a** random angle, **b** random angle and radius

3.6.4 Filtration and Identification

For convenience, assume that the circular image is converted into a sequence. Let the observed image $Z = \{z_n\}$ be an additive mixture Eq. 3.31 of the informative image $X = \{x_n\}$ defined by the autoregressive model Eq. 3.28 and white Gaussian noise $\Theta = \{\theta_n\}$.

$$z_n = x_n + \theta_n \tag{3.31}$$

The parameters of the model Eq. 3.28 and noise dispersion σ_θ^2 in observation model Eq. 3.31 are unknown and, possibly, vary. In the latter case, this variation is assumed to be fairly smooth. It is required to evaluate the informative image

X using observations Z. To solve this problem, we apply adaptive pseudo-gradient analogue of Kalman filter [6] provided by Eqs. 3.32–3.33, where $\tilde{x}_n$ is the extrapolated estimate and $\hat{x}_n$ is the corrected estimate.

$$\hat{x}_n = \tilde{x}_n + s_n(z_n - \tilde{x}_n) = \tilde{x}_n + s_n\Delta_n \tag{3.32}$$

$$\tilde{x}_n = \rho_n\hat{x}_{n-1} + r_n\hat{x}_{n-T} - \rho_n r_n\hat{x}_{n-T-1} \tag{3.33}$$

Variable parameter vector $\bar{\alpha}_n = (\rho_n, r_n, s_n)$ of this algorithm is calculated using a pseudo-gradient procedure Eq. 3.34:

$$\bar{\alpha}_{n+1} = \bar{\alpha}_n - \mu_n\bar{\beta}_n, \tag{3.34}$$

where $\bar{\alpha}_{n+1}$ is the next vector approximation after $\bar{\alpha}_n$ and

$$\bar{\beta}_n = \nabla[\Delta_n^2] = \nabla[(z_n - \tilde{x}_n)^2] \tag{3.35}$$

is the pseudo-gradient [33] of the quality functional $J(\bar{\alpha}_n) = M[\Delta_n^2] = M[(z_n - \tilde{x}_n)^2]$, that is, a random vector, the mean of which makes an acute angle with the gradient $\nabla J(\bar{\alpha}_n)$, μ_n is the sequence of positive coefficients affecting values of the steps of the procedure in Eq. 3.34. Thus,

$$\begin{aligned}
\rho_{n+1} &= \rho_n + h_n\Delta_n(\hat{x}_n - r_n\hat{x}_{n-T-1}), \\
r_{n+1} &= r_n + h_n\Delta_n(\hat{x}_{n-T} - \rho_n\hat{x}_{n-T-1}), \\
s_{n+1} &= s_n + h_n\Delta_n\Delta_{n-1}.
\end{aligned} \tag{3.36}$$

Figure 3.33 shows examples of the application of the described algorithm. Figure 3.31a shows the images simulated using model Eq. 3.27. These images were distorted by white Gaussian noise with variance $\sigma_\theta^2 = 1$ (Fig. 3.31b). The result of the filtering is shown in Fig. 3.31c. In both examples, as a result of processing the noise/signal ratio decreased by about 16 times.

Note that during the filtering process the image model is also identified, since the procedures in Eq. 3.36 actually estimate the parameters of model Eq. 3.28. A simpler identification of the model can be obtained by directly estimating the coefficients of radial and circular correlation. These coefficients characterize the structure of the image. From them it is possible to determine which correlation prevails, radial or circular. This can be used in various applications.

Let us consider, for example, images of the facies of biological fluids. The first image in Fig. 3.23 shows the facies of a healthy person (physiological type of facies). It has pronounced radial cracks (dark lines). Therefore, in Eq. 3.28 corresponding to this image, the radial correlation coefficient must be large. The second image in Fig. 3.23 shows the facies of a sick person. Here, the radial structure is particularly badly damaged. Therefore, it can be expected that the radial correlation coefficient will be small. Indeed, for the first image, estimates $a = 0.71$ and

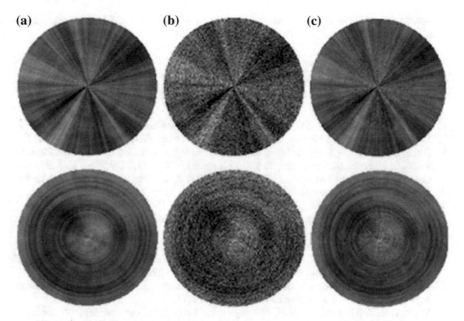

Fig. 3.33 Filtration of circular images: **a** informative images, **b** noisy images, **c** filtered images

$b = 0.89$ were obtained, and for the second image, estimates $a = 0.28$ and $b = 0.97$ were obtained. Thus, estimates of the parameters of the facies model provide an opportunity to make a conclusion about the general condition of the patient.

3.7 Conclusions

In this chapter, new approaches to solving problems of modeling and statistical analysis of multidimensional data sequences, which can be presented in the form of varying images, are considered. Description of images in the form of random fields, which are specified on an integer two-dimensional or multidimensional grid and on a circle, is taken as the basis. The synthesis and analysis of doubly stochastic model, which is a stochastic combination of autoregressive models, is performed. It is shown that the doubly stochastic model enables to describe spatially heterogeneous images and to form on its basis various algorithms for processing of these images. On the basis of the developed mathematical apparatus, a family of algorithms for estimation, filtering, and segmentation of multispectral satellite images and their time sequences is obtained. Their effectiveness and computational complexity are analyzed. Special attention is paid to adaptive algorithms of image processing, when there is uncertainty in the description of the data to be processed.

Acknowledgements The reported study was funded by the Russian Fund for Basic Researches according to the research projects № 16-41-732041 and № 18-47-730009.

References

1. Gonzalez R.C., Woods R.E.: Digital Image Processing, 4th edn, Pearson Education (2017)
2. LeCun, Y., Bengio, Y., Hinton, G.E.: Deep learning. Nature **521**(7553), 436–444 (2015)
3. Zeiler, M.D., Fergus, R.: Visualizing and understanding convolution network. In: Fleet, D., Pajdla, T., Schiele, B., Tuytelaars, T. (eds.) Computer Vision – ECCV 2014: 13th European Conference, Zurich, Switzerland, Proceedings, Part I, LNCS, vol. 8689, pp. 818–833. Springer International Publishing, NY (2014)
4. Bouman C.A.: Model Based Imaging Processing. Purdue University (2013)
5. Dudgeon, D.E., Mersereau, R.M.: Multidimensional Digital Signal Processing. Signal Processing Series. Prentice-Hall, Englewood Cliffs, New York (1984)
6. Krasheninnikov, V.R., Vasil'ev, K.K.: Multidimensional image models and processing. In: Favorskaya, M., Jain, L.C. (eds.) Computer Vision in Control Systems-3, ISRL, vol. 135, pp. 11–64. Springer International Publishing, Switzerland (2018)
7. Vasil'ev, K.K., Krasheninnikov, V.R.: Statistical analysis of multidimensional images sequences. Moscow: Radiotechnics (in Russian) (2017)
8. Vasil'ev, K.K., Dement'ev, V.E., Andriyanov, N.A.: Representation and processing of multispectral satellite images and sequences. Procedia Comput. Sci. **126**, 49–58 (2018)
9. Vasil'ev, K.K., Dement'ev, V.E., Luchkov, N.V.: Analysis of efficiency of detecting extended signals on multidimensional grids. Pattern Recognit. Image Anal. **22**(2), 400–408 (2012)
10. Shalygin, A.S., Palagin, Y.I.: Applied Methods of Statistical Modeling. Mashinostroenie, Mechanical Engineering Leningrad (1986)
11. Woods, J.W.: Two-dimensional Kalman filtering. In: Huang, T.S. (ed.) Two-Dimensional Digital Signal Processing I: Linear Filters. TAP, vol. 42, pp. 155–205. Springer, Berlin, Heidelberg, New York (1981)
12. Soifer, V.A. (ed.): Computer Image processing. Part I: Basic Concepts and Theory. VDM Verlag Dr. Muller E.K. (2009)
13. Jensen, J.: Introductory Digital Image Processing: A Remote Sensing Perspective. 4th edn. Pearson Education (2001)
14. Paliy, D., Foi, A., Bilcu, R., Katkovnik, V.: Denoising and interpolation of noisy Bayer data with adaptive cross-color filters. In: SPIE-IS&T Electronic Imaging (2007)
15. Zharrg, X.-P., Desai, M. D.: Adaptive denoising based on SURE risk. EEE Signal Process. Lett. **10**, 265–26 (1998)
16. Belyaeva O.V., Paschenko O.B., Filippov M.V.: Fast bilateral filtering of aerial photographs based on parallel decomposition into spatial filters. Math. Softw. Provis. Comput., Comput. Syst. Netw. **94**, 1–8 (in Russian) (2017)
17. Favorskaya, M.N., Levtin, K.: Early smoke detection in outdoor space by spatio-temporal clustering using a single video camera. In: Tweedale, J.W., Jain, L.C. (eds.) Recent Advances in Knowledge-based Paradigms and Applications. AISC, vol. 234, pp. 43–56. Springer International Publishing Switzerland (2014)
18. Jähne, B.: Digital Image Processing, 6th edn. Springer, Berlin, Heidelberg (2005)
19. York Pratt, W.K.: Digital Image Processing. PIKS inside. 3rd edn. Wiley, New York (2001)
20. Favorskaya, M., Jain, L.C., Buryachenko, V.: Digital video stabilization in static and dynamic scenes. In: Favorskaya, M.N., Jain, L.C. (eds.) Computer Vision in Control Systems-1, ISRL, vol. 73, pp. 261–309. Springer International Publishing, Switzerland (2015)
21. Krasheninnikov, V.R.: Correlation analysis and synthesis of random field wave models. Pattern Recognit. Image Anal. **25**(1), 41–46 (2015)
22. Gimel'farb, G.L.: Image Textures And Gibbs Random Fields. Springer, Berlin (1999)

23. Favorskaya, M.N., Jain, L.C., Savchina, E.I.: Perceptually tuned watermarking using non-subsampled shearlet transform. In: Favorskaya, M.N., Jain, L.C. (eds.) Computer Vision in Control Systems-4, ISRL, vol. 136, pp. 41–71. Springer International Publishing, Switzerland (2018)
24. Serra, J. (ed.): Image Analysis and Mathematical Morphology, vol. 2: Theoretical Advances. Academic Press, London (1988)
25. Vizilter, Y.V., Pyt'ev, Y.P., Chulichkov, A.I., Mestetskiy, L.M.: Morphological image analysis for computer vision applications. In: Favorskaya, M.N., Jain, L.C. (eds.) Computer Vision in Control Systems-1, ISRL, vol. 73, pp. 9–58. Springer International Publishing, Switzerland (2015)
26. Vasil'ev, K.K., Dement'ev, V.E., Andriyanov, N.A.: Doubly stochastic models of images. Pattern Recognit. Image Anal. 25(1), 105–110 (2015)
27. Vasil'ev, K.K., Dement'ev, V.E., Andriyanov, N.A.: Application of mixed models for solving the problem on restoring and estimating image parameters. Pattern Recognit. Image Anal. 26 (1), 240–247 (2016)
28. Vasil'ev, K.K., Dement'ev, V.E., Andriyanov, N.A.: Filtration and restoration of satellite images using doubly stochastic random fields. CEUR Work. Proc. 1814, 10–20 (2018)
29. Vasil'ev, K.K., Dement'ev, V.E., Andriyanov, N.A.: Development of the filtering algorithm for doubly stochastic images based on models with multiple roots of characteristic equations. In: International Conference on Pattern Recognition and Artificial Intelligence, pp. 643–647 (2018)
30. Dement'ev, V.E., Andriyanov, N.A.: Application of mixed models of random fields for the segmentation of satellite images. CEUR Work. Proc. 2210, 219–226 (2018)
31. Krasheninnikov, V.R., Subbotin, A.U.: Doubly stochastic models of cylindrical images. Radiotechnics 6, 5–8 (2018) (in Russian) (2018)
32. Krasheninnikov, V.R., Kuvayskova, Yu.E., Subbotin, A.U.: Autoregressive models of random fields on the circle. In: V International Conference on Information Technology and Nanotechnology, pp. 150–156 (in Russian) (2019)
33. Habibi, A.: Two-dimensional Bayesian estimate of images. Proc. IEEE 60(7), 878–883 (1972)

Chapter 4
Matrix Approach to Solution of the Inverse Problems for Multimedia Wireless Communication Links

Mikhail Sergeev and Nathan Blaunstein

Abstract The goal of this chapter is the analysis of a principal difference between the direct and inverse problems related to the applied electrodynamics, and radio, acoustic and optical wave physics, as well as, the relations of direct and inverse problems with each other. We use the concept of matrix or tensor equations technique for direct problem derivation, which deals with finding of the wave field for known distribution of sources of wave radiation. The same matrix approach is presented for the inverse problem resolving, which deals with determination and localization of the radiated sources' distribution a few limited cases of canonical objects and media. The strict analytical solution of such problems is obtained only for special cases of sources distribution and signal reconstruction. New methodology of how to identify the shape, form, structure, and type of material and define the parameters of media via knowledge on the wave field spatial, temporal, and spectral distribution arrived at the wave receiver and recorded by the detector, localizing any source and eliminating image speeding, has been created.

Keywords Direct problem · Inverse problem · Hadamard principal ·
Tikhonov's approach · Least squares method · Pseudo-inverse matrix
singular-value decomposition · Moore-Penrose matrix · Singular regularization ·
Levenberg–marquardt algorithm · Iteration algorithm

M. Sergeev
Saint-Petersburg State University of Aerospace Instrumentation, 67, B. Morskaia Str, 190000
Saint-Petersburg, Russian Federation
e-mail: mbs@gmail.com

N. Blaunstein (✉)
Ben-Gurion University of the Negev, POB 653 1, Ben Gurion St, Beer Sheva 84105, Israel
e-mail: nathan.blaunstein@hotmail.com

M. N. Favorskaya and L. C. Jain (eds.), *Computer Vision in Advanced Control
Systems-5*, Intelligent Systems Reference Library 175,
https://doi.org/10.1007/978-3-030-33795-7_4

4.1 Introduction

In this chapter, methods of using the matrix-vector systems of linear equations will be considered regarding a solution of the inverse problems occurring in the optic and radio communication, wired and wireless, presented in form of photos, pictures, or in other forms [1–3]. It should be notice that the corresponding solution models of the direct problems lie on the base of all considering inverse problems. Moreover, according to Hadamard's principle proposed in the eightieth century [1] and then by Tikhonov [4, 5] in twentieth century: the simpler selected model of the direct problem, more stable the solution of the inverse problem can be obtained.

Below we show that the most of the inverse problems can be declared via System of Linear Algebraic Equations (SLAE). Most strict and strong results in the theory of inverse problem solutions from the beginning were obtained exactly for SLAE. Then, we will present an example of reconstruction of video image with obstructions. That is, "spreading" along the photo recorded by the receiver in the optical or radio wireless communication channel.

The chapter is organized as follows. Section 4.2 discusses the direct and inverse problems in the communication theory. Least squares method and pseudo-inverse matrix are considered in Sect. 4.3. Singular-value decomposition based on Moore-Penrose matrix and singular regularization are presented in Sects. 4.4 and 4.5, respectively. Section 4.6 provides Levenberg–Marquardt algorithm for non-linear equations solution, while iteration algorithms for inverse problems solution and solution of convolutional integral equation and Wiener's filtering are discussed in Sects. 4.7 and 4.8, respectively. In Sect. 4.9, we consider elimination of images spreading in multimedia communication links. Inverse problem of source localization, micro-strip sensors reconstruction, and signal analysis is analyzed in Sects. 4.10–4.12, respectively. Section 4.13 concludes the chapter.

4.2 Direct and Inverse Problems in the Communication Theory

As a main object of investigations, let us consider SLAE written by Eq. 4.1, where $\mathbf{x} = (x_1, x_2, \ldots x_N)^T$ is the column vector of dimension N of the input values, $\mathbf{y} = (y_1, y_2, \ldots y_M)^T$ is the column vector of dimension M of the output values, $\mathbf{A} = \{a_{i,j}\}$ is the matrix of dimension $M \times N$, where $M \geq N$.

$$\mathbf{Ax} = \mathbf{y} \tag{4.1}$$

All values introduced in Eq. 4.1 can also be complex values. It is obvious that Eq. 4.1 describes the so-called direct problem usually formulated in the communication theory, where for the known input data (described by vector $\mathbf{x}$) it is required to find the output data (described by vector $\mathbf{y}$).

The inverse problem is a problem of definition of the input data (i.e., vector **x**) via the measured output data (i.e., vector **y**):

$$\mathbf{x} = \mathbf{A}^{-1}\mathbf{y}. \tag{4.2}$$

Definition of the inverse operator (matrix) $\mathbf{A}^{-1}$ is the main aim of the solution of the inverse problem.

According to Hadamard's definitions, the direct problem is a well-posed problem if the following three conditions are valid:

- The solutions are valid.
- The solution is unique.
- The solution is stable.

Stability means that deviations from the single solution are small during small deviations from the initial data of the problem. This smallness can be stated with help of the norm $\|\mathbf{x}\|$. According to Euclid definition of the norm, its length equals: $\|\mathbf{x}\| = \sqrt{\mathbf{x}^H\mathbf{x}}$ [4–6], where the above symbol ^{H} defines the conjugate transpose or Hermitian transpose of a matrix **x**. Then, the term small vector **x** determines the vector with small norm $\|\mathbf{x}\|$. In such definitions, we can determine the norm of operator (or matrix) **A** by Eq. 4.3 that defines the maximum value among all the various unit vectors.

$$\|\mathbf{A}\| = \sup_{\|\mathbf{x}\|=1} \|\mathbf{A}\mathbf{x}\| = \sup_{\|\mathbf{x}\|} \frac{\|\mathbf{A}\mathbf{x}\|}{\|\mathbf{x}\|} \tag{4.3}$$

Thus, the inverse problem (Eq. 4.2) is related to the direct one described by Eq. 4.1 via computation of vector **x** using knowledge of given or measured vector **y**. It should be noted that the theory of inverse problem solutions gives the exact results only when this problem can be deduced from SLAE that is when the vector A is simply a linear operator (matrix) [4–6].

As an example of the direct problem, we can consider a definition of the form of the optical or the radio pulse at the output of the transmitter consisting the linear filter [6]:

$$y(t) = \int_0^t h(\tau)x(t-\tau)d\tau, \tag{4.4}$$

where $h(\tau)$ is the impulse response function of the linear filter.

In spectral form, this relation can be presented as:

$$y(\omega) = \mathbf{H}(\omega)x(\omega). \tag{4.5}$$

Here,

$$H(\omega) = \int\limits_{0}^{\infty} h(\tau) \exp(i\omega\tau) d\tau \qquad (4.6)$$

is the transfer function of the filter.

Unfortunately, in practical applications the inverse problems are mostly unstable, i.e., Hadamard definition (introduced above, is not valid, and the full solution becomes incorrect. This means that small deviations of components of the output vector **y** lead to non-small deviations of those of the input vector **x**, even though the direct problem is initially and always has been stable. The reason of such stability of the direct problem is due to stability of the direct problem, namely, the existence of input noises together with the input signal, even after averaging by a filter of the device, can ultimately be seen in the output signal [4–6].

Thus, the formal solution of the inverse problem in the frequency domain (see Eq. 4.2) can be written in the same manner in the time domain provided by Eq. 4.7.

$$x(\omega) = \frac{y(\omega)}{H(\omega)} \qquad (4.7)$$

Finally, the following procedure—Inverse Fourier Transform (IFT), allows us to obtain the desired solution [5, 6]:

$$x(t) = \frac{1}{2\pi} \int\limits_{-\infty}^{\infty} x(\omega) \exp(-i\omega t) d\omega. \qquad (4.8)$$

The function $y(t)$ is defined as the result of experiment and, therefore, it definitely includes some errors. In this case, the experimental function $y(\omega)$ has a lot of high-frequency components, even including situations where $H(\omega)$ limits to zero and the solution of the inverse problem will be incorrect.

The objective estimation of the accuracy of the obtained solution $\hat{x}$ in situation, when the exact solution **x** is unknown, can be carried out by introducing a special value called the residual defined by Eq. 4.9.

$$\Phi = \|\mathbf{y} - \mathbf{A}\hat{\mathbf{x}}\|^2 \qquad (4.9)$$

It was found [2–6] that the smaller the values of the residual, the stricter the solution of the inverse problem. On the other words, the minimum of the residual denoted by $\min_{\hat{\mathbf{x}}} \Phi(\hat{\mathbf{x}})$ can be achieved for $\mathbf{x} = \hat{\mathbf{x}}$. This statement is seen as elementary in the least squares method and relates to the maximum likelihood estimation in statistics [4–6]. However, simple achievement of the minimum value of the residual (Eq. 4.9) does not guarantee the stability (according to Hadamard's principle) of the solution of the inverse problem.

In [4], Tikhonov introduced a fundamental definition of the regularized algorithm, which became a key statement in the theory of the solution of ill-posed inverse problems. It is based on the introduction of a special smoothing functional, which limits possible variations of the desired solutions, i.e., it decreases a class of possible solutions. A simple condition of such stabilization is written by Eq. 4.10.

$$\|\hat{\mathbf{x}}\|^2 \leq C < \infty \tag{4.10}$$

In this case, searching the minimum of the residual in Eq. 4.9 according to the calculus of variations leads to the minimization of the smoothing functional:

$$\Phi_\alpha = \|\mathbf{y} - \mathbf{A}\hat{\mathbf{x}}\|^2 + \alpha\|\hat{\mathbf{x}}\|^2 = \min_{\hat{\mathbf{x}}}. \tag{4.11}$$

Here, α plays a role of non-defined Lagrange product and is called in the literature the parameter of regularization. In this case, the desired solution fully depends on the parameter α, that is $\hat{\mathbf{x}} = \hat{\mathbf{x}}_\alpha$.

It is clear that for $\alpha = 0$, the task of solution estimation is definitely non-stabilized, but for $\alpha \to \infty$ the problem becomes fully stable. In [4], Tikhonov proposed the algorithm of selection of such a parameter α, for which a residual of the solution is similar to the variance (σ^2) of the measurements:

$$\Phi_\alpha = \|\mathbf{y} - \mathbf{A}\hat{\mathbf{x}}_\alpha\|^2 = \sigma^2. \tag{4.12}$$

It was proved in [2, 3] that the solution of Eq. 4.12 is unique and unified. In practical applications, one difficulty was found: the variance of noises during measurements (σ^2) is not always known from the experiment. Therefore, the parameter of regularization is often determined by ad hoc method.

As was found in [4], a problem based on the algorithm of regularization was stated as Tikhonov well-posed, also called conditionally well-posed. By introducing some distortions into the ideal model of the problem (Eq. 4.1), one can obtain small deviations from the exact solution, but the solution itself will remain enough stable.

4.3 Least Squares Method and Pseudo-Inverse Matrix

Let us present a smoothing functional (Eq. 4.11) in another form during definition of the norm according to Euclidean metric [4–6]:

$$\|\mathbf{y} - \mathbf{A}\hat{\mathbf{x}}\|^2 + \alpha\|\hat{\mathbf{x}}\|^2 = (\mathbf{y} - \mathbf{A}\hat{\mathbf{x}})^H(\mathbf{y} - \mathbf{A}\hat{\mathbf{x}}) + \alpha\hat{\mathbf{x}}^H\hat{\mathbf{x}}$$
$$= \left(\mathbf{y}^H - \hat{\mathbf{x}}^H\mathbf{A}^H\right)(\mathbf{y} - \mathbf{A}\hat{\mathbf{x}}) + \alpha\hat{\mathbf{x}}^H\hat{\mathbf{x}}.$$

The extremum of this functional is achieved by setting the gradient of $\hat{\mathbf{x}}^H$ to zero, that is,

$$\mathbf{A}^H(\mathbf{y} - \mathbf{A}\hat{\mathbf{x}}) + \alpha\hat{\mathbf{x}} = 0.$$

This equation can be rewritten by Eq. 4.13.

$$\left(\mathbf{A}^H\mathbf{A} + \alpha\mathbf{I}\right)\hat{\mathbf{x}} = \mathbf{A}^H\mathbf{y} \tag{4.13}$$

Here $\mathbf{I}$ is the diagonal zero matrix (identity matrix). There is a square non-zero matrix in the left-side of this equation and, therefore, its solution can be expressed as:

$$\hat{\mathbf{x}} = \mathbf{A}^+\mathbf{y}, \tag{4.14}$$

where the matrix equals

$$\mathbf{A}^+ = \left(\mathbf{A}^H\mathbf{A} + \alpha I\right)^{-1}\mathbf{A}^H \tag{4.15}$$

and is called pseudo-inverse matrix to matrix $\mathbf{A}$. It is important to know that always $\mathbf{A}^+\mathbf{A}\hat{\mathbf{x}} = \hat{\mathbf{x}}$. The pseudo-inverse matrix can be constructed even for each regular, not only rectangular matrix.

Equation 4.14 presents a solution of the least square method with regularization according to Tikhonov method [4–6]. In the case of $\alpha = 0$, this solution is simply the least square method that gives for the squared matrix $\mathbf{A}$ its inverse matrix $\mathbf{A}^{-1}$.

4.4 Singular-Value Decomposition and Moore-Penrose Matrix

Usage the pseudo-solution (Eq. 4.14) is not always effective even for the correct selection of the regularization parameter defined in Sect. 4.2 because during computations of pseudoinverse matrix in form of Eq. 4.15, a lot of computations should be done, during which computational noises have tendency to create a cumulative effect quickly, and finally, destroy the correct desired solution. To obey this effect, a singular-value decomposition of the matrix $\mathbf{A}$ is usually used, presented as a product of three matrices:

$$\mathbf{A} = \mathbf{U}^H\mathbf{S}\mathbf{V}, \tag{4.16}$$

where $\mathbf{U}$ and $\mathbf{V}$ are the unitary matrices of dimensions of $(M \times M)$ and $(N \times N)$, respectively. This means that $\mathbf{U}^H\mathbf{U} = \mathbf{I}$ and $\mathbf{V}^H\mathbf{V} = \mathbf{I}$, where $\mathbf{I}$ is the identity matrix introduced in Sect. 4.2.

In algebraic theory, columns of matrices $\mathbf{U}$ and $\mathbf{V}$ are called the left and right singular vectors, respectively. Matrix $\mathbf{S} = \{s_i\delta_{i,j}\}$ is the quasi-orthogonal matrix consisting of the singular values s_i, which are strictly positive and can usually be obtained one-by-one in decreasing of each value manner, $\delta_{i,j}$ are the Kronecker symbols. Notice that the non-zero eigenvalues of matrices $\mathbf{AA}^H$ and $\mathbf{A}^H\mathbf{A}$ are similar to the eigenvalues of matrix $\mathbf{A}$ that is $s_i = \sqrt{\lambda_i}$, where λ_i are the eigenvalues of matrix $\mathbf{A}$. The rank of any matrix $\mathbf{A}$ equals the rank of diagonal matrix $\mathbf{S}$.

Introduction the decomposition of Eq. 4.16 in Eq. 4.15 for $\alpha = 0$ yields:

$$\mathbf{A}^+ = \mathbf{V}^H\mathbf{S}^+\mathbf{U},$$

where

$$\mathbf{S}^+ = \left\{\frac{1}{s_i}\delta_{i,j}\right\}. \tag{4.17}$$

In such a form, the pseudo-inverse matrix is called Moore-Penrose matrix.

Use of a regular-value decomposition allows to estimate relatively simply the norms (Eq. 4.3) of all matrices in Eq. 4.16. Thus, because $\|\mathbf{V}^H\mathbf{V}\| = \|\mathbf{V}\|^2 = 1$, then

$$\|\mathbf{V}^H\mathbf{V}\| = \sup_{\|\mathbf{x}\|}\frac{(\mathbf{x}^H\mathbf{V}^H\mathbf{V}\mathbf{x})}{(\mathbf{x}^H\mathbf{x})} = \|\mathbf{V}\|^2 = 1,$$

and $\|\mathbf{V}\| = 1$ and $\|\mathbf{U}\| = 1$. In the same manner, the norm of diagonal matrix $\mathbf{S}$ can be found:

$$\|\mathbf{S}\|^2 = \sup_{\|\mathbf{x}\|}\frac{\sum_j x_j^2 s_j^2}{\sum_j x_j^2} \leq \max_j\left(s_j^2\right), \tag{4.18}$$

from which we obtain:

$$\|\mathbf{S}\| = \max_j\left(s_j\right) = s_{\max}. \tag{4.19}$$

From the decomposition of Eq. 4.16, it follows that:

$$\|\mathbf{A}\| \leq \|\mathbf{U}^H\|\|\mathbf{S}\|\|\mathbf{V}\| \leq \max_j\left(s_j\right) \text{ and } \|\mathbf{A}\| = \max_j\left(s_j\right) = s_{\max}. \tag{4.20}$$

For Moore–Penrose matrix, we obtain the same relations with the singular values:

$$\|\mathbf{A}^+\| = \max_j \left(s_j^{-1}\right) = 1/s_{\min}. \tag{4.21}$$

Let us now estimate the accuracy of the solution of the inverse problem using Moore-Penrose matrix. If we denote deviations of the solutions and the initial data as $\|\Delta\mathbf{x}\|$ and $\|\Delta\mathbf{y}\|$, respectively, then Eq. 4.14 yields the constraint:

$$\|\Delta\mathbf{x}\| \leq \|\mathbf{A}^+\|\|\Delta\mathbf{y}\|. \tag{4.22a}$$

From initial Eq. 4.1, we get the following constraint:

$$\|\mathbf{y}\| \leq \|\mathbf{A}\|\|\mathbf{x}\|. \tag{4.22b}$$

Producing the left and the right sides of Eqs. 4.22a and 4.22b and combining the last two inequalities, allow us to write:

$$\frac{\|\Delta\mathbf{x}\|}{\|\mathbf{x}\|} \leq \|\mathbf{A}\|\|\mathbf{A}^+\|\frac{\|\Delta\mathbf{y}\|}{\|\mathbf{y}\|} = cond(\mathbf{A})\frac{\|\Delta\mathbf{y}\|}{\|\mathbf{y}\|}, \tag{4.23}$$

where

$$cond(\mathbf{A}) \equiv \|\mathbf{A}\|\|\mathbf{A}^+\| = \frac{\max s}{\min s} = \frac{s_{\max}}{s_{\min}}$$

is called the conditional number of matrix $\mathbf{A}$. Conditional number demonstrates the accuracy of the inverse problem with respect to measurements. Only for the orthogonal transform, when $cond(\mathbf{A}) = 1$ is the error of the solution, not increased. Finally, it should be noticed that a solution of the least squares problem is more effective using Moore-Penrose matrix.

4.5   Singular Regularization

As follows from the constraint (Eq. 4.23), to increase accuracy of the pseudo-solutions one needs to decrease, first of all, the conditional number of Moore-Penrose matrix $\mathbf{A}$. To do that, one should simply decrease a set of singular values, for example, by introducing some small limit value τ to accounting only for those singular values, for which $s_i \geq \tau$. In this case, in Moore–Penrose matrix should be changed $\mathbf{S}^+$ on [5, 6]:

$$\mathbf{S}^+ \to \mathbf{S}^+ = \left\{ \frac{\chi(s_i - \tau)}{s_i} \delta_{i,j} \right\}, \tag{4.24}$$

where $\chi(s_i - \tau)$ is the Heaviside step function: $\chi(s_i - \tau) = 1$ if $s_i \geq \tau$, and $\chi(s_i - \tau) = 0$ if $s_i < \tau$. The limit value τ according to Eq. 4.23 should be taken not less than

$$\tau_0 = \frac{\|\Delta \mathbf{y}\|}{\|\mathbf{y}\|} s_{max}.$$

In this case, we have guaranteed that an error of solutions of the inverse problem will be enough small if $\frac{\|\Delta \mathbf{x}\|}{\|\mathbf{x}\|} \leq 1$.

More smoothing regularization can be proposed by taken regularization according to Eq. 4.11 proposed by Tikhonov [5, 6]. Accounting for the singular-value decomposition (Eq. 4.12) the pseudo-inverse matrix $\mathbf{S}^+$ defined by Eq. 4.24 should be changed on:

$$\mathbf{S}^+ \to \mathbf{\Sigma}^+ = (\mathbf{S} + \alpha \mathbf{S}^+)^+ = \left\{ \left(s_i + \frac{\alpha}{s_i} \right)^{-1} \delta_{i,j} \right\}. \tag{4.25}$$

The obtained decomposition for Moore-Penrose matrix (Eq. 4.17) already considers Tikhonov regularization. The norm of the matrix $\mathbf{\Sigma}^+$ can be estimated as:

$$\|\mathbf{\Sigma}^+\| = \min_i \left\{ \left(s_i + \frac{\alpha}{s_i} \right)^{-1} \right\} = \frac{1}{2\sqrt{\alpha}}. \tag{4.26}$$

Selection of the regularization parameter will be effective if according to Eq. 4.26 the following constraint will be achieved:

$$\frac{\|\Delta \mathbf{y}\|}{\|\mathbf{y}\|} \frac{s_{max}}{2\sqrt{\alpha}} \leq 1. \tag{4.27}$$

In this case, the following constraint should be valid: $s_{min} \leq \sqrt{\alpha} \leq s_{max}$.

4.6 Levenberg–Marquardt Algorithm for Non-linear Equations Solution

The Levenberg–Marquardt method is valid for solution of systems of non-linear equations, which can be written in the form of Eq. 4.28, where $\mathbf{F}(\mathbf{x})$ is any vector function of vector $\mathbf{x}$.

$$\mathbf{y} = \mathbf{F}(\mathbf{x}) \tag{4.28}$$

The aim of this method is to transfer general equations (Eq. 4.18) to a system of linear equations. For this purpose, let us suppose that the initial approximate vector $\mathbf{x}_0$ for a vector $\mathbf{x}$ is known. Then, the solution will be presented in the following manner:

$$\mathbf{x} = \mathbf{x}_0 + \Delta\mathbf{x}.$$

In assumption of a small error $\Delta\mathbf{x}$, we get according to [6]:

$$\Delta\mathbf{y} = \mathbf{y} - \mathbf{F}(\mathbf{x}_0) = \mathbf{F}(\mathbf{x}) - \mathbf{F}(\mathbf{x}_0) = \mathbf{A}\Delta\mathbf{x}. \tag{4.29}$$

Here, $\mathbf{A} = \frac{\partial \mathbf{F}(\mathbf{x}_0)}{\partial \mathbf{x}_0} = \left[\frac{\partial F_i(\mathbf{x}_0)}{\partial \mathbf{x}_{0j}}\right]$ is the Jacobian's matrix, $i = 1 \ldots M$ and $j = 1 \ldots N$ are the numbers of the corresponding vectors' elements. System of equations is the desired SLE for definition of the error $\Delta\mathbf{x}$.

The solution of t Eq. 4.28 can be found using Moore-Penrose pseudo-inverse matrix, that is,

$$\Delta\widehat{\mathbf{x}} = \mathbf{A}^+ \Delta\mathbf{y}. \tag{4.30}$$

Found in such a manner the solution of Eq. 4.14 allows correction of an initial approximate vector $\mathbf{x}_0$, i.e., $\mathbf{x}_0 = \widehat{\mathbf{x}} - \Delta\widehat{\mathbf{x}}$. Now exchanging $\mathbf{x}_0$ by $\widehat{\mathbf{x}}$, one can obtain a new approximation of the Jacobian matrix. Then, using Moore-Penrose matrix (Eq. 4.20), we repeat an iteration procedure. This iteration procedure (Eq. 4.30) is called in literature Levenberg-Marquardt algorithm. It can notice that Moore-Penrose matrices allow to use a procedure of regularization at each step of iteration. The rule to stop the iteration procedure is the following:

- When solution $\widehat{\mathbf{x}}$ becomes to be non-changing.
- When residual $\Phi = \left\|\mathbf{y} - \mathbf{F}(\widehat{\mathbf{x}})\right\|^2$ for Eq. 4.28 becomes to be stable or achieves a given level (for example, a value of variance of the accuracy of measurements of the vector $\mathbf{y}$).
- When the current solution becomes similar to desired solutions obtained earlier.

4.7 Iteration Algorithms for Inverse Problems Solution

Often, it is very complicated to make the operation of differentiation of the Jacobian matrices according to Levenberg-Marquardt algorithm described above. In this case the iteration algorithms are usually used.

Let us introduce a definition of the conjugate operator (matrix) $\mathbf{A}^H$. The conjugate matrix is that, for which the following equation is valid:

$$\mathbf{A}^H \mathbf{y} \cdot \mathbf{x} = \mathbf{y} \cdot \mathbf{A}\mathbf{x} \tag{4.31}$$

for arbitrary vectors $\mathbf{x}$ and $\mathbf{y}$.

Using for Eq. 4.31 the conjugate operator, we finally get:

$$\mathbf{A}^H \mathbf{y} = \mathbf{A}^H \mathbf{A}\mathbf{x}$$

or $\mathbf{A}^H \mathbf{y} - \mathbf{A}^H \mathbf{A}\mathbf{x} = 0$, which gives a vivid equation:

$$\mathbf{x} = \mathbf{x} + \lambda(\mathbf{A}^H \mathbf{y} - \mathbf{A}^H \mathbf{A}\mathbf{x}) \tag{4.32}$$

for arbitrary parameter λ. This parameter defines the degree of convergence of the current solution obtained during the iteration procedure.

Let us rewrite the latter equation in the following form by introducing the unit operator (matrix) $\mathbf{I}$, that is,

$$\mathbf{x} = \lambda\mathbf{A}^H \mathbf{y} + (\mathbf{I} - \lambda\mathbf{A}^H \mathbf{A}\mathbf{x})\mathbf{x}.$$

Here we should add an additional operator called the limited operator, which can be defined as [6]:

$$\mathbf{L}\mathbf{x} = \begin{cases} \mathbf{x}, & x_i \in (a,b) \\ 0, & x_i \notin (a,b) \end{cases}, \ a \leq b. \tag{4.33}$$

Accounting for Eq. 4.33, we finally get from Eq. 4.32 the following equation:

$$\mathbf{x} = \lambda\mathbf{A}^H \mathbf{y} + (\mathbf{I} - \lambda\mathbf{A}^H \mathbf{A})\mathbf{L}\mathbf{x}. \tag{4.34}$$

The iteration solution of Eq. 4.29 can be presented as follows:

$$\mathbf{x}^{(k+1)} = \lambda\mathbf{A}^H \mathbf{y} + (\mathbf{I} - \lambda\mathbf{A}^H \mathbf{A})\mathbf{L}\mathbf{x}^{(k)}. \tag{4.35}$$

To obtain a successful solution of Eq. 4.30, it is necessary to take precisely the initial approximation $\mathbf{x}^{(0)}$ and to take correctly the convergence parameter λ. Initially, the convergence can be obtained only if $\mathbf{A}^H$ is the conjugate operator of matrix $\mathbf{A}$.

4.8 Solution of Convolutional Integral Equation and Wiener's Filtering

The convolutional integral equation was introduced earlier in Sect. 4.1 for the simplest element of communication system (a linear filter) by Eq. 4.4. We present it again for the convenience of understanding of the matter, that is,

$$y(t) = \int h(\tau)x(t - \tau)d\tau.$$

Its formal solution (Eq. 4.5) in presence of errors during measurements leads to significant errors in solutions of an inverse problem. Therefore, Tikhonov's technique of regularization is usually used [5, 6], according to which a general residual should be minimized (see also Eq. 4.11):

$$\|\mathbf{y} - \mathbf{Ax}\|^2 + \alpha\|\mathbf{x}\|^2 = \min_{\mathbf{x}}. \tag{4.36}$$

In the case of integrated functions $f(t)$, its norm can be determined via the scalar product by Eq. 4.37, where $\hat{f}(\omega)$ denotes the Fourier transform of the function $f(t)$.

$$\|f\|^2 = \int\limits_{-\infty}^{\infty} f^*(t)f(t)dt = \frac{1}{2\pi} \int\limits_{-\infty}^{\infty} \hat{f}^*(\omega)\hat{f}(\omega)d\omega = \frac{1}{2\pi}\|\hat{f}\|^2 \tag{4.37}$$

Equation 4.37 is known as the Perceval's theorem [6]. Using Eq. 4.37, we can rewrite a general residual (Eq. 4.36) in the vivid form:

$$\frac{1}{2\pi} \int\limits_{-\infty}^{\infty} [\hat{y}(\omega) - H(\omega)\hat{x}(\omega)]^*[y(\omega) - H(\omega)x(\omega)]d\omega + \alpha\frac{1}{2\pi} \int\limits_{-\infty}^{\infty} \hat{x}^*(\omega)\hat{x}(\omega)d\omega$$
$$= \min_{x(\omega)}.$$

Taking the derivative of $\hat{x}^*(\omega)$ and limiting it to zero, we will finally get the following equation:

$$-H^*(\omega)[\hat{y}(\omega) - H(\omega)\hat{x}(\omega)] + \alpha\hat{x}(\omega) = 0. \tag{4.38}$$

Spectrum of the defined solution of Eq. 4.38 can be found as [6]:

$$\hat{x}(\omega) = \frac{\hat{y}(\omega)H^*(\omega)}{H(\omega)H^*(\omega) + \alpha} = \frac{\hat{y}(\omega)}{H(\omega)}R_\alpha(\omega), \tag{4.39}$$

where $R_\alpha(\omega) = \dfrac{H(\omega)H^*(\omega)}{H(\omega)H^*(\omega) + \alpha}$ is the regularization product according to Tikhonov [4].

It can be noticed that for $\alpha = 0$, we get $R_\alpha(\omega) = 1$, and then it can be found the formal solution (Eq. 4.7) of the inverse problem. Thus, a final solution can be written by use of the inverse Fourier transform of the inverse problem and can be presented by Eq. 4.40.

$$x(t) = \frac{1}{2\pi} \int\limits_{-\infty}^{\infty} \frac{\hat{y}(\omega)}{H(\omega)} R_\alpha(\omega) \exp\{-i\omega t\} d\omega \qquad (4.40)$$

Such presentation is known as Wiener's filtering with regularization [6]. We should outline that selection of the regularization parameter α essentially influences on the accuracy of solution of the inverse problem (such as the problem of reconstruction of images in optics) by rejection of so-called blur, which usually is called the clutter noise in radio and optical physics [4–6].

4.9 Elimination of Images Spreading in Multimedia Communication Links

We notice that all the approaches discussed above for defining solutions of the inverse problems in wireless communication links, relate to the problem of reconstruction of images in wireless multimedia. Thus, usually observed classical inverse problem in photography takes place, when some image should be reconstructed due to its spreading, namely, due to incorrect initial statement of photographic apparatus (see Fig. 4.1a). A simplest model of such a spreading of image is the operation of 2D integral convolution [6, 7]:

$$y(\boldsymbol{\rho}) = \iint A(\boldsymbol{\rho}') x(\boldsymbol{\rho} - \boldsymbol{\rho}') d^2 \boldsymbol{\rho}'. \qquad (4.41)$$

Here x is the initial image, and y is the spreading image. The matrix A introduced in Sect. 4.2, describes so-called Point Spread Function (PSF), which is supposed to be known. In Fig. 4.1 this function overlaps 6 points. Such problem can be met in optical image processing, for example, in occurrence of so-called "blurring" (as shown in Fig. 4.1a).

An inverse problem deals with reconstruction of the initial image x. As was mentioned above, such problem is solved using 2D Fourier transform, after that the initial equation in the spatial spectral domain can be converted to SLAE in the following form [6, 7]:

(a)   (b)

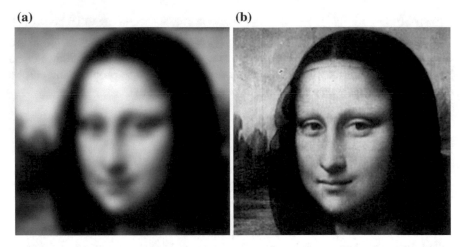

Fig. 4.1 Image of the Mona Liza portraiture with: **a** "spreading", **b** after reconstruction with rejection of the "blurring"

$$\hat{\mathbf{y}} = \hat{\mathbf{A}}\hat{\mathbf{x}}, \tag{4.42}$$

where $\hat{\mathbf{y}}$, $\hat{\mathbf{x}}$, and $\hat{\mathbf{A}}$ are the spatial spectra of the corresponding matrices.

Definition of the matrix $\hat{\mathbf{x}}$ is a solution of the inverse problem, which should be solved with regularization, as it is done using Eqs. 4.41 and 4.42. After that, by use of the inverse Fourier transform (Eq. 4.39), the initial image **x** can be finally reconstructed (see Fig. 4.1b). As illustrates Fig. 4.1b, after all procedures described above a picture fully obtains its initial natural view.

4.10   Inverse Problem of Source Localization

If the source is placed at the point 0 of a plane (x, y) as shown in Fig. 4.2, then for definition of its place it is enough to measure difference in time delays of signals, received at points 1–4 arbitrary distributed at the plane (see Fig. 4.2, points 1–4). Such a sort of problems occurs in definition of position of the source of radiation located above the terrain. Similar method is used in nondestructive testing of electronic plates based on acoustic emission occurring around the places of the bad contacts during mechanical deformation of these plates.

Let us consider existence of four receiving points, as shown in Fig. 4.2. This is minimal amount of points that enough for definition of the real position of the source. If the waves speed, let say, acoustic waves, is **v**, if this amount of receivers is really enough—should be checked. If **r** is the position of the source, and $\mathbf{r}_1, \mathbf{r}_2, \mathbf{r}_3, \mathbf{r}_4$ are the positions of the receiving points, then the delays of radiation arrival $t_{i,j}$ at the point of receiving can be determined as [6, 7]:

Fig. 4.2 The geometry of the
problem of positioning of the
source of radiation

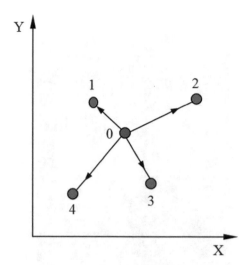

$$t_{1,2} = \frac{|\mathbf{r} - \mathbf{r}_1|}{c} - \frac{|\mathbf{r} - \mathbf{r}_2|}{c}, t_{1,3} = \frac{|\mathbf{r} - \mathbf{r}_1|}{c} - \frac{|\mathbf{r} - \mathbf{r}_3|}{c}, t_{1,4} = \frac{|\mathbf{r} - \mathbf{r}_1|}{c} - \frac{|\mathbf{r} - \mathbf{r}_4|}{c}.$$

$$(4.43)$$

Written in such a manner, a system of three equations allows precisely estimate
position of the source and correct additionally a waves speed. The above system
(Eq. 4.43) is pure non-linear and can be solved by one of the methods described in
Sect. 4.1.

4.11   Inverse Problem of Micro-Strip Sensors Reconstruction

For production of the multi-layered micro-strip detectors the existing takes place a
problem of mutual interactions between layered structures of charges collection.
Such problem takes place, for example, during performance of the Project
"ATLAS" at the Large Hadron Collider (LHC) in European Organization for
Nuclear Research (CERN) in Geneva (Swiss) [8], outer view on which is shown in
Fig. 4.3a.

During performance of such testing, it was supposed that a top detector (see
Fig. 4.3b) has ideal parallel structure. The placed below detector is also ideal, but
due to non-ideal technology of mechanical equipment can be shifted at distance
h along the axis OZ and turn on the angle α around the axis OY with respect to the
topside detector. It is necessary for constructing the algorithm of estimation of these
parameters according to numerous measurements of crossing responds (along axis
OZ) on the quasi-point vertical affections. Knowledge of these parameters allows

(a)

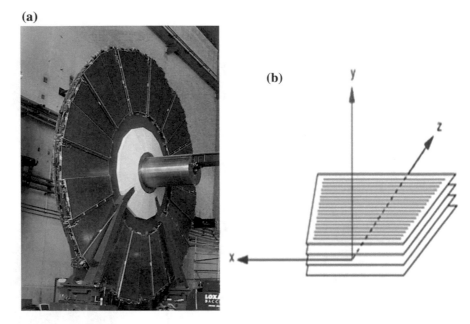

Fig. 4.3 To problem of micro-strips in the project "ATLAS" (Geneva): **a** outer view, **b** scheme of micro-strip detectors arrangement

for sufficiently increasing the accuracy of tomographic measurements of trajectories of all possible particles registered by LHC.

If coordinates of quasi-point interactions on the top layer of micro-strips are known exactly (X_0, Z_0), then the local coordinates of this interaction in the layer, lower to the current, can be found by Eq. 4.44.

$$X = X_0 \cos \alpha - (Z_0 + h) \sin \alpha, Z = (Z_0 + h) \cos \alpha + X_0 \sin \alpha, \tag{4.44}$$

This coordinates' transform is a superposition of the linear shift and the rotation. Having numerous measurements of only one transverse coordinate

$$Z_j = (Z_{0j} + h) \cos \alpha + X_{0j} \sin \alpha \tag{4.45}$$

for different interactions (X_{0j}, Z_{0j}), it can be found the parameters (h, α). For following layers the problem is solved in the same manner.

The solution of such a formulated inverse problem can be obtained with help of the method of least squares mentioned in Sect. 4.1. The accuracy of the solution of inverse problem increases with increase of number of independent measurements. In Fig. 4.4, an example of distribution of responses for 50 independent interactions is shown.

In Fig. 4.4, by points the position of point interactions is shown (see top layer), and by squares – responses on these interactions in the lower layer. Achieving in

Fig. 4.4 To solution of the inverse problem of particles' collection

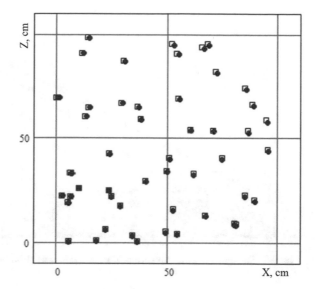

this case accuracy of the solution of the inverse problem according to estimation of parameters defined by (h, α) is not less than 10^{-7} cm and 10^{-7} angle degrees, respectively. At the same time, use only two independent interactions, the accuracy decreases approximately at four orders.

4.12   Inverse Problem for Signal Analysis

In radio physical applications, the important role plays the spectral analysis. A standard task of spectral analysis is usually solved using Fourier transform, for which were performed the corresponding algorithms called Fast Fourier Transform (FFT). However, often enough the equidistance of frequencies or time rights cannot be achieved. Moreover, in this case the condition of orthogonality of basic functions is not valid, and one cannot be used. The task of spectral function evaluation can be decided, as the inverse problem that can be considered as the solution of SLAE.

A lot of examples regarding radio tomography are discussed in literature [7, 8]. Based on knowledge obtained there, let us consider a time-dependent signal $S(t)$ and will relate it with its spectral function $S(\omega_j)$ given at the discrete map of frequencies ω_j via the well-known Eq. 4.46.

$$S(t) = \int_{-\infty}^{\infty} S(\omega) \exp\{i\omega t\} d\omega \approx \sum_{j} S(\omega_j) \exp\{i\omega_j t\} \Delta\omega_j \qquad (4.46)$$

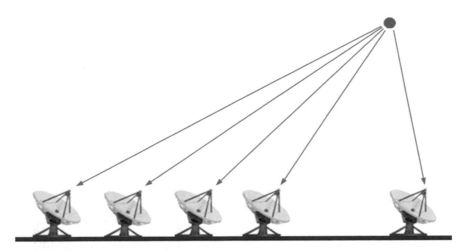

Fig. 4.5 The antenna array for localization of the source position

Spectral values of $S(\omega_j)$ determine the non-known parameters for the inverse problem.

If discretization of time-samples of the initial signal $S_m = S(t_m)$ can be done, then the task of the spectral evaluation and estimation can be deduced to SLAE solution:

$$\mathbf{y} = \mathbf{Ax},\tag{4.47}$$

where $\mathbf{y} = \{S_m\}$, $\mathbf{x} = \{S(\omega_j)\}$, and $\mathbf{A} = \{\exp\{i\omega_j t_m\}\}$. Definition of vector $\mathbf{x}$ is definitely a solution of the inverse problem of spectral estimation.

Let us notice that a correction of the formulated above inverse problem limits to unit if it realizes the discrete Fourier transform, that is, only in the cases of usage of the orthogonal expansion. In general case, this is not so trivial and it is necessary to introduce the regulated algorithms. Such a sort of solutions is usually used in the problems of extra-resolution and localization of the sources of electromagnetic radiation with usage of limited receiving antenna arrays, as shown in Fig. 4.5. The same is related to computation of antenna directivity diagrams. Let us consider this aspect more in details.

If $J(x)$ is the current distribution along some antenna aperture, then the field distribution along the angle θ is defined by the following integral:

$$E(\theta) = \frac{1}{L}\int\limits_{-\infty}^{\infty} J(x)\exp(ikx\cos\theta)dx,\tag{4.48}$$

where L is a linear dimension of the antenna aperture. The written integral in Eq. 4.48 defines Directivity Diagram (DD) of the antenna and is really a solution of the direct problem.

The inverse problem related to estimation of the current $J(x)$, which provides with the given DD. If in the integral (Eq. 4.48) we pass to the discrete values, the problem under consideration yields the SLAE solution of the same form as above, that is,

$$\mathbf{y} = \mathbf{A}\mathbf{x}.$$

Here $\mathbf{y} = \{E(\theta_m)\}$ is the vector of the measured values of directivity diagram $\mathbf{x} = \{J(x_n)\}$ is the vector of non-known values of current along the antenna aperture. Matrix $\mathbf{A}$ equals $\mathbf{A} \equiv \{\exp(ikx_n\cos\theta_m)\}$, and k is the wavenumber. The method of the solution of obtained SLAE can be selected from those described in Sect. 4.1.

An inverse problem formulated above can be completed, if we put a problem of recognition of the full DD according to its recovering fragment measured over some segment of angles θ. For this is enough to put the obtained values of the current inside the integral from Eq. 4.48 and to calculate it over full range of observation angles.

4.13   Conclusions

Matrix-vector systems of linear equations allows to find a solution of the inverse problems occurring in the optic and radio communication, wired and wireless, presented in form of signals and images. We develop a mathematical apparatus to solve the inverse problems via SLAE approach. In order to enforce calculation process, we apply singular-value decomposition based on Moore-Penrose matrix. Also, it was shown effectiveness of Levenberg–Marquardt algorithm application for non-linear equations solution. We show that Wiener's filtering with regularization essentially influences on the accuracy of solution of the inverse problem, for example, reconstruction of blurred images. Some practical examples regarding to inverse problems, viz. source localization, micro-strip sensors reconstruction, and signal analysis, are analyzed.

References

1. Horadam, K.J.: Hadamard Matrices and Their Applications. Princeton, New Jersey, Princeton University Press (2007)
2. Balonin, N., Sergeev, M.: Expansion of the orthogonal basis in video compression. Front. Artif. Intell. Appl. **262**, 468–474 (2014)

3. Sergeev A.M., Blaunstein N.S.: Orthogonal matrices with symmetrical structures for image processing. Informatsionno-upravliaiushchie sistemy [Information and Control Systems] **6**, 2–8 (in Russian) (2017)
4. Tikhonov, A.N., Goncharsky, A.V., Stepanov, V.V., Yagola, A.G.: Numerical methods for the solution of ill-posed problems. Springer Science + Business Media Dordrecht (1995)
5. Tikhonov, A.N., Leonov, A.S., Yagola, A.G.: Nonlinear Incorrect Problems. Nauka, Moscow (in Russian) (1995)
6. Yakubov, V.P., Shipilov, S.E.: Inverse problems of radio physics, Tomsk: Editors NTL, (in Russian) (2016)
7. Blaunstein, N., Yakubov, V.P. (eds.): Electromagnetic and Acoustic Wave Tomography: Direct and Inverse Problems in Practical Applications. CRC, Taylor & Frances Group, Boca Raton, FL (2019)
8. Kawamoto, T., Di, G.B.: New small wheel technical design report. http://cdsweb.cern.ch/record/1552862/. Accessed 4 Aug 2019

Chapter 5
Authentication and Copyright Protection of Videos Under Transmitting Specifications

Margarita N. Favorskaya and Vladimir V. Buryachenko

Abstract A lot of video content shared over the Internet have some problems, among which are the authentication and multimedia protection. The chapter provides a detailed overview of the transmitting specifications of video content, such as H.264/AVC, H.264/SVC, and H.265/HEVC in the sense of a watermarking scheme. Possible attacks against video sequence or single frames are classified as the intentional or accidental attacks. We study a selection of the relevant regions in the semi I-frames under formulated restrictions and conditions caused by the transmitted specifications. We simulate a sequence of I-frames in the original video sequence by a criterion of significant motion between three neighboring frames. The relevant regions are selected through a creation of the motion map, saliency map, textural map, blue component map, and joint map. Invariance to global Rotation, Scaling, Translation (RST) attacks is provided by feature-based approach. The novelty is in that we do not embed watermark in these stable regions, but coordinates of descriptors as they were in the host frame. This allows us to avoid the corresponding matches between descriptors in the host and watermarked frames. In order to provide invariance to RST attacks in the stable regions, we use exponential moments calculated on a unit circle. After compensation of the global RST attacks, we check other pairs of corresponding feature points consequently in order to find the possible local geometric distortions.

Keywords Video watermarking · Transmitting specification · Authentication · Copyright protection · Multilevel protection · Digital wavelet transform · Digital Shearlet transform · Digital Hadamard transform · Geometric attacks · Permutation attacks

M. N. Favorskaya (✉) · V. V. Buryachenko
Institute of Informatics and Telecommunications, Reshetnev Siberian State University of Science and Technology, 31, Krasnoyarsky Rabochy ave., Krasnoyarsk 660037, Russian Federation
e-mail: favorskaya@sibsau.ru

V. V. Buryachenko
e-mail: buryachenko@sibsau.ru

© Springer Nature Switzerland AG 2020
M. N. Favorskaya and L. C. Jain (eds.), *Computer Vision in Advanced Control Systems-5*, Intelligent Systems Reference Library 175,
https://doi.org/10.1007/978-3-030-33795-7_5

5.1 Introduction

Rapid development of Internet technology made the distribution of multimedia data much easier. That leads to a growth of the illegal operations such as duplication, modification, forgery, and copyright violation. This motivates a development of video authentication and copyright protection. The optimal tradeoff between imperceptibility for Human Visual System (HVS), payload, and robustness to attacks is the main goal of the watermarking techniques. The major applications of digital video watermarking are the following:

- Authentication. Fragile, semi-fragile, and robust watermarks are the commonly used entities for checking the integrity of images and videos. The scheme is sensitive to the content-changing manipulations.
- Copyright protection. This is the main issue in the digital data delivery networks. Watermark used for copyright protection identifies the copyright owner. Robust watermark that includes information about the sender and receiver of the delivered video is embedded into video signal.
- Broadcast monitoring. Watermark is embedded prior to transmission and extracted by the monitoring site that is set up within the transmission area. Each broadcasting video ought to be checked on a watermarking presence.
- Copy control. In this case, a watermark indicates whether a video content is copyrighted. Watermark is embedded into the header that prevents copying of data. Such watermark can be removed only with significant degradation of videos. This technique is a widely used application in video watermarking.
- Fingerprinting. The real-time applications of video streaming can be enforced by a fingerprinting policy. A watermark in a view of customer ID is embedded into videos to track back any user breaking the license agreement.
- Ownership identification. Client identification and verification facilitates the prevention, detection, and prosecution of the illegal data use, for example in e-markets.
- Video tagging. Tag assignment during image or video annotation can be considered as a watermarking process.
- Enhanced video coding. High Efficiency Video Coding (HEVC) extended variety applications, among which are broadcast of high definition TV signals over satellite, terrestrial transmission systems, video content acquisition, editing systems, security applications, Internet and mobile network video, and real-time conversational applications. However, a watermarking of HEVC encoded videos is a difficult task and requires special decisions because of the redundancy caused by codec application.

Our contribution deals with a development of the authentication and copyright protection of videos represented in some formats. Usually a watermarking process begins from selection of frames, suitable for embedding. Our method supports the detection of I-frames and selection the best regions for embedding using the joint map, which excludes moving and salient regions and involves high textural regions with prevailed blue component. Invariance to the main types of attacks regarding the compressed videos is provided by a feature-based approach for embedding with the original procedures.

The remainder of the chapter is organized as follows. Section 5.2 gives an overview of the transmitting specifications and possible attacks. Section 5.3 presents the related work in video watermarking. Section 5.4 provides the details for selecting relevant regions in I-frames. Proposed video watermarking method, which supports a protection against typical image processing, geometric, and permutation attacks, is discussed in Sect. 5.5. Experimental studies are considered in Sect. 5.6. Section 5.7 concludes the chapter.

5.2   Overview of Transmitting Specifications and Possible Attacks

Rapid development of video coding technologies allowed to extend the application areas principally from the multimedia messaging, video telephony, and video conferencing to the mobile TV, wireless and wired Internet video streaming, Standard Definition TV (SDTV), High Definition TV (HDTV), Full HD (applied in HDTV and movies written on Blu-Ray Disc and HD-DVD optical storage media), and Ultra High Definition TV (UHDTV). New technologies are appearing all the time, for example Picture In Picture (PIP). Such development would not be possible without standardization introduced since 1993:

- Coding of Moving Pictures and Associated Audio for Digital Storage Media at up to About 1.5 Mbit/s—Part 2: Video, ISO/IEC 11172-2 (MPEG-1 Video), ISO/IEC JTC 1, Mar. 1993.
- Generic Coding of Moving Pictures and Associated Audio Information—Part 1: Systems, ITU-T Rec. H.222.0 and ISO/IEC 13818-1 (MPEG-2 Systems), ITU-T and ISO/IEC JTC 1, Nov. 1994.
- Generic Coding of Moving Pictures and Associated Audio Information—Part 2: Video, ITU-T Rec. H.262 and ISO/IEC 13818-2 (MPEG-2 Video), ITU-T and ISO/IEC JTC 1, Nov. 1994.
- Narrow-Band Visual Telephone Systems and Terminal Equipment, ITU-T Rec. H.320, ITU-T, Mar. 1993.
- Video Coding for Low Bit Rate communication, ITU-T Rec. H.263, ITU-T, Version 1: Nov. 1995, Version 2: Jan. 1998, Version 3: Nov. 2000.
- Coding of audio-visual objects – Part 2: Visual, ISO/IEC 14492-2 (MPEG-4 Visual), ISO/IEC JTC 1, Version 1: Apr. 1999, Version 2: Feb. 2000, Version 3: May 2004.
- Advanced Video Coding for Generic Audiovisual Services, ITU-T Rec. H.264 and ISO/IEC 14496-10 (MPEG-4 AVC), ITU-T and ISO/IEC JTC 1, Version 1:

May 2003, Version 2: May 2004, Version 3: Mar. 2005, Version 4: Sept. 2005, Version 5 and Version 6: June 2006, Version 7: Apr. 2007, Version 8 (including SVC extension): Consented in July 2007.

Hereinafter, a short description of the transmitting specifications is discussed in Sect. 5.2.1, while Sect. 5.2.2 provides the analysis of possible attacks.

5.2.1 Transmitting Specifications

Variety of devices is involved into transmission. This causes a necessity to support a scalability of video coding standards. However, the prior international video coding standards, such as H.262 MPEG-2 Video (Generic Coding of Moving Pictures and Associated Audio Information—Part 2: Video, ITU-T Rec. H.262 and ISO/IEC 13818-2 (MPEG-2 Video), ITU-T and ISO/IEC JTC 1, Nov. 1994), H.263 (Video Coding for Low Bit Rate communication, ITU-T Rec. H.263, ITU-T, Version 1: Nov. 1995, Version 2: Jan. 1998, Version 3: Nov. 2000), and MPEG-4 Visual (Coding of audio-visual objects—Part 2: Visual, ISO/IEC 14492-2 (MPEG-4 Visual), ISO/IEC JTC 1, Version 1: Apr. 1999, Version 2: Feb. 2000, Version 3: May 2004) included only several tools for the scalability modes. It should be noted that H.264/AVC (Advanced Video Coding) provides the coding efficiency improved significantly compared to all prior standards. Joint Video Team of ITU-T VCEG and ISO/IEC MPEG standardized the following technique called as Scalable Video Coding (SVC) extension of H.264/AVC standard, which can adapt video bit stream to the end-user devices or network conditions [1]. Consider the properties of main video coding standards from a viewpoint of a watermarking.

Conceptually, H.264/AVC (Advanced Video Codec) involves two layers called as Network Abstraction Layer (NAL) and Video Coding Layer (VCL). NAL is organized in the units, each of one starts with a one-byte header showing the type of payload data, while the remained bytes store the content. VCL NAL units include the coded slices and non-VCL NAL units contain additional information about parameter sets and Supplemental Enhancement Information (SEI), which assist the decoding process. VCL of H.264/AVC is similar to the prior video coding standards, such as H.261, MPEG-1 Video, H.262 MPEG-2 Video, H.263, or MPEG-4 Visual, but involves useful features of flexibility and adaptability. In spite the pictures are partitioned into smaller coding units, H.264/AVC supports the traditional concept of subdivision into macroblocks and slices. The macroblocks cover a rectangular picture area of 16×16 luma samples and provide 8×8 samples of each of the two chroma components for video in 4:2:0 chroma sampling format. The samples of macroblock are predicted spatially or temporally, and the resulting prediction residual signal is coded. The macroblocks are organized into three basic slices parsed independently of other slices:

- I-slice: intra-picture predictive coding using spatial prediction from neighboring regions. H.264/AVC uses several directional spatial intra-prediction modes. Prediction signal is generated using the preceding neighboring samples of the blocks. For the luma component, the intra-prediction is applied to 4×4, 8×8, or 16×16 blocks, whereas for the chroma components, it is always based on a macroblock basis.
- P-slice: intra-picture predictive coding and inter-picture predictive coding with one prediction signal for each predicted region. The P-slices and B-slices represent a motion-compensated prediction with multiple reference pictures using variable block sizes. In P-slices, one distinct reference picture list is applied and the used reference picture can be independently chosen for each 16×16, 16×8, or 8×16 macroblock partition or 8×8 submacroblock. Additionally to the macroblocks and submacroblocks, the submacroblocks can be further split into 8×4, 4×8, or 4×4 blocks. Then the reference picture is replicated at the decoder through a reference index parameter.
- B-slice: intra-picture predictive coding, inter-picture predictive coding, and inter-picture bipredictive coding with two prediction signals combined with a weighted average to form the region prediction. In B-slices, two distinct reference picture lists including the previous and posterior reference pictures are yielded, and the prediction method can be selected between list 0, list 1, or biprediction list for each 16×16, 16×8, or 8×16 macroblock partition or 8×8 submacroblock. In the bipredictive mode, the prediction signal is formed by a weighted sum of the prediction signals from lists 0 and 1.

It should be noted that the luma signal of motion-compensated macroblocks cannot contain blocks smaller than 8×8 and is coded using 4×4 or 8×8 transform. For the chroma components, a two-stage transform, consisting of 4×4 transforms and Hadamard transform of the resulting Digital Cosine (DC) coefficients, is applied. In intra 16×16 mode, similar hierarchical transform is used for the luma component. All inverse transforms are specified by the exact integer operations, thus inverse-transform mismatches are avoided. A flexible partitioning of a picture into slices supports a concept of slice groups. Macroblocks of a picture can be arbitrarily partitioned into the slice groups via a slice group map.

H.264/AVC supports two methods of entropy coding known as Context-based Adaptive Variable Length Coding (CAVLC) that uses variable-length codes and Context-based Adaptive Binary Arithmetic Coding (CABAC) that utilizes arithmetic coding and more sophisticated algorithm for employing statistics, which permits to increase typical bit rate on 10–15% relative to CAVLC. Unfortunately, a block-based coding causes often the blocking artifacts. For their reduction, an adaptive deblocking filter is employed, within the motion compensated prediction loop.

H.264/SVC as an extension of H.264/AVC introduces notation of layers within the encoded stream. A base layer provides the lowest temporal, spatial, and quality representation of the video stream, while the enhancement layers encode additional information concerning the higher resolution, frame rate, and quality. A decoder

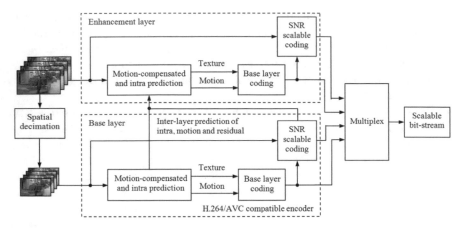

Fig. 5.1 The structure of SVC encoder

produces a video stream with certain desired characteristics, using the base layer and only the subsequent enhancement layers. During the encode process, a particular layer is encoded using reference only to lower level layers. Thus, the encoded stream can be truncated at any arbitrary point and at the same time remain a valid stream. Structure of SVC encoder with two layers is depicted in Fig. 5.1. H.264/SVC bit stream contains a Base layer (Layer 0) and one or more Enhancement layers (Layers 1, 2, etc.).

H.264/SVC provides three types of scalability called as temporal, spatial, and quality expressed by Signal to Noise Ratio (SNR). A bit stream provides temporal scalability, when the set of corresponding access units is partitioned into a temporal base layer and one or more temporal enhancement layers. This is a strategy of temporal data package based on the hierarchical prediction structures. Therefore, the temporal scalability is not suitable for watermarking.

Spatial scalability in SVC supports the traditional approach of multilayer coding used in H.262 MPEG-2 Video, H.263, and MPEG-4 Visual. Each layer having the own identifier corresponds to the supported spatial resolution. The motion-compensated prediction and intra-prediction are implemented for single-layer coding. For spatial scalability, SVC provides three types of prediction techniques appropriate for a watermarking, viz. the inter-layer motion prediction, inter-layer residual prediction and inter-layer intra-prediction. The inter-layer motion prediction means that the motion vector of marcoblock in the base layer will be upsampled as the motion vector of the unknown marcoblock in the enhancement layer. However, the motion vectors in the base layer are not appropriate for watermarking because they may be discarded by enhancement decoding. The inter-layer residual prediction is employed for all inter-coded macroblocks in the base layer and enhancement layer. Herewith, the residual signal in the base layer is upsampled as the prediction for the residual signal of the enhancement layer macroblock. This means that the residual signal of the base layer can be hided or encrypted. The residual signals of P- and B-slices in the base layer are encrypted and then written into bit steam. Here we have the end-to-end coding from the base layer to the enhancement layer. The inter-layer intra-prediction is used for the enhancement layer macroblock, while the base layer is intra-coded. Thus, the encrypted intra-coded block can be detected in the enhancement layer. In other words, lower resolution frames are encoded as the base layer, while the decoded and up-sampled base layer frames are used in the prediction of the higher-order layers. Additional information for reconstruction is encoded from the enhancement layer. Sometimes, re-using motion information can increase encoding efficiency.

It should be noted that H.264/SVC standard supports any resolution, cropping and dimensional aspect relation between two consecutive layers. For example, a certain layer may use Standard-Definition (SD) resolution (4:3 aspect), while the next layer supports High-definition (HD) resolution (16:9 aspect). Also the most common dyadic configuration adopts 2:1 relation between the neighbour layers. At the same time, H.264/SVC assumes non-dyadic ratios [2]. This solution demands the inclusion of a new algorithm class called Extended Spatial Scalability (ESS), where any relation between consecutive layers is supported.

The quality scalability is a special case of the spatial scalability with the identical picture sizes for the base and enhancement layers supported by the general concept for a spatial scalable coding—Coarse-Grain quality Scalable (CGS) coding. CGS uses the same inter-layer prediction mechanisms as for the spatial scalable coding but without the upsampling operations and inter-layer deblocking for the intra-coded reference layer macroblocks. The inter-layer intra- and residual-prediction are directly performed in the transform domain. In order to increase a flexibility of bit stream adaptation and error robustness, the Medium-Grain quality Scalability (MGS) and Fine-Grain quality Scalability (FGS) coding were implemented.

Let us compare video surveillance applications for H.264 AVC and SVC. Suppose that a camera serves multiple streams with 1280×720 and 720×480 resolutions. This makes camera more expansive but allows a stream with 1280×720 resolution to direct at the control device (for example, display) while another stream is decoded and displayed on a mobile phone. It should be noted that steps to decode, resize, and re-encode are already expensive. As shown in Fig. 5.2a, D1 stream is decoded and resized to Common Intermediate Format or Common Interchange Format (CIF) resolution to feed the video analytics. CIF resolution video is temporally decimated to achieve 7 frames per second (fps) and re-encoded in order to be available to the first-response over a wireless link. Using H.264 SVC codec, the full stream with 1280×720 resolution is stored on the Network Video Recorder (NVR), which can easily create D1 (or CIF) stream to free up disk space after a specified period. CIF stream is directed from NVR to CIF decode and analytics. Additional low quality stream can be organized for the mobile phone or another device with low capabilities (see Fig. 5.2b). Evident advantages are the following:

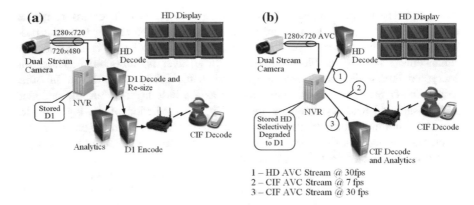

Fig. 5.2 Video surveillance applications [3]: **a** H.264/AVC, **b** H.264/SVC

- Reduced network bandwidth.
- Flexible storage management.
- Removing the expensive stages of decode and re-encode.
- High definition video available on NVR for archive.

Appearance of High Definition (HD), Full HD, and Ultra HD formats required the high efficiency video coding technology with higher capabilities than H.264/SVC. In 2010, the ITU-T Video Coding Expert Group (VCEG) and ISO/IEC Moving Picture Experts Group (MPEG) formed the Joint Collaborative Team on Video Coding (JCT-VC) for design the High Efficiency Video Coding (HEVC) format [4]. The HEVC or H.265/HEVC achieved significant improvement in coding efficiency, especially in the perceptual quality higher 50% bit rate compared to H.264/AVC [5]. Such improvement of the coding efficiency was obtained through the intensive computation complexity.

Typical HEVC video encoder structure is depicted in Fig. 5.3.

H.265/HEVC has a flexible partitioning scheme for large coding block sizes using Coding Tree Unit (CTU) [7]. The biggest block (64 × 64 pixels) is recursively split into smaller Coding Units (CUs) with the depth 0 of CTU to the depth 3 (CU with 8 × 8 pixels) that can be encoded in the intra-frame or inter-frame modes [8]. H.265/HEVC includes three types of units: the Coding Unit (CU), Prediction Units (PUs) for inter-prediction and intra-prediction, and Transform Unit (TU) for residual coding. For each of eight PU partitioning modes, the best matching block of the current PU respect to the reference frames is searched. This process results in a residual (prediction error) and a motion vector of the PU. The CUs, PUs, and TUs encapsulate the Coding Blocks (CBs), Prediction Blocks (PBs), and Transform Blocks (TBs), respectively, as well as, the associated syntax. Difference between this motion vector and motion vector of a neighbour encoded PU is encoded. Residual of CU is transformed to DCT domain using squared Residual Quad-Tree (RQT), which is evaluated from depth 0 (32 × 32 pixels) to depth 3 (4 × 4 pixels). Then these transform coefficients are quantized and entropy encoded.

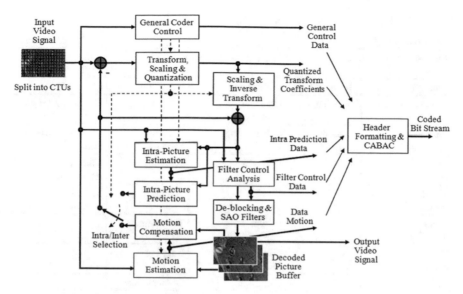

Fig. 5.3 Typical HEVC video encoder [6]

5.2.2 Possible Attacks

Attacks in the broadcast channel may be intentional or accidental. Intentional attacks are directed on the distortions of a part of video or a single frame. These attacks are categorized into common image processing and geometric attacks, while the accidental attacks are concerned to the common processing attacks. Classification of the possible attacks against videos is depicted in Table 5.1.

Frame dropping means a dropping one or several frames randomly from the watermarked video sequence. Frame averaging removes a dynamic aspect. Frame swapping means a switching of the order of frames randomly. If the numbers of the dropped frames, averaged frames, or swapped frames are large, then quality of the watermarked video sequence will rapidly be decreased. Some attacks referred to the common processing attacks, such as MPEG/JPEG compression, color distortions, contrast distortions, and noise adding, can be applied to a whole video or its single frame. A copy attack is applied to fake a watermarked image by collecting and analyzing a set of regions with the same watermark [9]. Single frame may be geometrically distorted by a rich set of techniques including Rotation, Scaling, Translation (RST), flipping, cropping, and local random bending attacks. Composition attacks are such that combine different types of the attacks simultaneously. Also the geometric attacks can be the global and local ones. Glossy copying, JPEG compression, change of frame rate, and change of resolution are the typical accidental attacks concerning the common image processing attacks.

It should be noted that any manipulation with videos is very easy process. At the same time, a restoration of the distorted watermarked video is a big problem

Table 5.1 Classification of the possible attacks

Intentional attacks				Accidental attacks
Against video sequence		Against single frame		Against video sequence
Common	Geometric	Common	Geometric	Common
Frame dropping	Cropping	Median filtering	Cropping	Glossy copying
Frame averaging	Random bending	Sharpening	Random bending	Change of frame rate
Frame swapping		Copying	Rotation	Change of resolution
MPEG compression		JPEG compression	Scaling	MPEG compression
Color distortions		Color distortions	Translation	
Contract distortions		Contrast distortions	Flipping	
Noise adding		Noise adding	Composition	

because the parameters of distortions are unknown. Nowadays, all existing blind watermarking techniques cannot prevent the most of known common image processing and geometric attacks. Usually, a watermarking technique is invariant more or less to the restricted types of the attacks. Full restoration of a watermark is possible under a non-blind watermarking scheme if an image or frame was not cropped or strongly distorted by the ordinary manipulations with intensities.

Some approaches against geometric distortions are recommended, among which are the following ones [10]:

- Spread spectrum modulation. This approach is divided into the additive and multiplicative schemes. Additive spread spectrum scheme supposes that the watermark bits are uniformly spread over the host image. Multiplicative spread spectrum scheme spreads the watermark bits according to the host contents. This approach became popular due to its robustness against the common image processing attacks and some geometric attacks. However, the interference effect of the host image causes a decoding performance degradation of a watermark under attacks.
- Template insertion. This is a common resynchronization technique, in which an additional watermark (called as template) is inserted into the host image. This template is applied as a reference for detection and compensation of geometric deformations, for example affine transform. The template is easily detectable in frequency domain with following removal.
- Invariant domain. An affine-invariant domain using Fourier-Mellin transform, generalized Radon transform, singular value, histogram shape, polar harmonic transforms, and invariant moments are the typical solutions to make a watermark

invariant to the geometric distortions. In this sense, the invariant moments are the best approach for full watermark reconstruction that significantly increases the robustness of a watermark to the geometrical attacks. The drawback is a high computational complexity of the invariant moments.

- Feature-based embedding. Such method is a relatively new resynchronization method based on an image content analysis. At present, Speeded-Up Robust Feature (SURF) detectors as the geometrically invariant features are employed. In this case, the watermark detection can be done without synchronization error. Nevertheless, the feature-based embedding is sensitive to an image modification regarding the color and contrast changes, and a watermark capacity is limited by the local regions for embedding.

The promising way to choose a method for embedding is to analyze a payload and expected attacks for the practical application.

5.3 Related Work

The issues of authentication and Copyright Protection (CP) of video appeared in 1990s. Fruitful ideas from that time could be found in [11–13]. Swanson et al. [11] proposed a scene-based multiscale watermark representation based on the combined static and dynamic representation of a watermark, masking with HVS assumptions, and DWT application along temporal axis of a scene. These authors introduced so called temporal low-pass frames (without motion) and temporal high-pass frames (with motion), according to which the watermark was a composite of the static and dynamic components. Also DWT was applied into each frame, thus we can consider this approach as 3D DWT application. This approach was extended by Piper et al. [13], who used the spread-spectrum scheme proposed by Cox et al. [14], application of a scalable watermark that should be detected in every resolution and quality layer [15], and so on.

One of the main challenges for a watermarking of the rate-scalable compressed video is that the most stable frames for embedding are I-slices of the base layer. Many video watermarking algorithms use this content. At the same time, the enhancement layers add information in a video stream and also ought to be protected by a watermark. In the ideal case, all layers require a corresponding protection. Lin et al. [16] proposed to embed one watermark in the base layer and another watermark in the enhancement layer(s). Such approach satisfies need for the temporal scalability because the enhancement layer does not alter the frames encoded in the base layer. However, it is difficult to provide a watermarking for the spatial and quality forms of scalability because the decoder merges the base and enhancement layers and all embedded watermarks can be distorted under Internet attacks in an unpredictable view. For supporting the scalability modes, Lin et al. [16] placed the most significant elements of a watermark in the beginning of video stream, while the elements of lesser significance were put in the following frames.

Meerwald and Uhl were the first, who formulated three evident strategies of a watermark embedding using SVC spatial scalability, which are depicted in Fig. 5.4 [17].

The first embedding scenario (Fig. 5.4a) implements any robust video watermarking schemes but cannot provide a control over the resulting bit-stream that makes difficult to detect a watermark in the compressed domain. In this case, lossy compression and downsampling of the full-resolution video have an impact on the embedded watermark. Van Caenegem et al. [18] proposed a robust watermarking scheme resilient to H.264/SVC but without scalability. The third embedding scenario (Fig. 5.4c) is too complex for implementation because of the inter-layer prediction structure of H.264/SVC, which requires a drift compensation to minimize error propagation [19]. These algorithms employ usually three types of embedding, viz. the motion compensated intra frames or residual frames, motion vector, or modifying the encoded bit streams. The compression domain embedding modifies a video coding pipeline with inserting the watermark modules within it [20]. Therefore, these schemes are always dependent on the given video coding algorithms and provide less flexibility.

The second embedding scenario (Fig. 5.4b) is the most promising for watermarking, when the integrated H.264/SVC video encoding and watermarking offer control over the bit-stream. Noorkami and Mersereau [21] embedded a watermark in non-zero quantized residual coefficients. Park and Shin [22] supposed an authentication method based on combined encryption and watermarking for H.264/SVC encoding. It should be noted that many watermarking algorithms did not consider a scalable bit-stream and the bit-rate overhead. Hereinafter, Park and Shin [23] studied an authentication scheme based on a fragile watermarking and a copyright protection scheme based on a non-blind watermarking for H.264/AVC

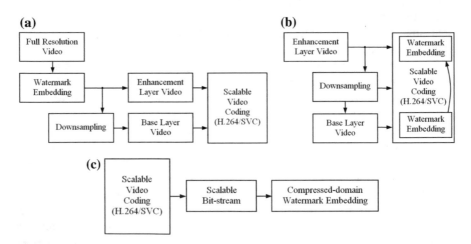

Fig. 5.4 Three scenarios for watermark embedding using H.264/SVC: **a** watermark embedding before video encoding, **b** integrated watermark embedding and coding, **c** compressed-domain embedding after encoding

and H.264/SVC. The proposed authentication scheme used the prediction modes and residuals, while the proposed CP scheme utilized DCT domain of uncompressed I-frames because I-frames in H.264/SVC include the most important information and are transmitted to all content user. Chang et al. [24] evaluated the error propagation in intra-coded frames using Discrete Cosine Transform (DCT)/ Discrete Sine Transform (DST) coefficients and modulating intra-prediction modes. It should be noted that the robust watermarking is usually used for a copyright protection, while the fragile or semi-fragile watermarks are adopted for authentication of video integrity.

Video watermarking algorithms cannot ignore the temporal dimension and motion information of the host video sequences. In the literature, they are categorized as mentioned below:

- Frame-by-frame watermarking. Frame-by-frame scheme is defined as the extension of image watermarking algorithms. This approach was the initial attempt due to its simplicity in implementation. Many similar algorithms are available in the literature [14, 25]. However, the frame-by-frame watermarking schemes often suffer from flickering, instability, compression in hybrid video coding, collusion, frame dropping, and de-synchronization.
- 3D decomposed watermarking. 3D variants of many transforms using in the image watermarking are suitable for video watermarking, for example 3D DFT [26]; 3D DCT [27], and multi-resolution 3D DWT [28], providing the imperceptibility and robustness of a watermark. These 3D decomposition-based methods solve the issues of temporal de-synchronization, video format conversion, and video collusion but their sensitivity to flickering and fragility remains because a motion is not considered.
- Motion compensated decomposition. These schemes prevent the collusion and compression attacks and also remove flickering. Algorithms work in the uncompressed domain [29] or joint compression domain [30]. Object motion within the frames are tracked and compensated followed by the transform. Watermark embedding is done on the transform coefficients before or after the quantization process in the intra frames or prediction frames, respectively. Motion Compensated Temporal Filtering (MCTF) provides a better framework for this type of video watermarking.
- Bit stream domain watermarking. The watermarking is executed on partially decoded bit streams [20]. This type provides a low computational cost that leads to the real-time watermarking. It prevents the decoding and re-encoding data loss, compared with the joint compression domain watermarking schemes. Bit-stream is partially decoded by entropy decoding followed by the de-quantization process. Watermark embedding is performed on the transform coefficients. The main disadvantage is appearance of errors due to the embedding modification that causes distortions.

- Motion vector based watermarking. This is a special case of bit stream domain watermarking. In any video encoder, a motion vector is always preserved and encoded with higher priority that makes such embedding robust to compression attacks. However, any small change in the motion vector causes a significant distortion in the host video. Thus, a careful balance ought to be specified [31]. Methods based on the higher magnitude, motion estimation mode selection, texture, and phase angle are used for selection of the motion vectors to be embedded.

For the scalable watermarking, system should be able to detect watermarks in all scalable bit-streams. If a quality of bit-stream decreases, the correlation between the watermark and watermarked signal may also decrease. For the watermark detection, the adaptive (for each layer) thresholds are usually applied. Shi et al. [32] proposed a scalable and credible watermarking algorithm towards SVC for CP system. This algorithm used a combination of the frequency mask, contrast mask, luminance adaption, and temporal mask in order to find the best watermark locations in the base and enhancement layers. In spite of heuristic aspect of the suggested masking, they succeeded to achieve false alarm rate values close to zero.

Bhowmik and Abhayaratne [33, 34] designed a non-commercial software framework for evaluating robustness of image and video watermarking to the content adaptation attacks. This software was called Watermark Evaluation Bench for Content Adaptation Modes (WEBCAM) and it is emulated a heterogeneous communication system, in which the content was encoded using the scalable coders to produce scalable bit streams. In such manner, a simulation of optical, wired or wireless networks was implemented for JPEG2000-based content adaptation [33].

Recently appeared H.265/HEVC standard is worthy of investigation, particularly in a watermarking aspect. Similar to H.264/AVC, the digital watermarking algorithms in HEVC are classified into three approaches in accordance with the embedding position: the prediction, transform and quantification, and entropy encoding. Embedding in the stage of prediction in its turn is represented by the intra prediction mode, the size of PU, and motion vector.

Embedding in the stage of prediction:

- Intra prediction mode. Yang et al. [35] formulated the mapping rule between the prediction modes and hiding information using a probability distribution of the statistically optimal and sub-optimal prediction modes. A hiding algorithm for H.265/HEVC based on the angle differences of intra prediction mode was proposed by Wang et al. [36]. They created the mapping table between the 35 types of prediction modes and corresponding angle values. For the encoding process, 4×4 luma blocks were utilized. Fast intra prediction algorithm with the enhanced intra mode grouping was proposed by Park and Jeong [37]. This algorithm converted each PU into transform (Hadamard instead of DCT) domain to classify a major directionality that provided the candidate modes.

Fig. 5.5 The angular
intra-prediction modes

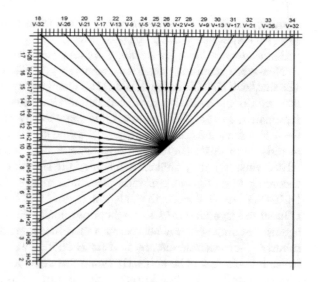

Then the intra mode group was update with certain conditions. This algorithm
provided a time saving for more than 25% in average with a slight visual quality
loss.

– The size of PU.

In order to improve a compression efficiency, the pixel values of PU should be
predicted close to the original pixels values. In HEVC [38], 33 angular prediction
modes for both luma and chroma and two non-directional prediction modes (DC,
Planar) are defined. Prediction block sizes are chosen from 4 × 4 up to 64 × 64
elements. DC intra-prediction is the simplest mode in HEVC. PU pixel is equaled to
the mean value of available neighboring pixels. Planar intra-prediction mode
implies 2D linear interpolation of neighboring pixels values. Angular
intra-prediction modes are the linear interpolations of pixel values in the corresponding 33 directions as it depicts in Fig. 5.5.

– Motion vector.

Van Pham et al. [39] developed a low complexity out-of-the-loop information
hiding algorithm for a video pre-encoded according to H.265/HEVC standard.
Information was hidden in the compressed domain by modifying the syntax elements of video stream. It should be noted that the luma and chroma Coding Blocks
(CBs) form the syntax elements. For this case, only low-complex entropy decoding
and encoding is required. The main challenge was to determine the optimal type
and amount of bit stream elements for embedding. Motion vectors and DCT
coefficients involved in a bit stream contained the hiding information. Also, Van
Pham et al. [39] proposed a selection strategy for distribution of hidden information
across the blocks and frames in the bit stream because modifying the syntax

elements causes a mismatch in coding information between encoder and decoder and as a consequence the propagated errors.

Embedding in the stage of transform and quantification.

Non-zero Quantized Transform Coefficients (QTCs) can be used for embedding. The first such algorithms employed QTCs changing by simple arithmetic operations that led to a sensitivity of Quantization Parameter (QP) and non-robustness. Hereinafter, Jo et al. [40] improved this approach by modifying non-zero QTCs of the 4×4 luma TBs of intra predicted I-frame during embedding process. In order to resist error drift phenomenon (this appears due to the feedback loop in H.265/ HEVC encoder), they differently modified the significant bits of 4×4 luma block according to its direction of intra prediction modes. Chang et al. [41] suggested DCT/DST-based data hiding algorithm for H.265/HEVC intra-coded frames without propagating errors to neighboring blocks. However, this scheme is fragile against re-compression attack because of drift compensation. Also, it is known a robust digital watermarking method for H.265/HEVC based on a repetition Bose–Chaudhuri–Hocquenghem (BCH) syndrome code technique without intra-frame distortion drift [42]. They improved the robustness to channel errors by applying BCH encoding respect to work [39], but the embedding capacity remained low.

Embedding in the stage of entropy encoding.

In work [43], CABAC as the entropy coding was proposed with a novel concept of Constant Bitrate Information Bit (CBIB) applied to Motion Vector Difference (MVD) in H.265/HEVC. Suggested CABAC-based video steganography algorithm was based on the codeword reservation and substitution rule for the encoding of MVDs.

Gaj et al. [44] studied a prediction mode based H.265/HEVC video watermarking resisting the re-compression attacks, as well as, common image processing attacks while maintaining a decent visual quality. The compressed domain watermarking scheme was proposed for H.265/HEVC bit stream that allowed them to handle a drift error propagation both for intra- and inter-prediction processes.

5.4 Selecting Relevant Regions in Semi I-Frames

As it follows from the literature review represented in Sect. 5.3, three scenarios for watermark embedding using H.264/SVC are possible: watermark embedding before video encoding, integrated watermark embedding and coding, and compressed-domain embedding after encoding. We simulate a sequence of I-frames in the original video sequence by a criterion of significant motion between three neighboring frames $(t - 1)$, t and $(t + 1)$. The selecting relevant regions are under formulated the restrictions and conditions caused by the transmitted specifications.

In existing video compression standards, the basic coding unit is Group Of Pictures (GOP). The first frame in each GOP is the intra-coded frame called as I-slice (in the terms of H.264/AVC). The following frames are the predicted frames and bi-directionally frames called as P-slice and B-slice, respectively. Hereinafter, we will support the terms I-frame, P-frame, and B-frame in our discussion regarding the original video. It should be noted that P-frame and B-frame were introduced for the motion estimation and motion compensation in order to eliminate further temporal redundancy of video contents.

It is well-known that nowadays Convolutional Neural Networks (CNNs) have wide distribution in many practical applications including the motion detection in videos [45], salient object detection [46], and texture analysis [47] due to their high accuracy results after long-time learning. Use of three CNNs with different structures or multi-functional CNN creation as a preprocessing step is a non-reasonable decision in the sense of the computational and time-consuming costs. In our task, we do not need high accuracy estimates. Our goal is to provide the fast computed approximate maps for following analysis.

The steps of procedure for selecting the relevant regions, including a creation of the motion map, saliency map, textural map, blue component map, and joint map are discussed in Sects. 5.4.1–5.4.5, respectively.

5.4.1 Motion Map

Block Matching Algorithm (BMA) is a simple and non-consuming in time (relative to optical flow and other accurate motion estimates) method that finds any motion in scene. The basic BMA is referred as Full Search (FS) strategy. Some modifications of FS such as Exhaustive Search (ES), Three-Step Search (TSS), four-step search, conjugate direction search, dynamic window search, cross-search algorithm, two-dimensional logarithmic search, Diamond Search (DS), and adaptive rood pattern search differ by the searching algorithm [48]. First, a current frame is divided into the non-crossed blocks with sizes 16×16 pixels, which are defined by the intensity function $I_t(x, y)$, where (x, y) are the coordinates, t is the discrete time moment. Second, for each block in small neighborhoods $-S_x < d_x < +S_x$ and $-S_y < d_y < +S_y$ the most similar block in following frame $I_{t+1}(x + d_x, y + d_y)$ is detected.

Let us compare the most popular modifications such as ES, TSS, and DS. Modification ES is very close to FS strategy, when a similarity is computed in each point of a search area. This leads to the best matching estimates but complexity is the highest in respect to other modifications. Modification TSS is executed by three steps, for which neighborhood sizes are decreased in scales 4, 2, and 1. After each step, a reference point (point associated with maximum similarity measure) is chosen. Modification TSS is a popular algorithm thanks to its simplicity and high productivity. Its drawback connects with missing of small motion in a scene. Modification DS uses two patterns called as Large Diamond Search Pattern (LDSP) and Small Diamond Search Pattern (SDSP). The difference between LDSP and SDSP is in that a similarity is computed in points (x, y) relative the reference point under conditions $|x| + |y| = 2$ and $|x| + |y| = 1$, respectively. Pattern LDSP is applied until the best value is computed in a point differed from the reference point,

(a)   **(b)**

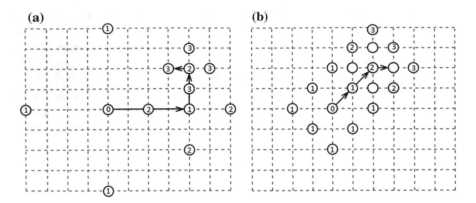

Fig. 5.6 Examples of search patterns: **a** pattern TSS, **b** pattern DS

otherwise pattern LDSP is changed by pattern SDSP. Modification DS can find global minimum exactly because a search pattern is not large. Modification DS has Peak Signal to Noise Ratio (PSNR) value closed to PSNR value provided by modification ES under lesser computational cost. Example, which demonstrates a sequence of the applied patterns for modifications TSS and DS, is depicted in Fig. 5.6.

Similarity is determined by minimization of the error functional according to some metrics, such as Sum of Absolute Differences (SAD) (Eq. 5.1), Sum of Squared Differences (SSD) (Eq. 5.2), and Mean of Squared Differences (MSD) (Eq. 5.3), where n is the number of the analyzed surrounding blocks.

$$SAD\left(d_x, d_y\right) = \sum_{x=1}^{N} \sum_{y=1}^{N} |I_{t+1}(x, y) - I_t(x + d_x, y + d_x)| \tag{5.1}$$

$$SSD\left(d_x, d_y\right) = \sum_{x=1}^{N} \sum_{y=1}^{N} (I_{t+1}(x, y) - I_t(x + d_x, y + d_x))^2 \tag{5.2}$$

$$MSD\left(d_x, d_y\right) = \frac{1}{n \times n} \sum_{x=1}^{N} \sum_{y=1}^{N} (I_{t+1}(x, y) - I_t(x + d_x, y + d_x))^2 \tag{5.3}$$

Computational estimates are the following ones: $O(255 \cdot w \cdot h) \approx O(w \cdot h)$ for modification ES, $O(25 \cdot w \cdot h) \approx O(w \cdot h)$ for modification TSS, and $O(4 \cdot (k + 1) \cdot w \cdot h) \approx O(w \cdot h \cdot k)$ for modification DS, where w and h are the width and height of frame, respectively, k is the number of steps for pattern LDSP. The complexity of all these algorithms is characterized by the proportional dependence from the frame sizes. Number of calculations for modification TSS is decreased in 9 times respect to modification ES. Modification DS has not such advantage because a step number of pattern LDSP is unknown.

Due to independence of the neighborhood block matching, we may organize a parallel data processing. For this purpose, we share an area of processed frame between the cores of processor and apply standard OpenMP for C, C++, or FORTRAN program parallelization [49]. Real-time mode can be achieved with 4 threads for processing of frames with the sizes 2560 × 1440 pixels by modification TSS and 1920 × 1080 pixels by modification DS with average frequency 30 frames per second. Modification ES is not suitable for the real-time processing of frames even with the sizes 640 × 360 pixels.

The detected motion area is excluded from I-frame. As a result, we obtain a motion map *MM*.

5.4.2   Saliency Map

In spite of different saliency maps exist [50], the intensity, color, and contrast saliency maps provide the main contribution. Thus, we consider a creation of these three maps, which are jointed in a common saliency map by linear fusion.

Simplified mathematical model called as DIVision of Gaussians (DIVoG) was proposed in [51] as a fast algorithm for intensity and color maps representation. DIVoG algorithm includes three steps:

- Step 1. The Gaussian pyramid U is constructed from a set of bottom-up layers $U_1, U_2, ..., U_n, n = 5$. The first layer U_1 has a resolution $w \times h$ pixels. Then a layer resolution is decreased by $\left(w/2^{n-1}\right) \times \left(h/2^{n-1}\right)$ and derived via down-sampling using 5 ×5 Gaussian filter. Thus, U_n is a top layer of pyramid.
- Step 2. The reversed Gaussian pyramid D is generated, starting with a layer U_n, in order to derive its base D_1. Thus, the pyramid layers are derived via up-sampling using 5 ×5 Gaussian filter.
- Step 3. The minimum ratio matrix $m_{n_{ij}}$, performing the pixel by pixel division of U_n and D_n in each layer, is computed by Eq. 5.4.

$$m_{n_{ij}} = \min\left(D_{n_{ij}}/U_{n_{ij}}, U_{n_{ij}}/D_{n_{ij}}\right) m_{n-1_{ij}} \quad n \geq 1 \qquad (5.4)$$

Then an element $sm_{1_{ij}}$ of saliency map SM_{In}, normalized in the range [0 ... 1], is formed by Eq. 5.5.

$$sm_{1_{ij}} = \left(1 - m_{1_{ij}}\right) \in SM_{In} \qquad (5.5)$$

The elements sm_{1_R}, sm_{1_G}, and sm_{1_B} of saliency map SM_{Cl} are normalized in the range [0...255] with following conjunction into a single salience color map SM_{Cl}

and calculated by Eq. 5.5 in similar manner. To avoid a division by zero in Eq. 5.4, we assume that all values of intensity and color are positive numbers.

The contrast map is the most complex and involves responses of the texture energy contrast, texture gradient contrast, and color contrast. The measures introduced by Laws [52] are one of the fast ways to estimate the texture energy contrast and texture gradient contrast. These measures are computed by applying 2D convolution kernels generated from the set of 1D convolution kernels of length three and five provided by Eq. 5.6, where L is the mask of average gray level, E is the mask of edge features, S is the mask of spot features, W is the mask of wave features, R is the mask of ripples features for kernels with dimensions 3 and 5.

$$
\begin{aligned}
L_3 &= \begin{bmatrix} 1 & 2 & 1 \end{bmatrix} & E_3 &= \begin{bmatrix} 1 & 0 & -1 \end{bmatrix} & S_3 &= \begin{bmatrix} 1 & -2 & 1 \end{bmatrix} \\
L_5 &= \begin{bmatrix} 1 & 4 & 6 & 4 & 1 \end{bmatrix} & E_5 &= \begin{bmatrix} -1 & -2 & 0 & 2 & 1 \end{bmatrix} & S_5 &= \begin{bmatrix} -1 & 0 & 2 & 0 & -1 \end{bmatrix} \\
W_5 &= \begin{bmatrix} -1 & 2 & 0 & -2 & 1 \end{bmatrix} & R_5 &= \begin{bmatrix} 1 & -4 & 6 & -4 & 1 \end{bmatrix}
\end{aligned}
\tag{5.6}
$$

The texture energy contrast CE_i in area Ω_i is provided by Eq. 5.7, where $D(e_i, e_j)$ is the Euclidian distance between texture energies e_i and e_j in the central region Ω_i and the surrounding regions Ω_j, respectively, w_{ij}^{en} is the spatial constraint of texture energy.

$$
CE_i = \sum_{j \in \Omega_j} D(e_i, e_j)^2 w_{ij}^{en}
\tag{5.7}
$$

L^2-norm of function $D(e_i, e_j)$ is the convolutions between the original frame I (x, y), where (x, y) are the coordinates, and one of masks M_k determined by Eq. 5.6.

$$
D(e_i, e_j) = \left| \sum_{(x,y) \in \Omega_i} |I(x, y) * M_k(x, y)| - \sum_{(x,y) \in \Omega_j} |I(x, y) * M_k(x, y)| \right|_2
\tag{5.8}
$$

Spatial constraint w_{ij}^{en} enhances the effect of nearer neighbor. This constraint is an exponential function provided by Eq. 5.9, where α is the weighted coefficient controlling a spatial distance, en_i and en_j are the normalized texture in the central region Ω_i and the surrounding regions Ω_j, respectively.

$$
w_{ij}^{en} = e^{-\alpha \|en_i - en_j\|^2}
\tag{5.9}
$$

It should be noted that Eqs. 5.7–5.9 are used for each Gaussian pyramid level. Texture gradient contrast can be estimated in similar manner using Eqs. 5.7–5.9 but instead of the masks denoted in Eq. 5.6 the gradient masks are applied.

Color contrast includes tremendous saliency information of a given frame. Fu et al. [53] proposed the combined saliency estimator taking the advantages of color contrast and color distribution. The main idea is to compute the color contrast $\hat{C}_i$

saliency in area Ω_i using Eq. 5.10, where $D(c_i, c_j)$ is Euclidian distance between colors c_i and c_j in the central region Ω_i and the surrounding regions Ω_j, respectively, w_{ij}^c is the spatial constraint of color contrast. (Given frame in RGB-color space ought to be transformed in a color-opponent space with dimension L for lightness and a and b for the color-opponent dimensions—Lab-color space.)

$$\hat{C}_i = \sum_{j \in \Omega_j} D(c_i, c_j)^2 w_{ij}^c \tag{5.10}$$

Spatial constraint w_{ij}^c is computed similarly by Eq. 5.9. Human attention has higher probability to fall onto the center area of a frame. This fact is considered by the color distribution CD_i in Eq. 5.11 that defines a color contrast CC_i.

$$CC_i = \hat{C}_i \cdot CD_i \tag{5.11}$$

All three normalized contrast estimators join in a single map as a saliency contrast map SM_{Cn}.

Normalized saliency maps based on intensity, color, and contrast are fused to the saliency map SM_m linearly with empirical coefficients k_{In}, k_{Cl}, and k_{Cn}, respectively:

$$SM_m = k_{In} \cdot SM_{In} + k_{Cl} \cdot SM_{Cl} + k_{Cn} \cdot SM_{Cn}. \tag{5.12}$$

A visibility of saliency map can be enforced by multiply on coefficient k_G calculated by Eq. 5.13, where $d(p_i, p_j)$ is Euclidean distance from ith pixel p_i to a definite foreground or background jth pixel p_j, k is parameter, $k = 0.5$, Ω_F and Ω_B are the areas of foreground and background, respectively.

$$k_G = \exp\left(-k \frac{\min_{j \in \Omega_F}\left(d(p_i, p_j)\right)}{\min_{j \in \Omega_B}\left(d(p_i, p_j)\right)}\right) \tag{5.13}$$

The main idea of Eq. 5.13 is to give more weight to pixels, which are closer to the definite foreground region. The obtained saliency map ought to be inversed for our task (called as SM) because the watermark embedding into salient object can provide artifacts of visibility.

5.4.3 Textural Map

Textural map can be created by different ways respect to the goal – recognition, clustering, reconstruction, or watermarking [54, 55]. The earliest statistical approaches were based on the moments of 1st–4th orders (average, dispersion, homogeneity, smoothness, entropy, skewness, and kurtosis), Haralick texture features [56], Galloway texture features [57], Tamura texture features [58], among other. Haralick et al. [56] suggested a set of 14 textural features with the main parameters—homogeneity, contrast, and entropy, which were extracted from a co-occurrence matrix. Galloway [57] introduced five original features of run-length statistics, which were extracted from a gray-level image. At present, these run-length statistics have a historical meaning. Tamura features included the coarseness, directionality, regularity, contrast, line-likeness, and roughness. First three features are more significant respect to remaining ones. MPEG-7 standard had employed the regularity, directionality, and coarseness as the texture browsing descriptor [59]. It should be noted that the Tamura and MPEG-7 texture descriptors are non-invariant to scaling. The following propositions were connected with building a gray-level co-occurrence matrix [60], calculation of spectral features based on 2D wavelet transform, and Gabor transform, which is invariant to scaling.

We have interest for two techniques, viz. fast computed Local Binary Patterns (LBPs) and approximates of fractal estimation, which provide more accurate detection of textural areas due to the homogeneity analysis. Among numerous modifications of LBP, we found measures of texture classification.

LBP describes a unique encoding of the central pixel with position c regarding its local neighborhood containing P pixels with a predefined radius value R. LBP is calculated by Eq. 5.14:

$$LBP_{P,R} = \sum_{p=0}^{P-1} s(g_p - g_c)\, 2^p, \tag{5.14}$$

where $g(\cdot)$ is the gray-scale value of pixel, $g(\cdot) \in [0, \ldots, 255]$ and

$$s(\cdot) = \begin{cases} 1, & s(\cdot) \geq o \\ 0, & s(\cdot) < 0 \end{cases}.$$

A uniformity measure U returns the number of bitwise 0/1 and 1/0 transitions in LBP. LBP means the uniform LBP if $U \leq 2$. Pietikäinen et al. [61] created 58 uniform patterns in $(8, R)$ neighborhood. Ojala et al. [62] introduced the rotation-invariant variance measure VAR provided by Eq. 5.15.

$$VAR_{P,R} = \frac{1}{P} \sum_{p=0}^{P-1} (g_p - \mu)^2, \tag{5.15}$$

where

$$\mu = \frac{1}{P} \sum_{p=0}^{P-1} g_p.$$

These two measures $LBP_{P,R}$ and $VAR_{P,R}$ might be used for texture classification. As Teutsch and Beyerer mentioned [63], the gradient magnitude is estimated using these parameters by Eq. 5.16.

$$G(LBP_{P,R}) = \begin{cases} \sum_{r=R_1}^{R_n} \sqrt{VAR_{P,r}} & \text{if } LBP_{P,R} \text{ is uniform} \\ 0 & \text{else} \end{cases} \qquad (5.16)$$

Liu et al. [64] suggested four LBP-like descriptors: two local intensity-based descriptors, such as Central Intensity LBP (CI-LBP) and Neighborhood Intensity LBP (NI-LBP), and two local difference-based descriptors, such as Radial Differences LBP (RD-LBP) and angular differences LBP (AD-LBP). Using this simple technique, we may find the high textural areas in frame for a watermark embedding.

However, the textural regions may be homogeneity, for example, visualization of sky with small clouds. The homogeneity regions are bad regions for data embedding. They can be detected using fractal analysis.

The self-similarity and irregularity of textures can be defined by fractal theory. For images with ideal fractal structure, fractal dimension D is computed in a view of Eq. 5.17, where parameters r and N_r are estimated in dependence of the chosen method.

$$D = \lim_{r \to 0} \frac{\log(N_r)}{\log(1/r)} \qquad (5.17)$$

However, it is difficult to compute D using Eq. 5.17 directly. Also, the most texture images are not the ideal fractal images [65]. Some approximates for calculation of fractal dimension were proposed since 1990s, for example, the reticular cell counting approach [66] and the probability modification [67]. The most famous is Box Counting Dimension (BCD) method [68], which can be applied to the patterns with and without self-similarity.

In BCD method, an image with resolution $M \times M$ pixels can be considered as a 3D surface with (x, y) projection on an image plane, where the coordinate z denoted a gray level. An image is divided into boxes with $s \times s \times s'$ pixels, where s' is a height of each box, $1 < s < M/2$ and s is an integer. The parameter r is defined as $r = s/M$. These boxes are indexed with (i, j) in the 2D space. If the minimum and maximum grayscale levels in the (i, j)th grid fall into the kth and lth boxes, respectively, then the contribution of n_r in the (i, j)th grid is defined by simple counting:

$$n_r(i,j) = l - k + 1. \qquad (5.18)$$

The parameter N_r shows the contributions on $n_r(i, j)$ as their summation:

$$N_r = \sum_{i,j} n_r(i,j). \tag{5.19}$$

The parameter N_r can be computed for different sizes of the partitioned boxes r in order to be convinced in the fractal properties of the test texture. However, if a texture image is noisy, then the fractal dimension can be estimated from the least-square linear fit of $\log(N_r)/\log(1/r)$ using different values of parameter r.

Also, some methods for fractal estimation of the color images are developed [69].

Fractal dimension measures a space occupation. However, often the textures with different structures have the identical values of fractal dimension. Lacunarity reflects a property of texture fullness. Lacunarity as a special distribution of specific gap size l along the texture was introduced [70]. Unlike fractal dimension, a lacunarity is a scale dependent measure. The earlier algorithms computed a lacunarity using the analysis of the mass distribution in the deterministic or random set. One of the most famous and simple algorithms, based on Differential Box Counting (DBC) method, was proposed by Du and Yeo [71]. For each $l \times l$ gliding box, the relative height of a column $h_l(i, j)$ is calculated by Eq. 5.20, where i and j are the image coordinates, v and u are the maximum and minimum pixel values inside this box, respectively.

$$h_l(i,j) = [v/l] - [u/l] \tag{5.20}$$

Probability distribution $P(H, l)$ is computed by Eq. 5.21, where H is each relative height, $\delta(\cdot)$ is a Kronecker's delta.

$$P(H,l) = \sum_{i,j} \delta(h_l(i,j), H) \qquad \delta(x,y) = \begin{cases} 1, & x = y \\ 0, & x \neq y \end{cases} \tag{5.21}$$

Probability dense function $Q(H, l)$ and, finally, the lacunarity $\Lambda(l)$ for a box size l are defined by Eqs. 5.22–5.23, respectively.

$$Q(H,l) = P(H,l) \bigg/ \sum_{\forall H} P(H,l) \tag{5.22}$$

$$\Lambda(l) = \sum H^2 \cdot Q(H,l) \bigg/ \left(\sum H \cdot Q(H,l) \right)^2 \tag{5.23}$$

It should be noted that a low value of lacunarity indicates the homogeneity, while a high value of lacunarity specifies the heterogeneity structure of texture.

The approach, when a lacunarity is computed in terms of the local binary patterns, was proposed by Backes [72]. This approach is based on the gliding-box method and DBC method designed for binary images and gray scale images, respectively. The mass was calculated as a sum of 1 in LBP operator. The obtained results demonstrated the computational simplicity and accuracy.

Use of LBP approach, or fractal approach, or both approaches together allows to build textural map *TM*.

5.4.4   Blue Component Map

Methods of color image segmentation can be divided into four categories: feature-based techniques, region-based techniques, edge-based techniques, and hybrid techniques, which combine procedures concerning to the first three categories. Clustering, adaptive *k*-means clustering, and histogram thresholding are referred to the feature-base techniques, which neglect the spatial information among pixels sharing the same color. Split-merge methods, region growing methods, graph-theoretical methods, and neural network methods are belong to region-based techniques. Various gradient methods, Hough transform, and wavelet transform are related to the edge-based techniques.

For building a blue component map *BM*, we apply a hybrid technique, which extracts blue component in each pixel position with following morphological processing of "blue" regions in order to enlarge the detected "blue" regions.

5.4.5   Joint Map

Relevant regions for watermark embedding are detected from the joint map *JM*, where four maps have imposed each other:

$$JM = MM \cap SM \cap TM \cap BM. \tag{5.24}$$

Obtained joint map *JM* includes the preferable regions for embedding. However, we may exclude any map, for example *BM* or/and *SM*, if a total region for embedding will not be enough for watermarking procedure. Thus, Eq. 5.24 shows the ideal case but is only recommended.

Example of *JM* building is depicted in Fig. 5.7. The regions for embedding a watermark are shown in white color of Fig. 5.7e.

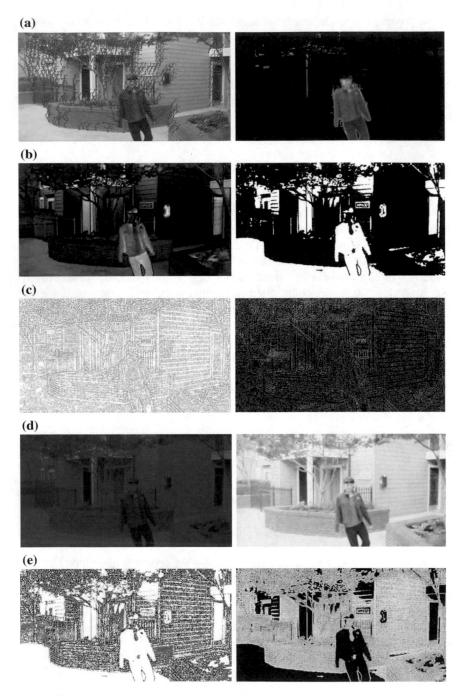

Fig. 5.7 Example of *JM* building: **a** motion vectors, motion map MM, **b** salient regions, salient map *SM*, **c** LBP-detected regions, textural map *TM*, **d** blue intensity regions, blue map *BM*, **e** inverted association of four maps, joint map *JM*

5.5 Proposed Video Watermarking Method

Proposed video watermarking method is robust to several types of typical Internet attacks, e.g. the common image processing attacks, global and local geometric attacks, and permutation attacks. Usually the common image processing attacks and geometric attacks are referred to a single frame, while the permutation attacks are directed against video. The common image processing attacks reduce watermark energy and include median filtering, sharpening, noise adding, and JPEG compression. Compensation of this type of attacks can be considered as a preliminary step that includes estimation of intensity equalization (in order to correct whitening attacks) and correction of contrast in simple cases.

Geometric attacks induce synchronization errors between the original and extracted watermarks and involve so called RST attacks, cropping, random bending, and composition. Detection of cropped regions (usually black or white) is an easily executed procedure but the cropping attacks remote the embedding data without possibility of any prediction. If the cropped regions are not large, then with some probability interpolation of the extracted watermark may improve a visibility of watermark. Global geometric synchronization is achieved by the content-based methods or invariant transforms. However, at present time any protecting method is useful for restricted types of attacks. Therefore, a combination of different protecting methods is a promising way for watermark defense.

Our method is based on a careful detection of relevant regions for embedding. The invariance to global RST attacks is provided by comparison of feature points' sets using SURF descriptors (Sect. 5.5.1). The novelty is that we do not embed watermark in these stable regions but embed the coordinate values of SURF descriptors as they were in the host frame. This allows us to avoid the corresponding matches between SURF descriptors in the host and watermarked frames and extend a volume of embedded information. In order to provide invariance to RST attacks in the stable regions, we use Exponential Moments (EMs) calculated on a unit circle (Sect. 5.5.2).

The proposed watermark embedding and extraction schemes follow the ordinary technique [73]. The algorithmic implementations are mentioned in Sects. 5.5.3 and 5.4, respectively.

5.5.1 Scale-Invariant Feature Points Extraction

Feature-based watermarking schemes are robust to the desynchronization attacks and sometimes to the common image processing attacks. In this research, we use Hessian-based detectors, e.g. multi-scale Gaussian model, which approximates Laplacian-of-Gaussian (LoG) using Difference-of-Gaussians (DoG) [74].

Let $I(x, y)$ be an image and $G(x, y, \sigma)$ be a variable-scale Gaussian filter defined by Eq. 5.25.

$$G(x, y, \sigma) = \frac{1}{2\pi\sigma^2} e^{-(x^2 + y^2)/2\sigma^2} \tag{5.25}$$

The scale function $L(x, y, \sigma)$ is computed as a convolution between Gaussian filter and image and provided by Eq. 5.26, where sign "*" is the convolution operation.

$$L(x, y, \sigma) = G(x, y, \sigma) * I(x, y) \tag{5.26}$$

The DoG function, $D(x, y, \sigma)$ is calculated as a difference between two nearby scales under consideration of multiplicative factor k (Eq. 5.27).

$$D(x, y, \sigma) = (G(x, y, k\sigma) - G(x, y, \sigma)) * I(x, y) = L(x, y, k\sigma) - L(x, y, \sigma) \tag{5.27}$$

As a result, we obtain a set of candidate feature points in the relevant regions. It is desirable to have the feature points distributed uniformly in frame. For this purpose, we impose a grid with large sizes cells (for example, 5×4 cells) and detect a restricted number of feature points in each cell.

5.5.2 Exponential Moments

Moment invariants are widely employed in watermarking schemes due to the invariance of their kernel functions to RST attacks and some intensity variations. We can mention Zernike and pseudo-Zernike moments [75], Fourier–Mellin moments [76], complex moments [77], wavelet moments [78], Polar Harmonic Transform (PHT) [79], exponent-Fourier moments [80], and Exponent Moments (EMs) [10]. Compared with other orthogonal moments, the PHT and EMs are characterized by better image reconstruction, lower noise sensitivity, lower computational complexity, and good invariant properties. However, EMs application is more suitable for our task due to higher simplicity respect to PHT.

Suppose that a function set $P_{n,m}(r, \theta)$ is defined in a polar coordinate system and contains the radial function $A_n(r)$ and Fourier factor in angle direction $\exp(jm\theta)$ provided by Eq. 5.28:

$$P_{n,m}(r, \theta) = A_n(r) \exp(jm\theta), \tag{5.28}$$

where

$$A_n(r) = \sqrt{2/r} \exp(j2n\pi r), \quad n, m = -\infty, \ldots, +\infty, \ 0 \le r \le 1, \ 0 \le \theta \le 2\pi.$$

The set of $P_{n,m}(r, \theta)$ is orthogonal on the unit circle

$$\int\limits_{0}^{2\pi} \int\limits_{0}^{1} P_{n,m}(r,\theta)P^*_{k,l}(r,\theta)r\,dr\,d\theta = 4\pi\delta_{n,k}\delta_{m,l},$$

where 4π is the normalization factor, $\delta_{n,k}$ and $\delta_{m,j}$ are the Kronecker symbols, $P_{k,l}{}^*$ (r,θ) is the conjugate of $P_{k,l}(r,\theta)$.

The 2D function $f(r,\theta)$ can be decomposed into the set of $P_{n,m}(r,\theta)$ using Eq. 5.29:

$$f(r,\theta) = \sum_{n=-\infty}^{n=+\infty} \sum_{m=-\infty}^{m=+\infty} E_{n,m}A_n(r)\exp(jm\theta), \qquad (5.29)$$

where $E_{n,m}$ is the EMs of order n with repetition m, $A_n{}^*(r)$ is the conjugate of $A_n(r)$. EMs have a view of Eq. 5.30.

$$E_{n,m} = \frac{1}{4\pi} \int\limits_{0}^{2\pi} \int\limits_{0}^{1} f(r,\theta)A^*_n(r)\exp(-jm\theta)r\,dr\,d\theta \qquad (5.30)$$

Due to a principle of the orthogonal functions, 2D function $f(r,\theta)$ can be reconstructed approximately by limited orders of EMs ($n \leq n_{\max}$, $m \leq m_{\max}$).

EMs are satisfied to RST invariant properties. Translation invariant is achieved by placement of the origin of local coordinate center in the coordinates of feature point. EMs magnitudes $|E_{n,w}(f)|$ are invariant respect to rotation transform. We may find an angle α using a phase shift $\exp(jm\alpha)$ of the $E_{n,m}(f)$. Further, EMs magnitudes are invariant to scaling if the computation area covers the same content because the EMs are defined on the unit circle. Last proposition is fulfilled in the neighborhood of stable feature point.

We embed the coordinates of feature point in EMs calculated on a unit circle of stable region. Also, if it is possible, we embed a number of current frames as a prevention of permutation attacks. During extraction, we choice three stable regions, which do not lie on a straight line. Average values of coordinates along OX and OY axes specifies coordinates of pseudo-center of image. Differences of pseudo-centers of the watermarked host and distorted images mean a transition values a_{31} and a_{32} in affine transformation matrix provided by Eq. 5.31, where (x, y) and (x', y') are the coordinates of feature points in the watermarked host and distorted image, respectively.

$$\begin{bmatrix} x' \\ y' \\ 1 \end{bmatrix} = \begin{bmatrix} a_{11} & a_{12} & 1 \\ a_{21} & a_{22} & 1 \\ a_{31} & a_{32} & 0 \end{bmatrix} \cdot \begin{bmatrix} x \\ y \\ 1 \end{bmatrix} \qquad (5.31)$$

After solving the transformation matrix in Eq. 5.31, we restore the distorted image globally using bilinear interpolation procedure [81].

After compensation of the global RST attacks, we check other pairs of corresponding feature points consequently in order to find the possible local geometric distortions. If difference exceeding the computational errors is detected, then we apply local bilinear interpolation as a particular case of local image warping [82].

Cropping areas cannot be resorted even with minimum probability. Such pixels are marked as unknown in the extracted watermarks.

5.5.3   Embedding Scheme

We enforced a conception of multilevel protection, when fragile, semi-fragile, and robust (textual or visual) watermarks are embedded in the same content. A fragile watermark is destroyed under all types of attacks, while generally a robust watermark ought to save its structure under the attacks. A semi-fragile watermark is destroyed partially under some attacks. A fragile watermark can be visible or invisible [83]. For this purpose, we apply Digital Hadamard Transform (DHT) with controlling visible degree by original function from sigmoid family.

The scheme with multiple watermarks for embedding in each I-frame includes the following steps:

- Step 1. Prepare fragile and textual or visual watermarks in a suitable view for embedding (binary view).
- Step 2. Chose the place for fragile watermark embedding in the host I-frame.
- Step 3. Segment the host I-frame into areas for fragile and textual or visual watermarks (Sect. 5.4).
- Step 4. Calculate the intensity interval for further compensation of common image processing attacks and put these parameters into secrete key.
- Step 5. Create code chain for fragile watermark and put it into secrete key. Each element in chain has size of 8×8 due to a frequency transform for embedding.
- Step 6. Embed a fragile watermark as a visual, semi-visual, or invisible element using DHT.
- Step 7. Estimate the sizes of the textual or visual watermark and chose a corresponding frequency transform, i.e. the digital wavelet transform or digital shearlet transform.
- Step 8. Apply Arnold's transform for the textual or visual watermark [73] including the corresponding parameters into a secrete key.
- Step 9. Calculate the stable regions in the constructed joint map JM and embed the coordinates of SURFs into EMs.
- Step 10. Create code chains for the textual or visual watermark (except of the stable regions) and put them into a secrete key. Each element in this chain has size of 8×8 due to a frequency transform for embedding.

- Step 11. Embed the textual or visual watermark into corresponding code chains using digital wavelet transform or digital shearlet transform.

It should be noted we may apply Arnold's transform to the watermarked host I-frames in order to increase its security during a transmission through unprotected Internet networks.

5.5.4 Extraction Scheme

Let us consider the extraction scheme under proposition of possible Internet attacks mentioned below:

- Step 1. Apply the inverse Arnold's transform for a whole I-frame if attack was detected.
- Step 2. Analyze intensity values of the transmitted I-frame and improve the intensity values and contract if it is required to use conventional methods of digital frame improvement.
- Step 3. Analyze transmitted I-frame on the possible cropping attacks. If full black or white areas are detected, then these areas are considered as the prohibited areas. Information in the cropped areas is fully lost.
- Step 4. Analyze a fragile watermark in I-frame. If a fragile watermark is normal, then go to Step 14.
- Step 5. Build the joint map *JM* for the extracted I-frame.
- Step 6. Extract feature points and calculate their exponential moments' values invariant to global RST attacks according to the joint map *JM*. Extract the coordinates of feature points in the host I-frame.
- Step 7. Compare the coordinates of three best correspondences of feature points, which do not lie on the straight line.
- Step 8. Calculate a center of mass between three feature points in the host and transmitted I-frames and define the translation shifts along OX and OY axes, respectively.
- Step 9. Put the center of the transmitted I-frame into a center of mass and solve Eq. 5.31 in order to find the rotation angle and scaling factor.
- Step 10. Execute a synchronization correction of the transmitted I-frame against the global RST attacks using a bilinear interpolation.
- Step 11. Analyze the remaining feature points in order to find possible local geometric attacks and compensate them applying a local warping technique [84].
- Step 12. Extract the textual or visual watermark.
- Step 13. Apply the inverse Arnold's transform to the textual or visual watermark if it was executed.
- Step 14. Restore the extracted textual or visual watermark if it is required using the conventional methods for digital image improvement.

Table 5.2 Description of videos used in experiments

Video	Screen	Resolution, pixels	Frames number	PSNR (original and watermarked video without attacks)	Video parameters
Sam_1.avi [85]		640 × 360	280	31.39	Fast motion, object of interest
EP_jugg.avi [86]		1280 × 720	720	30.96	Slow motion, static camera, object of interest
Gleicher4.avi [85]		640 × 360	420	32.12	Slow parallel motion, high jitters on video
Akiyo.mp4 [87]		352 × 264	250	34.85	Static camera, simple background, object of interest

(continued)

Table 5.2 (continued)

Video	Screen	Resolution, pixels	Frames number	PSNR (original and watermarked video without attacks)	Video parameters
Yuna_long.avi [85]		640 × 360	400	36.45	Fast motion, high jitters, object of interest
volley_10100153. mp4 [88]		270 × 480	480	32.87	Fast motion, several objects of interest, static camera
diving_1104.mp4 [88]		270 × 480	180	29.45	Complex scene structure, object of interest, static camera
bridge_close [87]		176 × 144	2001	28.45	Complex scene structure, slow motion

(continued)

Table 5.2 (continued)

Video	Screen	Resolution, pixels	Frames number	PSNR (original and watermarked video without attacks)	Video parameters
Bus [87]		176 × 144	150	33.15	Complex scene structure, fast motion, several objects of interest
Carphone [87]		176 × 144	382	32.81	Fast motion, complex scene structure, object of interest

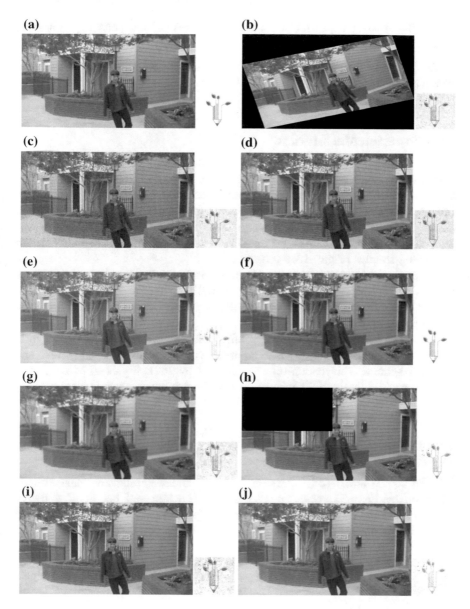

Fig. 5.8 Examples of different attacks on frame 70 of Sam1.avi video [85], left—watermarked image, right—extracted watermark: **a** no attacks, **b** rotate (15°), **c** salt & pepper noise, **d** Gaussian noise, **e** gamma correction (20% contrast increase), **f** median filtering, **g** blurring, **h** crop (25%), **i** JPEG compression, **j** scaling

- Step 15. Analyze the extracted numbers of frames and compensate permutation attacks if it is necessary.

The proposed extraction scheme allows to represent the textual or visual watermark in a view close to the original watermarks.

5.6 Experimental Studies

Proposed watermark embedding algorithm under some types of Internet attacks was tested using 10 video sequences. Their short descriptions are presented in Table 5.2. A watermark was implemented in I-frames of video sequence and then various negative effects were applied to these frames. Watermark recovery was carried out only on the basis of I-frame data, without using interframe information and coding features of the video sequence.

Some visual examples of attacks are shown in Fig. 5.8. They include the frame noise (Gaussian noise with a frequency of 0.01, noise like "Salt and pepper"), contrast correction, motion blur, median filtering, frame compression, RST attacks, and cropping [89].

The effectiveness of the watermark embedding algorithm was evaluated by a number of parameters. One way is to compare a quality of the original I-frame and extracted I-frame. This parameter is estimated by Peak Signal-to-Noise metric (PSNR) provided by Eq. 5.32, where MAX_I is the maximum possible pixel value of the frame.

$$PSNR = 10 \ \log_{10} \left[\frac{MAX_I^2}{MSE} \right] \tag{5.32}$$

Mean square error (MSE) between the host and watermarked image can be calculated by Eq. 5.33, where m and n are the width and height of a frame, respectively, I and I_w are the intensity values of the host I-frame and the extracted I-frame in coordinates (x, y), respectively.

$$MSE = \frac{1}{m \times n} \sum_{i=1}^{m} \sum_{j=1}^{n} [I(i,j) - I_w(i,j)]^2 \tag{5.33}$$

The higher is PSNR value, the better is a watermarked frame quality and lower losses.

In addition, we evaluate a quality of the restored watermark using Normalized Correlation Coefficient (NCC) metric [90]:

Table 5.3 Estimates of the watermarked videos properties after different attacks

Types of attacks	PSNR (original and watermarked images after attack), dB			NCC (original and extracted watermarks)		
	Sam_1	EP_jugg	Gleicher4	Sam_1	EP_jugg	Gleicher4
No attacks	31.39	30.96	34.85	1.000	1.000	1.000
Rotation (15°)	29.08	30.23	33.21	0.965	0.923	0.961
Salt & pepper noise (0.01)	31.17	30.44	34.47	0.985	0.996	0.998
Gaussian noise (0, 0.01)	31.11	29.87	34.67	0.983	0.995	0.991
Gamma correction (1.2)	30.37	28.46	33.21	0.783	0.986	0.977
Blurring filter (motion length 10, 45°)	28.41	29.21	33.93	0.966	0.912	0.972
Gamma correction, Gaussian noise, blurring	30.22	28.77	32.88	0.723	0.856	0.931
Median filtering (3 × 3)	31.78	30.21	34.45	0.967	0.981	0.978
Gaussian noise (0, 0.01) and Median filtering (3 × 3)	31.39	30.12	34.21	0.986	0.974	0.927
Scaling (1.15)	29.76	27.66	33.59	0.923	0.863	0.891
Crop black (25%)	27.67	27.21	31.56	0.607	0.745	0.723
JPEG compression	31.12	30.57	34.21	0.956	0.923	0.964

$$NCC = \frac{\sum_{i=1}^{k} \sum_{j=1}^{l} [[w(i,j) - \mu_w] \times [\bar{w}(i,j) - \mu_{\bar{w}}]]}{\sqrt{\sum_{i=1}^{k} \sum_{j=1}^{l} [w(i,j) - \mu_w]^2} \times \sqrt{\sum_{i=1}^{k} \sum_{j=1}^{l} [\bar{w}(i,j) - \mu_{\bar{w}}]^2}}, \quad (5.34)$$

where k and l are the watermark width and height, respectively, $w(i,j)$ and $\bar{w}(i,j)$ are the intensity value of the original and restored watermarks, respectively, μ_w and $\mu_{\bar{w}}$ are the mean values of the original and restored watermarks, respectively.

NCC value can be in the range $[-1, +1]$. Value close to 1 indicates a high degree of watermark correlation. Value close to 0 means the strong differences between the restored and original watermarks that is caused by the negative impact of attacks on a video sequence. Value close to -1 shows a high degree of watermark correlation, but when one of images has the inversed intensity values of pixels.

Quality assessment of the restored watermark and the types of attacks applied to I-frames of video sequences sam_1.avi [85], with 640 × 360 resolution EP_jugg. avi with 1280 × 720 resolution [86], and Gleicher4.avi with 640 × 360 resolution [85] are given in Table 5.3.

As one can see from Table 5.3, we obtained the dissimilar results because the video sequences used in the experiments had different quality. Thus, Gleicher4.avi video shows the best estimates of the restored watermark, as well as, the best PSNR values compared to other video sequences. That is explained by the simpler background and the lack of foreground objects moving fast. EP_Jugg.avi video has the areas poorly suited for watermark embedding, because there are very bright area, in the upper left corner and very dark area, at the bottom right corner. That causes the losses after various attacks. Proposed algorithm is robust to the most types of attacks. However, the geometric attacks, such as rotation and scaling, impact on the watermark significantly. At the same time, the algorithm is highly robust against the unintentional attacks and transmission attacks, such as noise, frame blur, and image compression. It should be noted that the most difficult cases occur, when several types of attacks take place.

5.7   Conclusions

In this research, we introduce classification of Internet attacks as intentional and accidental. Intentional attacks are directed on the distortions of a part of video or a single frame. These attacks are categorized into common image processing and geometric attacks, while the accidental attacks are concerned to the common processing attacks. It was found that global geometric attacks have the most significant effect on a quality of the restored watermark.

The proposed embedding algorithm based on the joint map of relevant for embedding regions and stable regions with the embedded feature points' coordinates shows good results and high resistance to such attacks as noise, color and brightness correction of the frame, and scaling. Also, we estimated available amount of embedded information in I-frame (without visible distortions). Thus, for frame with 640×360 resolution, the available amount of embedded information is about 20 Kb. As the size of a watermark increases, the recovery quality of the image, as well as, PSNR value, sharply decreases.

Acknowledgements   The reported study was funded by the Russian Fund for Basic Researches according to the research project № 19-07-00047.

References

1. Schwarz, H., Marpe, D., Wiegand, T.: Overview of the scalable video coding extension of the H.264/AVC standard. IEEE Trans. Circuits Syst. Video Technol. **17**(9), 1103–1120 (2007)
2. Segall, A. Sullivan, G.: Spatial scalability within the H.264/AVC scalable video coding extension. IEEE Trans. Circuits Syst. Video Technol. **17**(9), 1121–1135 (2007)
3. Oliver, M.: Tutorial: the H.264 scalable video codec (SVC). Available at https://www.eetimes.com/document.asp?doc_id=1275532&page_number=3. Assessed 26 Apr 2019

4. Bross, B., Han, W.-J., Sullivan, G.J., Ohm, J.-R., Wiegand, T.: High efficiency video coding (HEVC) text specification draft 9 (SoDIS). JCTVC-K1003. Available from http://phenix.it-sudparis.eu/jct/doc_end_user/current_document.php?id=6803. Accessed 26 Apr 2019
5. Ohm, J.-R., Sullivan, G.J., Schwarz, H., Tan, T.K., Wiegand, T.: Comparison of the coding efficiency of video coding standards–including high efficiency video coding (HEVC). IEEE Trans. Circuits Syst. Video Technol. **22**(12), 1669–1684 (2012)
6. Overview of the High Efficiency Video Coding (HEVC) Standard: Available at https://ieeexplore.ieee.org/document/6316136/figures#figures. Accessed 26 Apr 2019
7. Sullivan, G.J., Ohm, J.-R., Han, W.-L., Wiegand, T.: Overview of the high efficiency video coding (HEVC) standard. IEEE Trans. Circuits Syst. Video Technol. **22**, 1649–1668 (2012)
8. Kim, I.-K., Min, J., Lee, T., Han, W.-J., Park, J.: Block partitioning structure in the HEVC standard. IEEE Trans. Circuits Syst. Video Technol. **22**(2), 1697–1706 (2012)
9. Lu, C.S., Hsu, C.Y.: Near-optimal watermark estimation and its countermeasure: antidisclosure watermark for multiple watermark embedding. IEEE Trans. Circuits Syst. Video Technol. **17**(4), 454–467 (2007)
10. Wang, X.-Y., Wang, A.-L., Yang, H.-Y., Zhang, Y., Wang, C.-P.: A new robust digital watermarking based on exponent moments invariants in nonsubsampled contourlet transform domain. Comput. Electr. Eng. **40**(3), 942–955 (2014)
11. Swanson, M.D., Zhu, B., Tewfik, A.H.: Multiresolution scene-based video watermarking using perceptual models. IEEE J. Sel. Areas Commun. **16**(4), 540–550 (1998)
12. Zhu, W.W., Xiong, Z.X., Zhang, Y.Q.: Multiresolution watermarking for images and video. IEEE Trans. Circuits Syst. Video Technol. **9**(4), 545–550 (1999)
13. Piper, A., Safavi-Naini, R., Mertins, A.: Coefficient selection methods for scalable spread spectrum watermarking. In: Kalker, T., Cox, I., Ro, Y.M. (eds.) Digital Watermarking. LNCS, vol. 2939, pp. 235–246. Springer, Berlin (2003)
14. Cox, I.J., Kilian, J., Leighton, T., Shamoon, T.G.: Secure spread spectrum watermarking for multimedia. IEEE Int. Conf. Image Process. **6**, 1673–1687 (1997)
15. Lu, W.M., Safavi-Naini, R., Uehara, T., Li, W.Q.: A scalable and oblivious digital watermarking for images. Int. Conf. Signal Process. **3**, 2338–2341 (2004)
16. Lin, E., Podilchuk, C.I., Kalker, T., Delp, E.J.: Streaming video and rate scalable compression: what are the challenges for watermarking? In: SPIE Proceedings, vol. 4314, Security and Watermarking of Multimedia Content III, pp. 116–127 (2001)
17. Meerwald, P., Uhl, A.: Robust watermarking of H.264/SVC-encoded video: quality and resolution scalability. In: Kim, H.-J., Shi, Y., Barni, M. (eds.) Digital Watermarking. LNCS, vol. 6526, pp. 159–169. Springer International Publishing (2010)
18. Van Caenegem, R., Dooms, A., Barbarien, J., Schelkens, P.: Design of an H.264/SVC resilient watermarking scheme. In: Proceedings of SPIE, Multimedia on Mobile Devices, vol. 7542 (2010). https://doi.org/10.1117/12.838589
19. Noorkami, M., Mersereau, R.M.: A framework for robust watermarking of H.264 encoded video with controllable detection performance. IEEE Trans. Inf. Forensics Secur. **2**(1), 14–23 (2007)
20. Zhang, L., Zhu, Y., Po, L.M.: A novel watermarking scheme with compensation in bit-stream domain for H.264/AVC. In: IEEE International Conference on Acoustics, Speech and Signal Processing, pp. 1758–1761 (2010)
21. Noorkami, M., Mersereau, R.M.: Digital video watermarking in P-frames with controlled video bit-rate increase. IEEE Trans. Inf. Forensics Secur. **3**(3), 441–455 (2008)
22. Park, S.W., Shin, S.U.: Combined scheme of encryption and watermarking in H.264/scalable video coding (SVC). In: Tsihrintzis, G.A., Virvou, M., Howlett, R.J., Jain, L.C. (eds.) New Directions in Intelligent Interactive Multimedia, SCI, vol. 142, pp. 351–361. Springer, Berlin (2008)
23. Park, S.-W., Shin, S.U.: Authentication and copyright protection scheme for H.264/AVC and SVC. J. Inf. Sci. Eng. **27**, 129–142 (2011)

24. Chang, P.C., Chung, K.L., Chen, J.J., Lin, C.H., Lin, T.J.: An error propagation free data hiding algorithm in HEVC intra-coded frames. In: IEEE Asia-Pacific Conference on Signal and Information Processing, pp. 1–9 (2013)
25. Hartung, F., Girod, B.: Watermarking of uncompressed and compressed video. Signal Process. **66**(3), 283–301 (1998)
26. Deguillaume, F., Csurka, G., O'Ruanaidh, J.J., Pun, T.: Robust 3D DFT video watermarking. Secur. Watermarking Multimed. Contents SPIE **3657**(1), 113–124 (1999)
27. Lim, J.H., Kim, D.J., Kim, H.T., Won, C.S.: Digital video watermarking using 3D-DCT and intracubic correlation. SPIE Secur. Watermarking Multimed. Contents III **4314**(1), 64–72 (2001)
28. Xu, D.-W.: A blind video watermarking algorithm based on 3D wavelet transform. In: International Conference on Computational Intelligence and Security, pp. 945–949 (2007)
29. Meerwald, P., Uhl, A.: Blind motion-compensated video watermarking. In: IEEE International Conference on Multimedia and Expo, pp. 357–360 (2008)
30. Noorkami, M., Mersereau, R.M.: Compressed-domain video watermarking for H.264. IEEE Int. Conf. Image Process. **2**, 890–893 (2005)
31. Pei, W., Zhendong, Z., Li, L.: A video watermarking scheme based on motion vectors and mode selection. Int. Conf. Comput. Sci. Softw. Eng. **5**, 233–237 (2008)
32. Shi, F., Liu, S., Yao, H., Liu, Y., Zhang, S.: Scalable and credible video watermarking towards scalable video coding. In: Qiu, G., Lam, K.M., Kiya, H., Xue, X.Y., Kuo, C.C.J., Lew, M.S. (eds.) Advances in Multimedia Information Processing—PCM 2010. LNCS, vol. 6297, pp. 697–708. Springer, Berlin (2010)
33. Bhowmik, D., Abhayaratne, C.: A watermark evaluation bench for content adaptation modes. Available at http://svc.group.shef.ac.uk/webcam.html. Accessed 26 Apr 2019
34. Bhowmik, D., Abhayaratne, C.: A framework for evaluating wavelet based watermarking for scalable coded digital item adaptation attacks. In: SPIE Wavelet Applications in Industrial Processing VI, vol. 7248 (2009). https://doi.org/10.1117/12.816307
35. Yang, G., Li, J., He, Y., Kang, Z.: An information hiding algorithm based on intra-prediction modes and matrix coding for H. 264/AVC video stream. AEU Int. J. Electron. Commun. **65**(4), 331–337 (2011)
36. Wang, J.-J., Wang, R.-D., Xu, D.-W., Li, W.: An information hiding algorithm for H.265/HEVC based on angle differences of intra prediction mode. J. Softw. **10**(2), 213–221 (2015)
37. Park, M., Jeong, J.: Fast HEVC intra prediction algorithm with enhanced intra mode grouping based on edge detection in transform domain. J. Adv. Comput. Netw. **3**(2), 162–166 (2015)
38. Helle, P., Oudin, S., Bross, B., Marpe, D., Oguz Bici, M., Ugur, K., Jung, J., Clare, G., Wiegand, T.: Block merging for quadtree-based partitioning in HEVC. IEEE Trans. Circuits Syst. Video Technol. **22**(12), 1720–1731 (2012)
39. Van Pham, L., De Praeter, J., Van Wallendael, G., De Cock, J., Van de Walle, R.: Out-of-the-loop information hiding for HEVC video. In: International Conference on Image Processing, pp. 3610–3614 (2015)
40. Jo, K., Lei, W., Li, Z.: A watermarking method by modifying QTCs for HEVC. Comput. Perform. Commun. Syst. **1**(8–1), 16 (2016)
41. Chang, P.-C., Chung, K.-L., Chen, J.-J., Lin, C.-H., Lin, T.-J.: A DCT/DST-based error propagation-free data hiding algorithm for HEVC intra-coded frames. J. Vis. Commun. Image Represent. **25**(2), 239–253 (2014)
42. Elrowayati, A.A., Abdullah, M.F.L., Manaf, A.A., Alfagi, A.S.: Robust HEVC video watermarking scheme based on repetition-BCH syndrome code. Int. J. Softw. Eng. Its Appl. **10**(1), 263–270 (2016)
43. Jiang, B., Yang, G., Chen, W.: A CABAC based HEVC video steganography algorithm without bitrate increase. J. Comput. Inf. Syst. **11**(6), 2121–2130 (2015)
44. Gaj, S., Kanetkar, A., Sur, A., Bora, P.K.: Drift-compensated robust watermarking algorithm for H.265/HEVC video stream. J. ACM Trans. Multimed. Comput. Commun. Appl. **13**(1), 11.1–11.24 (2017)
45. Li, P., Wang, D., Wang, L., Lu, H.: Deep visual tracking: Review and experimental comparison. Pattern Recogn. **76**, 323–338 (2018)

46. Ji, Y., Zhang, H., Wu, J.Q.M.: Salient object detection via multi-scale attention CNN. Neurocomputing **322**, 130–140 (2018)
47. Zhang, H., Xue, J., Dana, K.: Deep TEN: texture encoding network. In: IEEE Conference on Computer Vision and Pattern Recognition, pp. 708–717 (2017)
48. Manikandan, L.C., Selvakumar, R.K.: A new survey on block matching algorithms in video coding. Int. J. Eng. Res. **3**(2), 121–125 (2014)
49. Slabaugh, G., Boyes, R., Yang, X.: Multicore image processing with OpenMP. IEEE Signal Process. Mag. **27**(2), 134–138 (2010)
50. Favorskaya, M., Buryachenko, V.: Fast salient object detection in non-stationary video sequences based on spatial saliency maps. In: De Pietro, G., Gallo, L., Howlett, R.J., Jain, L. C. (eds.) Intelligent Interactive Multimedia Systems and Services, SIST, vol. 55, pp. 121–132. Springer International Publishing Switzerland (2016)
51. Katramados, I., Breckon, T.P.: Real-time visual saliency by division of Gaussians. In: IEEE International Conference on Image Processing, pp. 1741–1744 (2011)
52. Laws, K.I.: Rapid texture identification. SPIE Image Process. Missile Guid. **238**, 376–380 (1980)
53. Fu, K., Gong, C., Yang, J., Zhou, Y., Gu, I.Y.-H.: Superpixel based color contrast and color distribution driven salient object detection. Signal Process. Image Commun. **28**(10), 1448–1463 (2013)
54. Favorskaya, M., Jain, L.C., Proskurin, A.: Unsupervised clustering of natural images in automatic image annotation systems. In: Kountchev, R., Nakamatsu, K. (eds.) New Approaches in Intelligent Image Analysis: Techniques, Methodologies and Applications. ISRL, vol. 108, pp. 123–155. Springer International Publishing, Switzerland (2016)
55. Favorskaya, M., Pyataeva, A., Popov, A.: Texture analysis in watermarking paradigms. Procedia Comput. Sci. **112**, 1460–1469 (2017)
56. Haralick, R.M., Shanmugum, K., Dinstein, I.: Textural features for image classification. IEEE Trans. Syst. Man Cybern. **3**(6), 610–621 (1973)
57. Galloway, M.M.: Texture analysis using gray level run lengths. Comput. Graph. Image Process. **4**(2), 172–179 (1975)
58. Tamura, H., Mori, S., Yamawaki, T.: Texture features corresponding to visual perception. IEEE Trans. Syst. Man Cybern. **8**(6), 460–473 (1978)
59. Manjunath, B.S., Salembier, P., Sikora, T.: Introduction to MPEG-7: Multi-media Content Description Language. Wiley, New York (2002)
60. Selvarajah, S., Kodituwakku, S.R.: Analysis and comparison of texture features for content based image retrieval. Int. J. Latest Trends Comput. **2**(1), 108–113 (2011)
61. Pietikäinen, M., Hadid, A., Zhao, G., Ahonen, T.: Local binary patterns for still images. In: Pietikäinen, M., Hadid, A., Zhao, G., Ahonen, T. (eds.) Computer Vision Using Local Binary Patterns, CIV, vol. 40, pp. 13–47. Springer-Verlag London Limited (2011)
62. Ojala, T., Pietikäinen, M., Mäenpää, T.: Multiresolution gray-scale and rotation invariant texture classification with local binary patterns. IEEE Trans. Pattern Anal. Mach. Intell. **24**(7), 971–987 (2002)
63. Teutsch, M., Beyerer, J.: Noise resistant gradient calculation and edge detection using local binary patterns. In: Park, J.I., Kim, J. (eds.) Computer Vision—ACCV 2012 Workshops. LNCS, vol. 7728, pp. 1–14. Springer, Berlin (2013)
64. Liu, L., Zhao, L., Long, Y., Kuang, G., Fieguth, P.: Extended local binary patterns for texture classification. Image Vis. Comput. **30**(2), 86–99 (2012)
65. Favorskaya, M., Petukhov, N.: Comprehensive calculation of the characteristics of landscape images. J. Opt. Technol. **77**(8), 504–509 (2010)
66. Gangepain, J.J., Roques-Carmes, C.: Fractal approach to two dimensional and three dimensional surface roughness. Wear **109**(1–4), 119–126 (1986)
67. Keller, J., Crownover, R., Chen, S.: Texture description and segmentation through fractal geometry. Comput. Vis. Graphics Image Process. **45**(2), 150–160 (1989)
68. Peitgen, H.O., Jurgens, H., Saupe, D.: Chaos and Fractals New Frontiers of Science, 2nd edn. Springer Science + Business Media Inc., Springer, New York (2004)

69. Casanova, D., Florindo, J.B., Falvo, M., Bruno, O.M.: Texture analysis using fractal descriptors estimated by the mutual interference of color channels. Inf. Sci. **346–347**, 58–72 (2016)
70. Mandelbrot, B.: The Fractal Geometry of Nature. WH Freeman and Company, New York (1982)
71. Du, G., Yeo, T.S.: A novel lacunarity estimation method applied to SAR image segmentation. IEEE Trans. Geosci. Remote Sens. **40**(12), 2687–2691 (2002)
72. Backes, A.R.: A new approach to estimate lacunarity of texture images. Pattern Recogn. Lett. **34**(13), 1455–1461 (2013)
73. Favorskaya, M.N., Jain, L.C., Savchina, E.I.: Perceptually tuned watermarking using non-subsampled Shearlet transform. In: Favorskaya, M.N., Jain, L.C. (eds.) Computer Vision in Control Systems-3, ISRL, vol. 136, pp. 41–69. Springer International Publishing Switzerland (2018)
74. Bay, H., Ess, A., Tuytelaars, T., Van Gool, L.: Speed-up robust features (SURF). Comput. Vis. Image Underst. **110**(3), 346–359 (2008)
75. Singhal, N., Lee, Y.-Y., Kim, C.-S., Lee, S.U.: Robust image watermarking using local Zernike moments. J. Vis. Commun. Image Represent. **20**(6), 408–419 (2009)
76. Shao, Z., Shang, Y., Zhang, Y., Liu, X., Guo, G.: Robust watermarking using orthogonal Fourier-Mellin moments and chaotic map for double images. Signal Process. **120**, 522–531 (2016)
77. Zhu, H., Liu, M., Li, Y.: The RST invariant digital image watermarking using Radon transform and complex moments. Digit. Signal Proc. **20**(6), 1612–1628 (2010)
78. Wang, X.-Y., Yang, Y.-P., Yang, H.-Y.: Invariant image watermarking using multi-scale Harris detector and wavelet moments. Comput. Electr. Eng. **36**(1), 31–44 (2010)
79. Yap, P.-T., Jiang, X., Kot, A.C.: Two-dimensional polar harmonic transforms for invariant image representation. IEEE Trans. Pattern Anal. Mach. Intell. **32**(7), 1259–1270 (2010)
80. Hu, H.-T., Zhang, Y.-D., Shao, C., Ju, Q.: Orthogonal moments based on exponent functions: exponent-Fourier moments. Pattern Recogn. **47**(8), 2596–2606 (2014)
81. Hurtik, P., Madrid, N.: Bilinear interpolation over fuzzified images: enlargement. In: IEEE International Conference on Fuzzy Systems, pp. 1–8 (2015)
82. Favorskaya, M., Buryachenko, V., Tomilina, A.: Structure-based improvement of scene warped locally in digital video stabilization. Procedia Comput. Sci. **112**, 1062–1071 (2017)
83. Favorskaya, M., Savchina, E., Popov, A.: Adaptive visible image watermarking based on Hadamard transform. In: IOP Conference Series: Materials Science and Engineering, MIST Aerospace, vol. 2018450, pp. 052003 (2018)
84. Favorskaya, M.N., Buryachenko, V.V.: Warping techniques in video stabilization. In: Favorskaya, M.N., Jain, L.C. (eds.) Computer Vision in Control Systems-3, ISRL, vol. 135, pp. 177–215. Springer International Publishing Switzerland (2018)
85. Grundmann, M.: Auto-directed video stabilization with robust L1 optimal camera paths video dataset. Available at http://www.cc.gatech.edu/cpl/projects/videostabilization/. Accessed 26 Apr 2019
86. EllenPage_Juggling Video: Available at http://www.youtube.com/watch?v=8YNUSCX_akk. Accessed 26 Apr 2019
87. Standard Videos for Image Processing: Available at http://trace.eas.asu.edu/yuv/. Accessed 5 May 2019
88. Safdarnejad, S.M., Xiaoming, L., Lalita, U., Brooks, A., Wood, J., Craven, D.: Sports videos in the wild (SVW): a video dataset for sports analysis. Available at http://cvlab.cse.msu.edu/project-svw.html. Accessed 5 May 2019
89. Zhou, X., Zhang, H., Wang, C.: A robust image watermarking technique based on DWT, APDCBT, and SVD. Symmetry **10**(3), 77–91 (2018)
90. Sachin, G., Vinay, K.: A RDWT and block-SVD based dual watermarking scheme for digital images. Int. J. Adv. Comput. Sci. Appl. **8**(4), 211–219 (2017)

Chapter 6
Object Selection in Computer Vision: From Multi-thresholding to Percolation Based Scene Representation

Vladimir Yu. Volkov, Mikhail I. Bogachev and Airat R. Kayumov

Abstract We consider several approaches to the multi-threshold analysis of monochromatic images and consequent interpretation of its results in computer vision systems. The key aspect of our analysis is that it is based on a complete scene reconstruction leading to the object based scene representation inspired by principles from percolation theory. As a generalization of the conventional image segmentation, the proposed reconstruction leads to a multi-scale hierarchy of objects, thus allowing embedded objects to be represented at different scales. Using this reconstruction, we next suggest a direct approach to the object selection as a subset of the reconstructed scene based on a posteriori information obtained by multi-thresholding at the cost of the algorithm performance. We consider several geometric invariants as selection algorithm variables and validate our approach explicitly using prominent examples of synthetic models, remote sensing images, and microscopic data of biological samples.

Keywords Object selection · Multi-threshold analysis · Percolation · Hierarchical structure · Adaptive thresholding · CLSM imaging · Z-stack

V. Yu.Volkov (✉) · M. I. Bogachev
Saint-Petersburg Electrotechnical University (LETI), 5, Prof. Popova str.,
197376 Saint Petersburg, Russian Federation
e-mail: vl_volk@mail.ru

M. I. Bogachev
e-mail: rogex@yandex.ru

V. Yu.Volkov
State University of Aerospace Instrumentation, 67, Bolshaya Morskaya,
190000 Saint Petersburg, Russian Federation

M. I. Bogachev · A. R. Kayumov
Kazan Federal University, 18 Kremlyovskaya st., 420008 Kazan, Russian Federation
e-mail: kairatr@yandex.ru

© Springer Nature Switzerland AG 2020
M. N. Favorskaya and L. C. Jain (eds.), *Computer Vision in Advanced Control Systems-5*, Intelligent Systems Reference Library 175,
https://doi.org/10.1007/978-3-030-33795-7_6

161

6.1   Introduction

Image analysis is a ubiquitous part of modern technology. Various applications range from remote observations of the Earth's surface to video analysis systems obtained from either optical, laser, radar, infrared or acoustic based sensors, Synthetic Aperture Radars (SAR), etc. Typical problems arising in image analysis are detection, extraction, selection, and localization of objects of different shapes from raw or preprocessed images. They in turn allow for the identification of objects, tracking them, matching and combining images from heterogeneous sensors, indexing and image recovery.

Common requirements for modern observation systems include the ability of easily merging data from complementary data sources, e.g. combination of satellite and airborne photography, measurements obtained using different physical principles with different resolution. For example, in Geographic Information Systems (GIS) information is typically considered at object-level, while raw remote sensing images of the Earth's surface are typically acquired at pixel-level in the form of digitized images. To achieve the goal of early integration of the observed information into GIS, an early transition from pixel-level to object-level representation is strongly encouraged. The above requirements as a part of the modern GEographic Object-Based Image Analysis (GEOBIA) paradigm strongly encourage the development of rather universal analysis concepts and data representation frameworks with an early transformation of the remote sensing imagery into object-based representation at early stages of the analysis [1, 2].

Another example is related to the biomedical applications, where new microscopic imaging technologies, such as fluorescent staining followed by Confocal Laser Scanning Microscopy (CLSM) became widespread and available to a broad research and laboratory community. Access to the technologies stimulated penetration into previously unexplored areas, including deep multi-layered analysis of three-dimensional biological structures without damaging their integrity and viability. In particular, CLSM with differential fluorescent staining revealed the internal organization of various complex three-dimensional structures from intracellular molecular organization to subpopulations of cells in consortia that allows obtaining dynamic pictures of their interaction, differentiation and adaptive response to stress conditions. Although new vision technologies do provide a deeper insight into the unknown depth of life, there are still certain limitations on the side of quantitative assessment and interpretation of their results [3–5]. To support the quantitative assessment and decision making, vision technologies should be appropriately complemented by sophisticated data processing, visualization and interpretation solutions. Common tasks to be solved include image analysis at various stages from preprocessing to selection, classification and quantification of objects with different coloring, shape and morphology. By now various image analysis technologies are available in the form of universal tools or focused solutions to problems implemented in a series of software tools [6–11]. Besides that, new methods of selection and localization of objects are intensively introduced into medical applications, such as magnetic resonance tomography of organs, the detection of cancer pathology, as well as, into the diagnosis of various diseases of the internal organs.

The variety and variability of shapes and textures of objects, as well as, intense non-stationary background make their extraction and analysis in vision system generally a complex problem. Typical limitations arise from low signal-to-background ratio in the area of interest, low quality images, excessive quantization, fuzzy boundaries of objects and structures that are typical for both natural and artificial structures such as rivers, roads, bridges, or buildings observed by remote sensing systems, as well as, biological structures observed by biomedical vision systems. Other common conditions are strongly non-Gaussian background probability distributions that also appear rather asymmetric, with the tails of the distributions following approximately log-normal or mixed (contaminated-normal) densities. Moreover, the properties of the background are structurally and often statistically similar to the areas of interest and thus can either imitate or mask the presence of objects. The above conditions strongly limit the performance of conventional methods like adaptive thresholding, as incorrect choice of thresholds often leads to the loss of useful objects at an early stage. Another problem is the low quality of the generated images, spots, blurred borders, as well as, internal speckle noise typical for SAR systems, coherent laser imaging systems etc. [12].

Given the overall variety and non-stationarity of the conditions, a common approach to general image analysis is the use of multi-step programmable algorithms, where the raw images undergo several consecutive steps of processing typically represented by the so-called pipelines. Following the initial adjustment using representative data samples, the algorithm parameters can be next applied to a large series of images obtained under more or less similar conditions this way allowing fast processing of large data sets and obtaining good overall statistics that often compensates algorithm optimization imperfections. Pipelines commonly consist of several steps including initial image preprocessing, illumination correction, image segmentation, object selection and identification, and, finally, estimation of the quantities of interest. Preprocessing is commonly performed at pixel level representation and resolves such problems as the inhomogeneous intensity correction that can be performed using dozen various methods, with a systematic review of them being represented, for instance, by a three-level classification scheme depicted in [13].

Further steps in the pipeline are usually based on a combination of rather universal algorithms. For particular application in biomedical imaging, dedicated software environments such as the widely used open-source Cell Profiler [14–17] and Fiji/ImageJ [18] have been developed, as well as, underwent a series of further consecutive advancements in the last decade.

At each step of the algorithm, several options are typically available, with many of them having at least one or more free parameters. On the one hand, this allows for a more accurate fine tuning of the algorithm parameters considering particular task requirements. On the other hand, the analysis results thus become strongly dependent on the adequate parametrization. While draft pipelines for many common problems are supplied as examples thus providing a certain first approximation to the optimized algorithm, the amount of effort that has to be invested into its adequate parametrization by an unexperienced used user remains significant. Moreover, excessive parametrization of the multi-step pipeline algorithms also leads to the increased probability of random error scenarios indicating by spurious results, since the number of parameter combinations increases to the point, where all combinations cannot be checked with a small trial cohort, as it is usually done prior to using the pipeline mode analysis for a large cohort. From this point of view, the use of data-driven approaches seems to be generally preferential to model-based solutions, since the latter require parameterization that is in turn often limited by quite little information about the object(s) of interest commonly available from the raw scene. Otherwise, the choice of the remaining free parameters should be based on clear objective criteria that are transparent and easily reproducible.

At the moment, segmentation of an image into individual objects is commonly based on such characteristics are intensity uniformity and color matching. Regional methods are mainly based on the assumption that neighboring pixels within the object area exhibit more or less similar values [19]. A detailed review of object segmentation methods that are rather universal for various image analysis applications not limited to remote sensing systems, microscopic and/or biomedical imaging [20]. It classifies the segmentation methods according to the objects they are dealing with, including the pixel-level thresholding and clustering methods, edge detection methods, region-level, and other classifiers, the latter including many non-parametric methods like machine learning, neural networks, fuzzy sets, etc. Without going deeper into details, we have to note that while the variety of available methods allows for a more specific choice of an optimal algorithm for a given problem, it also makes more complicated the reproducibility of the results, especially given the number of free parameters that can be controlled by the user and which are often subjectively if not arbitrary chosen.

For an efficient multi-scale analysis, the conventional image segmentation approaches often appears insufficient, since generalized hierarchical representation of objects no longer allows for an unambiguous assignment of a single pixel to a single object. In contrast, a single pixel often belongs to a hierarchy with fine resolution objects and structures being embedded into larger ones.

To partially overcome the above limitations and to better fit the requirements of merging data from different sources by early integration of imaging results into dedicated information systems preferably at object level the discussed approach has to be revisited. More generally, an early transition from pixel- to object level representation and unification of the following procedures at object level requires a common image analysis and reconstruction framework with a set of clear object selection rules and unified object representation that should be universal despite of the methods that have been used for their initial selection and quantification. Accordingly, the requirements to the reconstruction methodology include its universality based on common geometric principles, algorithmic transparency and simple transferability between solutions for imaging techniques based on different physical principles. Expansion from 2D to 3D imaging by different techniques ranging from ground penetrating radars to CLSM and tomographic systems further stipulate the necessity of a certain universal framework of 3D objects reconstruction from a set of multiple 2D layers with potentially broad application area [21, 22].

In this chapter, we therefore suggest an original approach to image analysis based on its multi-threshold processing. The results of multi-thresholding represented by a series of binary layers are next used to image segmentation and selection. Thus, one of the specific features of our approach is that segmentation and selection is performed based on a posteriori information after obtaining binary images for a series of potential thresholds. The keynote advantage is that no preliminary learning from a subset is required and the algorithm parameters are adjusted for each image and further for each object individually while at the cost of the overall analysis performance.

We start with rather simple approach where the best global threshold value is obtained from the maximum area occupied by objects of certain size and/or shape parameters that is tested for the entire range of potential thresholds. As the proposed approach requires good statistics and, thus, performs well only when a sufficient quantity of objects of interest exhibiting similar statistical properties are available, we next suggest an alternative approach based on the percolation properties of the base object. Despite of being well established and widely utilized in statistical physics and its various applications including structural characterization of complex objects and patterns in 2D, 3D or higher dimensional state space representations [23–27], the percolation based approach is quite rarely used in image analysis. Few notable exceptions include complex object selection and reconstruction hindered by non-stationary background noise reminiscent to speckle patterns commonly observed in vision systems with coherent lightning [28–30], as well as, several focused applications such as crack detection in concrete structures [31–33]. Regarding applications to topographic data, very recently a percolation based approach to characterize and distinguish between landscape patterns have been suggested [34].

The base object contains the entire image at zero threshold and loses some of its content when the threshold is increased, until breaking up into separate isolated objects at a certain point, leading to the gap in the maximum object size. Locations of such gaps are good guidelines for choosing thresholds, as they indicate critical points (phase transitions) between connected and disconnected image fragments. In the following, we generalize the percolation inspired approach by extrapolating the same principles from the base object to all newly formed separated objects at different thresholds. We suggest a universal percolation based framework to represent the results of image multi-thresholding in the form of 3D structures spanning through a series of binary layers. Based on these structures, further characterization is possible either (a) by selection of certain layers corresponding to threshold value optimized according to certain geometric criteria or (b) by using integrative parameters obtained from multi-threshold representation of the image, such as the percolation coefficient.

Finally, we show how the same conceptual framework can be used to represent and analyze and effectively characterize higher dimensional datasets exemplified by 3D multi-layer data obtained by CLSM microscopic imaging.

The remainder of this chapter is organized as follows. Section 6.2 presents a short literature review. The proposed methodology is discussed in Sect. 6.3. Implementation of the proposed methodology is given in Sect. 6.4, while its applications are presented in Sect. 6.5. Section 6.6 concludes the chapter.

6.2   Related Work

In the theory and practice of object recognition in remote sensing images, computer vision and data processing, two main approaches are commonly exploited. The discriminant approach is based on comparison with a given template. This approach commonly utilizes correlation and spectral methods that are characterized by weak robustness against possible distortions of individual elements of the object description.

The syntactic approach based on the structural features of the object of interest appears more promising, as it is associated with the analysis of their internal structure and the allocation of their local features. This approach typically includes transformation or reconstruction of the original image in order to be able to use the selection of objects according to given criteria. In most approaches, the reconstruction stage has the objective of the image segmentation, i.e. extraction of areas homogeneous with respect to some criterion. At this stage, the connectivity and homogeneity properties of the regions are the key properties used in the analysis. Next the regions are selected according to the specified features of the objects of interest. Common difficulties with the implementation of this approach are associated with the need to describe a variety of different objects by a finite system of rules. In addition, it requires substantially larger amount of calculations, especially when using a variety of geometric transformations. Conventional segmentation schemes use features that stand out from the original image and only indirectly take into account the properties of the objects of interest. In particular, the properties of the histogram of the original image, the properties of edges and contours are widely used. The results of the subsequent selection of objects are practically not used for segmentation.

An excellent overview of modern methods of image segmentation is given in [20], where four categories are distinguished, each of which is based on its key element. Such key elements are the pixels, edges, areas, and others. The first category includes methods of threshold processing and clustering, the second category provides the boundary detectors. The third category includes the farming areas, the method of the watershed, split and merge, level set and active contours. The remaining fourth category includes the use of wavelets, neural networks, fuzzy sets, etc.

Each method has its own way of accounting a priori information about the objects of interest. It is noted that methods based on domain properties, such as Fractal Net Evolution Approach (FNEA) and methods based on graph theory (graph methods), dominate the creation of compact object support domains and for acceptable scales. Methods using graphs represented by four basic algorithms, namely the best merge, Minimum Spanning Tree (MST), minimum average cut, and normalized cut [20].

There are two ways to shape feature areas. One of them ("bottom-up") is based on the merger of smaller objects into larger ones, which is used, in particular, the property of homogeneity. Another method ("top-down"), on the contrary, considers the original image as an initial single segment, followed by its fragmentation into separate parts (the minimum average cut, the normalized cut).

These approaches are very constructive for segmentation of images obtained by remote sensing systems (SAR, multi-spectral, hyperspectral, panchromatic systems, etc.). However, they also have significant limitations. First of all, it is their computational complexity associated with the solution of optimization problems, and the requirement of high processing power, because often the number of objects to be constructed is very large. In addition, the methods do not have optimality in the choice of the starting points in a sequence of iterations, often resulting in controversial decisions depending on changes of initial conditions.

These disadvantages may be somewhat reduced if use better the specifics of the selected objects, training, as well as, a combination of different methods to overcome the shortcomings of each of them [20, 35, 36]. Ultimately, all the methods under consideration are reduced to organizing image pixels into some multiscale hierarchical structures, which further allow the use of different criteria for the selection of objects. The challenge is to make such a structure more transparent and not so difficult to use.

Multi-threshold processing transforms the original monochrome image into a set of binary layers. In the case of a sufficiently large number of thresholds, it can be assumed that there is no loss of information with such a transformation. At the same time, binary image processing is easier and faster than gray level image processing.

Various applications of multi-threshold processing for image segmentation are considered in numerous works, only some of them can be specified [31–33, 37, 38]. Basically, multi-threshold segmentation is based on the histogram properties of the original image. In most cases, the last step is to select the optimal threshold. Properties of objects of interest and the results of their selection are not taken into account.

The expected object properties are required to implement the object selection. The key assumptions are the connectivity of the pixels in the area of the object of interest, and the isolation of one object from another. Typically, there is an acute lack of information about objects, except for the typical size and some assumptions about the area, perimeter, shape, and orientation. The original idea is to select and set the optimal threshold value based on the results of the selection of objects in multi-threshold framework to achieve the best selection based on a posteriori information. This approach was originally proposed in [39] for the selection of small-scale objects.

Further development of this idea is described in this chapter and includes the evaluation of certain geometric parameters of the object in binary images after multi-threshold processing and the corresponding selection of objects [20, 40]. The optimal threshold value is selected according to the extremum of the selected parameter. This geometric parameters are the area of the object, the ratio of the perimeter square to the area of the entire object, or the ratio of the square of the main axis to the area of the object.

In this chapter, we develop the idea of reconstructing a three-dimensional hierarchical structure of objects based on the multi-threshold analysis of the raw image. We separate objects from each other based on the percolation effect [41]. This effect is associated with the elimination of empty pixels that appear below the enhanced threshold from the object content, which ultimately leads to the breakup of the integrity of the object and the emergence of new isolated objects as its fragments. Thus, the objects of interest are represented in the form of 3D structures spanning through a series of binary layers. After 3D reconstruction, one can select the objects of interest using various criteria, such as their percolation properties, geometric characteristics, or texture parameters.

6.3 Methodology

Let a monochrome image $I(x, y)$, where I is the intensity and x, y coordinates of the pixels, is binarized by a fixed global threshold T. The result is a binary layer (slice) B_T: $\{B_T = 1$, if $I(x, y) \geq T$; $B_T = 0$, if $I(x, y) < T\}$, in which a subset of units represents the objects of interest (foreground), for example, buildings, structures, vehicles, waterfront, functional organs in medical CT images, or cells in the biological samples.

Objects of interest typically exhibit such common features as localization and connectivity of pixels, a certain geometric shape, isolation from other objects, and the difference in intensity from the background. A subset of zeros at $\{I(x, y) < T\}$ belongs to the background that differs from the foreground objects. Such qualitative (nonparametric) description of the objects of interest does not allow to suggest a statistical model and, thus, also to design an optimal selection algorithm. It is clear that background structures may also exhibit qualitatively similar characteristics, and the difference from useful objects may be purely quantitative. Under these conditions, the problem can be solved by applying certain restrictions to the selected objects. In particular, a restriction on the minimum intensity differences and on the minimum area of objects of interest can be introduced. It is also necessary to determine certain morphological properties of objects that can be described by geometric invariants.

If the pixel intensities of the objects of interest are generally higher than the background pixel intensities, the generally accepted method for selecting a global threshold is the Otsu method, which operates under fairly general conditions. It is based on the analysis of the intensity histogram of the raw image and requires the minimum sum of intragroup variances for subsets $\{I(x, y) \geq T\}$ and $\{I(x, y) < T\}$, respectively. Since in practical scenarios the objects of interest are blurred and the background is heterogeneous and noisy, the global threshold leads to overlapping distributions and inevitable errors.

Ideally, each object of interest requires its own threshold value, and such local thresholds can be formed by using local (moving or gliding) windows, which are slightly larger than the size of the objects of interest, and within which the background is considered homogeneous. In this case, one should first specify the area of interest (associated with the size of the expected object) and the reference (background) area, which surrounds the area of interest and is used to estimate the adaptive threshold. The above approach has been widely utilized in one-dimensional radar signals processing in order to detect point-like objects [42–44].

Examples of similar approach application to the analysis of 2D images are presented in [45]. Gliding window based methods require a priori knowledge of sizes of the objects of interest. In case of inconsistency of the controlled area with the size of the object of interest there are losses, which also entail the use of different windows for different objects. In addition, the use of the background window leads to the loss of resolution thus way making it more difficult to discriminate closely located objects.

Other ways of local thresholding in images include, for example, the use of local histograms [46], are characterized by high computational complexity with limited control over the results.

Alternative approaches are possible based on multi-threshold processing. With sufficiently small threshold step, the resulting binary layers stores all the information contained in the original image. At the same time, each binary layer is easily morphologically processed at low computational cost. Historically, the first developed methods focused on independent processing of each binary layer in order to select objects on a given basis.

One of the methods suggests setting a threshold for each category of objects of interest, which is selected according to a certain criterion [11, 39, 47]. In these cases, we use various parameters to describe the category of objects, such as the area of the object, its orientation, or invariant characteristics, such as the ratio of the perimeter square to the area, the compression ratio of the ellipse, as well as, other geometric or texture characteristics. As a rule, the category of objects was set by the corresponding range of features. In this case, each binary layer selects objects that meet the specified properties, and the binarization threshold for such objects is selected such that the maximum number of objects of this category (or their total area, respectively), taking into account the required preservation of the object shape is being selected. This process can be automated, resulting in adaptive methods for setting the global threshold.

However, this approach does not allow setting an individual threshold for each object of interest, if the objects belong to different categories. To select a local threshold, one has to establish links between individual layers, and each pixel belongs to the same, or to a new object that occurs in place of the previous one due to the fragmentation effect when the threshold level increases. It is required to establish links between pixels with the same coordinates on different binary layers. If one specifies a certain parameter that determines the specified relationships between pixels, a three-dimensional hierarchical structure based on a single binary (white) object can be formed, resulting at zero threshold value, and coinciding in area with the entire image.

To partially overcome the above limitations, we next propose the percolation based framework. Let us start with $T = 0$, when $I(x, y) > 0$ is satisfied for the entire image thus forming a binary layer containing one global isolated (base) object with area S_0, equals to the full size of the. The threshold T is then enhanced by ΔT, so that some pixels appear below the new threshold ΔT, and a new binary layer is created at the top that satisfies $I(x, y) > \Delta T$. If ΔT is relatively small and only a small fraction of the pixels is eliminated from the object, the global (base) isolated object remains connected, despite the reduction of its area $S_T < S_0$.

At each step, the ratio $S_{T+\Delta T}/S_T$ characterizes the fraction of connected pixels preserved within the isolated object. As long as this ratio is equal to or greater than one half, the object in the upper layer can be treated as an unambiguous *successor* of the object in the lower layer, which in turn appears its sole *predecessor*. Further additions to the threshold leave an increasing number of pixels below the new thresholds $T + k \Delta T$, where k is the integer number of the layer, $k = 0, 1, \ldots, K$, and the area of the object decreases.

Ultimately, the increase in the threshold leads to the formation of gaps in the original base object, and to the separation of isolated fragments from it. The ratio $K_p = S_{T+\Delta T}/S_T$ can be considered a characteristic of the persistence of the object as the threshold increases. At a certain critical point, corresponding to the *percolation threshold T_c*, the fraction of pixels with intensity below the threshold becomes large enough, and these pixels merge together forming a large global gap, limited only by the size of the image, which leads to fragmentation and destruction of the original base object. Such phase transition event is known as percolation [41]. Since percolation theory is well represented in the statistical physics literature and has been widely used for decades in a variety of fields ranging from diffusion modelling to forest fire propagation, the discussion of the underlying theoretical framework here can be limited to fairly simple examples that are relevant to understanding percolation related image analysis methodology.

Once the persistence parameter $K_p \geq 0.5$, when the object is fragmented into two equal parts, there will always be (taking into account that the boundary pixels fall out of the total area) its ambiguous successor in its upper layers until the destruction of the original object and the appearance of two or more new objects instead. Considering larger K_p values (above 0.5) increases the resolution of the analysis while also increases the number of objects (for smaller $K_p < 0.5$), a single isolated object in the lower layer may have more than one successor isolated object in the upper layer, leading to unambiguity. The fraction of pixels in an object

$P_c = S_{Tc}/S_0$ with intensity above the percolation threshold T_c provides an estimate to the percolation coefficient for that object.

The initial binary layer for the new object is formed at the moment of percolation of the predecessor, and then the new object accumulated further binary representations for different threshold T as long as the condition $I(x, y) > T_k$ is satisfied. In fact, the percolation coefficient characterizes the texture of the object surface. If the object has a flat top with a constant value of the intensity $I(x, y)$ = const, it completely disappears in one threshold step, and its percolation coefficient $P_c = 1$. In this case, the intensity value itself does not affect the value of the percolation coefficient, i.e. it appears invariant to the transformations like shift or rescaling of the image. If the object changes its intensity smoothly, i.e. there is a small intensity gradient, then the heredity between adjacent layers with increasing threshold will remain as long as $S_{T+\Delta T} \geq K_p \cdot S_T$.

To ensure that a given object has a single successor, the K_p value must remain between 0.5 and 1 in each of the top layers. All other separated fragments are treated as new objects. Each of them initially appears at the basic T_b threshold value. To increase detalization, one can set $K_p = 1$, which is the most stringent requirement, when the loss of even one pixel for the original object leads to the formation of a new object. In contrast, choosing $K_p = 0.5$ results in a significant reduction in the total number of three-dimensional objects spanning through multiple binary layers.

For illustration, Fig. 6.1 represents four isolated objects hindered by additive Gaussian noise characterized by different averages and standard deviations. As a consequence, they demonstrate different rates of area reduction with increasing threshold level thus leading to different percolation properties. The discrepancies in

Fig. 6.1 Four noise objects with different percolation properties

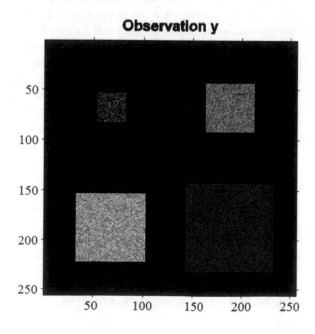

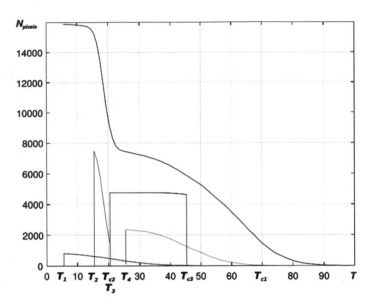

Fig. 6.2 Objects area variations as a function of threshold level

object characteristics allows distinguishing them from each other based on their properties such as area reduction after multi-thresholding, as indicated in Fig. 6.2. The upper line shows the total area exceeding specified threshold values.

At zero threshold, a single object with the area of the entire image $S_\Sigma(0)$ can be observed. As the threshold increases, as expected, the total area above the threshold decreases, as shown by the upper curve. One of the first significant area losses by the overall base object leads to the formation of the first regular isolated object at threshold T_1, which next with further enhancement of the threshold reduced its area rather slowly until nearly complete elimination indicated by its percolation coefficient being close to zero. The second isolated object with larger area separates at T_2 exhibiting the percolation coefficient of around one quarter. The appearance of the third object corresponds to threshold T_3 that is further destructed at T_{c3} with nearly no intermediate area loss being observed, leading to its percolation coefficient close to one. Finally, the fourth isolated object appears at T_4 and nearly completely eliminated at T_{c4} so that its percolation coefficient is close to zero like for the first object.

Figure 6.3 presents an illustration of the multi-threshold image analysis with a random noise field and two rectangular objects with different intensities hindered by the background noise.

If the image contains only random noise, i.e. the intensity is randomly distributed throughout the image, then the location of the pixels that appear below the increased threshold is also random (see the example in Fig. 6.3a). Theoretically it is well known [41] that for an infinite size of the image and pure background noise (invariant to its distribution) the percolation coefficient equals to $P_c \approx 0.593$, since

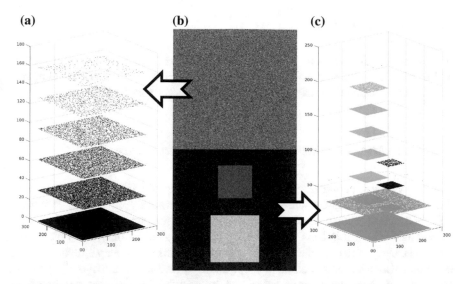

Fig. 6.3 Illustration of multi-threshold 3D representation of images containing: **a** random intensity field (noise), **b** two regular (rectangular) objects with superimposed (additive) noise, **c** 3D reconstruction. 3D reconstructions show how objects span through multiple binary layers. Changes from lower to upper layers show the fragmentation of the object with the enhancement of the threshold to the moment of reaching the percolation threshold and complete disappearance of the object

pixelized image exhibit the same topology as the square lattice [48]. For finite but sufficiently large images, percolation coefficient $P_c = S_{Tc}/S_0$ can be well estimated by averaging statistics over various noise configurations. The value of T_c can be obtained as the corresponding quantile of the distribution of the noise intensity. A detailed illustration of two rectangular objects hindered by background noise is shown in Fig. 6.3c, which represents separate 3D reconstructions of two objects with different percolation threshold values.

As a result of the application of the above approach, a three-dimensional hierarchical structure is formed containing all selected objects, in which every pixel no longer belongs to a single binary layer, but may correspond to several binary layers in the structure $k = 0, \ldots, K$. Based on this reconstruction, image segmentation and object selection can be easily generalized.

A similar 3D structure can be obtained by reducing the threshold from its highest level to zero [34]. This representation has the advantage that the most intensive objects that form isolated clusters are being selected first. Here, the percolation events are indicated by merging of two or more objects, which also corresponds to the destruction of the original objects and the emergence of new ones. The disadvantage of this approach is that all small-scale objects detected in the upper layers must be preserved until it is determined whether they are the successors of larger objects with small percolation coefficients and whether their base sizes satisfy the limits leading to the reduction of the overall algorithm performance.

6.4   Implementation

In this section, the invariant geometric criteria for the adaptation, object selection and adaptive thresholding, and adaptive thresholding based on invariant geometric criteria are considered in details in Sects. 6.4.1–6.4.3, respectively.

6.4.1   Invariant Geometric Criteria for Adaptation

Among various practical problems that can be solved by the proposed three-dimensional representation using multi-threshold processing, adaptive threshold based image segmentation and object selection is one of the most straightforward. In this case, each pixel will be assigned to only one object. For each selected objects that spans through multiple binary layers in a three-dimensional representation, the optimal T_{opt} threshold corresponding to the layer with its best flat representation can be selected by different geometric or texture criteria. Prominent examples of such criteria include areas or ranges of areas of the selected objects. It appears efficient when many similar objects have to be selected, the statistics is sufficient to obtain a reliable estimate of the optimal threshold value. The key disadvantage is that this approach is associated with the absolute values of areas of the objects of interest and, thus, cannot be considered as scale-invariant.

Invariant geometric metrics that largely overcome the above limitation include the elongation coefficient of the perimeter defined as $P_S = P^2/4\pi S$, where P is the perimeter of the object, S is its area [40]. More generally, this coefficient is known as a metric of the compactness of the domain [46]. Another geometric invariant that also characterizes compactness is the elongation coefficient of the main axis L defined as $P_L = \pi L^2/4S$ [49]. These metrics are normalized such that they are equal one for round (circular) objects, while increase in case of the elongation of either the edge or the object itself, respectively. If the objects of interest are compact, a single binary layer can be chosen to select each of them, which provides a minimum of appropriate evaluation of the specified measures. For a more detailed overview of geometric and texture based criteria we refer to [46].

6.4.2   Object Selection and Adaptive Thresholding

It is believed that the main property that distinguishes the object of interest from the background noise is the connectivity of adjacent pixels in the binary image B_T. Let us first consider the implementation of the global thresholding scheme.

In each binary layer, objects that satisfy the specified properties are selected, and the binarization threshold for such objects is selected such that the maximum

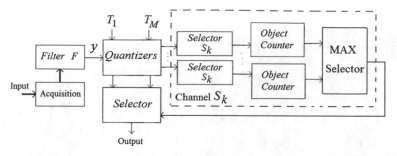

Fig. 6.4 The structure for adaptive object selection by area

number of objects of this category and/or their total area can be observed. This process can be automated, resulting in adaptive thresholding methods. The structure of the object selection algorithm by area is shown in Fig. 6.4.

Each channel is set to its own area range of isolated objects. Adaptive selection of objects by area can be implemented by applying a threshold against the results of multi-threshold processing such that the maximum number of objects of a given size and/or their maximum total area can be observed. In this case, a range of areas of the desired objects of interest can be chosen for selection. As a result of the adaptation, each channel will have its own threshold value, at which the corresponding objects will be selected in the best way.

In Fig. 6.5, a model of a monochrome noisy image on a 256×256 grid is presented, in which the pure signal (Fig. 6.5a) is hindred by the additive Gaussian noise (Fig. 6.5b), thus resulting in the noise image (Fig. 6.5c). The objects of interest have a small signal-to-noise ratio $d = 1.163$ in each signal pixel. The signal-to-noise ratio is determined as the ratio of the shift of the average to the mean square of the noise. Rectangular objects of 20×8, 20×16, 20×32, and 20×64 pixels are being considered, with the smallest area equals 160 pixels.

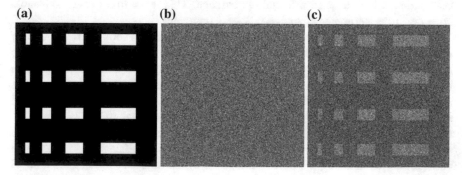

Fig. 6.5 Image model containing rectangular objects hindered by Gaussian noise: **a** initial image, **b** Gaussina noise, **c** image hindered by Gaussian noise

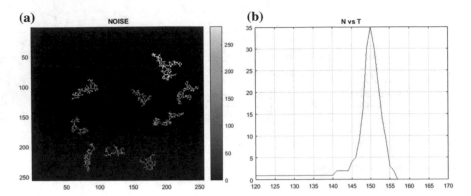

Fig. 6.6 Exclusion of small objects: **a** the result of binarization of the noise field, **b** the dependence of the number of objects on the threshold value

Let us first consider a purely noise field. At each binarization threshold value, related objects appear that contain a different number of pixels. Their number depends on the threshold value [39, 47]. Since the shape of the object of interest will be of interest in the future, small objects will be excluded from consideration, which can significantly reduce the noise level after binarization. The appearance of very large objects is unlikely as mainly fractured fragments containing some typical number of pixels that also depends on the threshold value can be observed. In Fig. 6.6, the results of binarization of the noise field after the exclusion of small objects up to 120 pixels are shown. The dependence of the number of such objects on the threshold value is shown in Fig. 6.6b.

Figure 6.7 shows the results of single-threshold selection of connected objects of rectangular shape, taking into account the removal of small objects. There are two types of distortion of the shape of objects: the loss of pixels in the area of the object and addition of extra pixels along its boundaries. At high threshold values required for a small number of false objects, useful objects mostly lose pixels. At small signal-to-noise ratios, useful objects undergo significant deformations of the boundaries, which acquire a fractal appearance. This leads to a rather noticeable increase in the perimeter of such connected fragments.

The optimal threshold should ensure acceptable preservation of the shape of useful objects. In particular, it is possible to require approximate equality of the number of pixels missing inside the object and the number of pixels adjacent to its border from outside. In this case, the optimal threshold will not correspond to the maximum of the selected objects of the specified area, but will be slightly shifted towards higher values. In Fig. 6.7a, the binarization threshold is $T = 130$, while the maximum number of objects is reached at $T = 109$ (Fig. 6.7d).

When the threshold is lowered (Fig. 6.7c), there are significantly more adjacent background pixels along the outer boundaries of the objects. Further reduction of the threshold leads to these adjacent pixels forming conglomerates (Fig. 6.7d). In the case of Fig. 6.7c, the number of useful objects may decrease. However, in the

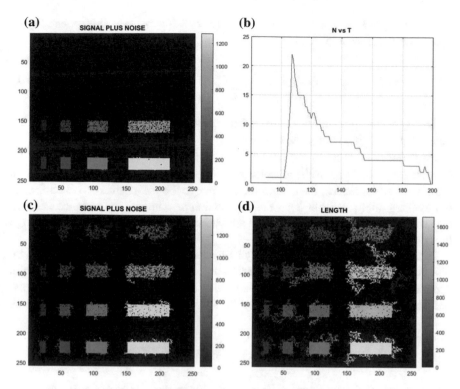

Fig. 6.7 The result of global threshold based selection of rectangular objects and dependence of the number of selected objects on the threshold value: **a** result of binarization using threshold $T = 130$, **b** dependence of the number of objects on the threshold value, **c** result of threshold reduction, **d** result of further threshold reduction

background area there are false objects, the area of which is more or less comparable to the area of useful objects in this case.

Assuming that the intensity values in the pixels of the image are mutually independent, and the background and the objects of interest are homogeneous, one can calculate the hit rate of the object of interest on a given area S including n pixels. If the binarization threshold is sufficiently high, a small number of background pixels that are adjacent to the object forming extended fractured edges can be neglected. Then the task of detecting the object of interest in noise is solved by fixing the k exceedances of the threshold of the n possible in the area S and the comparison statistics k with the account threshold m (the method of binary integration) [39, 50].

In pure form, the binary integration method can be implemented by summing the number of exceedances within the sliding window of specified dimensions. At each position of the sliding window statistics k is distributed according to the binomial law. The probability of reaching or exceeding the k_T threshold by k statistics is given by Eq. 6.1, where C_n^k are the binomial coefficients, p is the probability of exceeding the threshold in each pixel.

$$P(k \geq k_T) = \sum_{k=m}^{n} C_n^k p^k (1-p)^{n-k} \tag{6.1}$$

Let p_0 be the probability of the noise region and p_1 be the probability of the object region. It is assumed that $p_1 > p_0$.

At sufficiently large n, the binomial distribution can be approximated by the Gaussian one, and the decision statistics $dk = \sqrt{n}(p_1 - p_0)/\sqrt{p_0(1-p_0)}$ can be introduced as the ratio of the shift of the average to the mean square of the noise. In binary integration, statistics k has the expectation $m = np$ and the variance $\sigma^2 = np(1-p)$. Thus, both the expectation and the variance of the decisive statistics change in the area of the object.

In the case of selection of objects by area, statistics k is no longer represented by binomial distribution, since only connected objects are selected, and their number is significantly lower than C_n^k. By analogy with the case of binary integration, the probability of reaching or exceeding the threshold k_T by statistics k can be written as

$$P(k \geq k_T) = \sum_{k=k_T}^{n} B_n^k p^k (1-p)^{n-k}, \tag{6.2}$$

where B_n^k are the coefficients, whose values determine the number of connected objects consisting of k pixels over an area of n pixels. It is possible to calculate the exact values of these coefficients only for the one-dimensional model and for a small area of objects $n \leq 9$ [47].

The difficulties of the calculus of probabilities according to Eq. 6.1 interfere with the determination of the exact threshold value of the account k_T. However, this can be done through adaptation. For the adaptive setting of the threshold, selection of objects by area is used, taking into account the restrictions on the distortion of the shape of objects.

The simulation results are shown in Fig. 6.8, where 49 square objects of 16×16 pixels are placed on a standard Gaussian noise background (Fig. 6.8b). The signal-to-noise ratio (the relative shift of the expected value) in each pixel is $d = 1.163$. The dependence of the total number of connected objects on the threshold value is shown in Fig. 6.8c. At selection of objects by the area, the acceptable distortion of borders of objects is reached at the threshold values exceeding $T = 133$ (Fig. 6.8d). At lower values of the threshold, the shape of objects is significantly distorted by fractal noise, which destroys the boundaries.

The method of selection of objects by area provides a good selection of the shape of objects even at small signal/noise ratios, almost not inferior to the noise immunity method of binary integration (Fig. 6.8e). It is well seen that the latter distorts the shape of objects very significantly.

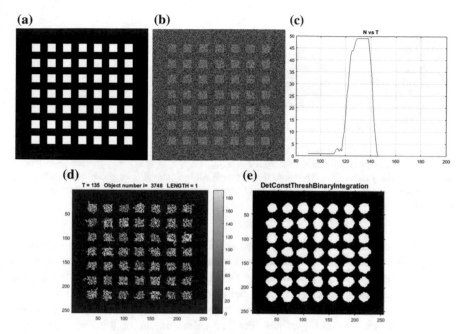

Fig. 6.8 The simulation results: **a** simulation of selection by area of square objects, **b** image with Gaussian noise, **c** dependence of the number of selected objects on the threshold value, **d** results of selection by area, **e** results of object detection by binary integration

In the detection task, the binary integration method can be used after pre-selection of objects. Since the selection of objects by area removes noise objects, it reduces the probability of false alarm. Consequently, the binarization threshold can be significantly reduced, which gives a gain in the probability of correct detection in the subsequent binary accumulation.

6.4.3   Adaptive Thresholding Based on Invariant Geometric Criteria

Figure 6.9 contains an example of the use of an adaptive global threshold for the selection of objects of circular shape hindered by background non-stationary noise based on the results of multi-threshold analysis. The optimization criterion here is the minimum of the coefficient of elongation of the perimeter of the object $P_S = P^2/4\pi S$, where P is the perimeter of the object, S is its area. This coefficient also known as the measure of the compactness of the domain [46] equals one for round (circular) objects. Any modifications of the object shape lead to its enhancement. Thus, for circular shaped objects hindered by additive noise this coefficient should be as close to one as possible to indicate minimum corruption by the noise.

(a) (b) (c) (d)

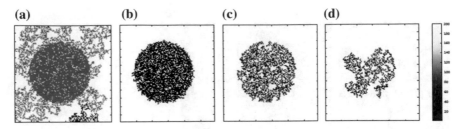

Fig. 6.9 Illustration of the use of multi-threshold image analysis containing a circular object against the background of additive noise: **a** $T < T_{opt}$, **b** $T = T_{opt}$, **c** $T > T_{opt}$ (before percolation), **d** $T > T_{opt}$ (after percolation). The intensity of the object indicates its elongation coefficient along the perimeter of P_S, which increases due to the fragmentation of the object with increasing threshold

At low thresholds, the object merges with the surrounding noise exhibiting strongly fractured edges due to adjacent structures formed by background pixels enhanced by positive noise values (Fig. 6.9a). With increasing the threshold, these adjacent fractured structures are being detached from the object (Fig. 6.9b). At the same time some fragments within the object are being eliminated due to the influence of negative noise values. Further enhancement of the threshold value leads again to the fracture of the object edges now due to the formation of adjacent fragments from inside the edge (Fig. 6.9c), and finally to its fragmentation once the percolation threshold is reached (Fig. 6.9d). Thus, both low and high threshold values lead to an enhancement of the elongation coefficient of the perimeter of the object due to noise impact. By minimizing this factor, we can choose the best binary layer to represent circular objects (if they are present in the image) with the least possible edge fracture by noise.

Although the formal minimization of the P_S coefficient is justified for round objects, it is also suitable for the allocation of other regular objects with a perimeter elongation factor close to one, for example, for square objects, where this coefficient is approximate equaled to 1.273. The effect of noise typically leads to a significant (one to two orders of magnitude) increase in P_S for such objects due to the elongation of their edges, that is well above minor discrepancies in the coefficients for different regular shapes of objects in the absence of noise. The advantage of this methodology is that it is simple, has only a few free parameters that are easy to interpret. To increase the detalization of the analysis, one can consider higher K_p values leading to a single object that is broken down into characteristic sub-objects of different sizes (e.g. for the separate selection of cells and cell nuclei from microscopic images).

Now let us consider two important parameters that determine the boundary conditions, namely the minimum base size of the object S_{bmin}, which is included in the initial 3D reconstruction, and the maximum coefficient P_S, which will be taken into account in multi-layer content analysis. A reasonable choice of both parameters allows one to speed up the algorithm considerably, as well as, eliminate

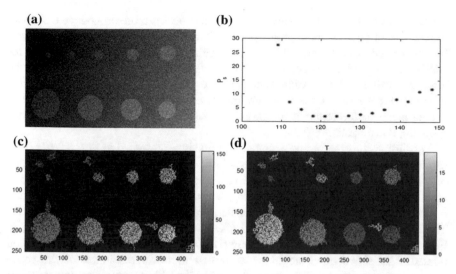

Fig. 6.10 Illustration of multi-threshold image analysis containing several round objects distorted by additive noise with non-stationary (gradient) component: **a** image hindered by Gaussian noise, **b** the graph at the top right illustrates the typical U-shaped $P_S(T)$ function for regular objects in noise, which has a minimum P_S value corresponding to the optimal T_{opt} threshold, **c** the result of the adaptive threshold processing procedure, which is performed independently for each selected object with the base area $S_b > 100$, **d** the result of the adaptive threshold processing procedure the shape parameter $P_S < 20$. The intensities of the objects in **c** determine the optimal threshold values of T_{opt}, and the intensity in **d** correspond to the values of the perimeter elongation factor for the selected objects

high-frequency noise represented by numerous small isolated objects, many of them containing only one or few pixels. However, appropriate selection of both parameters requires knowledge of the typical size of the object of interest, as well as, the image resolution, although in practice they are often adjusted during testing.

Figure 6.10 illustrates the adaptive procedure for setting the threshold value after multi-threshold analysis of an image containing several circular objects distorted by additive noise with a non-stationary (gradient) component. Figure 6.10 shows how the optimization procedure based on P_S minimization described above can be used to find the best local T_{opt} thresholds for a series of objects that are similar only in shape but have significantly different sizes. Due to the significant background gradient, the optimal thresholds in Fig. 6.10c are significantly different, and are between 60 and 160. In turn, the corresponding minimum P_S values achieved for different objects also vary from values just over one to almost 20. Despite the uniform shape of the objects, the results show a difference in the P_S coefficients obtained for round objects in the presence of noise. As can be seen from Fig. 6.10d, larger objects exhibit higher P_S values, despite the similarity of shape and noise characteristics, and normalization.

This fact is a consequence of the fracture of the edges of objects corrupted by noise. First, the edge of a large object is longer and in the same image resolution contains more finite elements (pixels), hence the effect is more pronounced. Second, the fracture of edges is more pronounced at low threshold values than at high ones. In multiscale processing, this will require additional P_S correction for similar-shaped objects depending on their size, as well as, depending on the threshold value and the signal-to-noise ratio. Such corrections require detailed research in this area in order to ensure invariance in multiscale selection of objects and to preserve the required resolution of objects in the image including corresponding discretization effects.

For further illustration, the example previously represented in Fig. 6.8 has been reconsidered with local adaptive thresholding based on minimizing P_S for each selected object. Results for the signal-to-noise ratio $d = 1.163$ shown in Fig. 6.11a have been obtained for $S_{bmin} = 150$, $P_{Smax} = 60$ and $K_p = 0.5$ while the adaptive threshold value as a function of the selected object number is shown in Fig. 6.11b. The results indicate that 11 out of 49 (22%) objects are missing. In contrast, in Fig. 6.11c, where similar results are shown for $K_p = 0.85$, no single object has been missed. The above effect could be attributed to lower resolution of the scene reconstruction at $K_p = 0.5$. In particular, when an object containing two or more

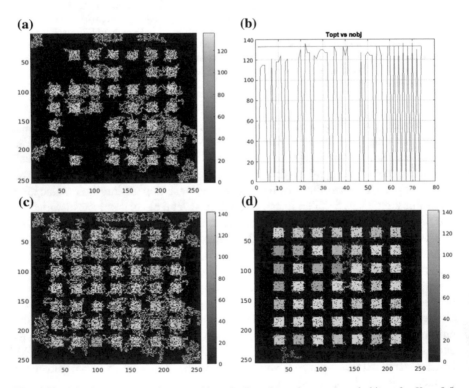

Fig. 6.11 Adaptive selection of square objects in Gaussian noise: **a** selected objects for $K_p = 0.5$, **b** adaptive threshold as a function of the selected object number, **c** selected objects for $K_p = 0.85$, **d** objects selected by global thresholding

sample squared objects disintegrated due to the increased threshold T, it eventually preserves more than half of the area of the previous object. In the limiting scenario, two objects are separated with slightly unequal areas, and the larger one is then considered as the successor of the previous (e.g. double square) object. In turn, the latter object is never considered as a separate one in the reconstruction. To avoid the above effect the analysis resolution could be simply increased by the enhancement of K_p. Importantly, the above problem arises only once true squared objects are separated by small gaps only. By increasing distance between the objects, one could achieve similar performance also for the smaller K_p values.

Figure 6.11d represents more accurate result obtained by the fixed global thresholding. The above result is not surprising, as small objects that are used for individual threshold adaptation in Fig. 6.11a, c provide each with considerably smaller statistics compared to the entire area above the threshold in Fig. 6.11d. Accordingly, in the absence of gradient global thresholding based approach still appears more reasonable.

Figure 6.12 illustrates the application of the algorithm inspired by [34] to the test image (Fig. 6.12a), where the largest object size (for normalization, actually its fraction from the total image size is shown) is obtained as the function of the threshold T (Fig. 6.12b). Figure 6.12 shows that, like for the landscape analysis, there are several characteristics gaps that can be also found as local peaks (outliers) in the derivative. By setting the threshold values at the corresponding maxima one can follow the stepwise decomposition of the base object into sub-objects. The first characteristic peak corresponds to the segmentation of the entire image into a global object and background (Fig. 6.12c).

Further peaks correspond to the detachment of further objects (Fig. 6.12d–f). The key advantage of this approach is that it, in marked contrast to the previously considered one, allows for an easy selection of large objects even when they are single, i.e. there is no need to accumulate sufficient statistics from multiple objects. However, a clear drawback of this approach that when the base object is split into two or several objects, only the largest of its successor objects is being tracked further, meaning that others are ignored.

Based on the generalization of the percolation based approach suggested above, the same raw image has been reconstructred into a hierarchy of objects shown in Fig. 6.13, where intensity denotes the individual percolation thresholds P_c of each object. Figure 6.13 shows that this approach leads to a more detailed decomposition of the image with reconstruction depth controlled by the persistence coefficient K_p. It should be emphasized that the problem of selection of objects based on percolation is more general than the problem of image segmentation, when the image is divided into non-intersecting fragments. In this case, individual pixels can belong to different selected objects at different threshold layers. Figure 6.13 illustrates this effect, where the highlight on an object is selected as a separate object.

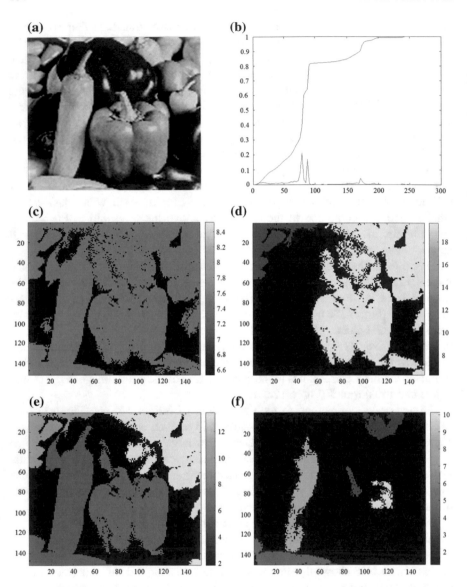

Fig. 6.12 Multi-threshold selection of largest objects for a series of thresholds. Only thresholds corresponding to significant changes in the maximum object size, i.e. the local peaks of the first differences, are being considered: **a** initial image, **b** function of threshold T, **c**–**f** selection with different thresholds

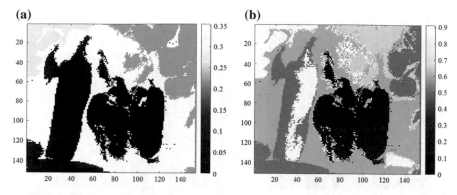

Fig. 6.13 An overview of the 3D representation of the entire scene based on the percolation concept using reconstruction depth: **a** $K_p = 0.6$, **b** $K_p = 0.8$. Individual objects (shown by different intensities) appear one above another, i.e., shades are selected separately from objects at different (higher) thresholds. Reconstruction depth is controlled by K_p parameter. Intensity denotes individual percolation coefficients P_c for each selected object

6.5   Applications

Finally, we show how the above formalism can be applied to real-world observational data from remote sensing of the Earth's surface, as well as, microscopic biomedical images.

Figure 6.14 contains the results of selection of objects on the area on the remote observation image (Fig. 6.14a). In Fig. 6.14b the dependence of the number of selected objects on the threshold value is shown. Image (Fig. 6.14b) obtained by setting the threshold to the maximum number of allocated connected objects ($N_{obj} = 30$ at $T = 101$). The intensity tint reflects the area values of the objects in pixels. With the increase of the threshold it is possible to increase the image resolution (Fig. 6.9d), but less intense objects disappear. If the objects are isolated, then after selection each object is localized, i.e. the coordinates of its center, as well as, other parameters of shape and texture are determined.

The disadvantage of selection by area is the need to set the area parameter in absolute values (in pixels), which is difficult in cases of changing the scale of the image. This method does not work well in the case of a non-uniform background, which can give false objects that are comparable in area to the objects of interest (Fig. 6.14e, f).

Figure 6.15 shows how the algorithm based on the global percolation curve can be used for the image segmentation. SAR remote sensing image is being processed where characteristic peaks corresponding to particular phase transitions correspond to the threshold values that in turn allow for an accurate selection of the agricultural land, river bed, and forest coverage, respectively.

Figure 6.16 shows a satellite photo image and the results of its reconstruction with particular P_c values depicted for each object. Figure 6.16 shows that,

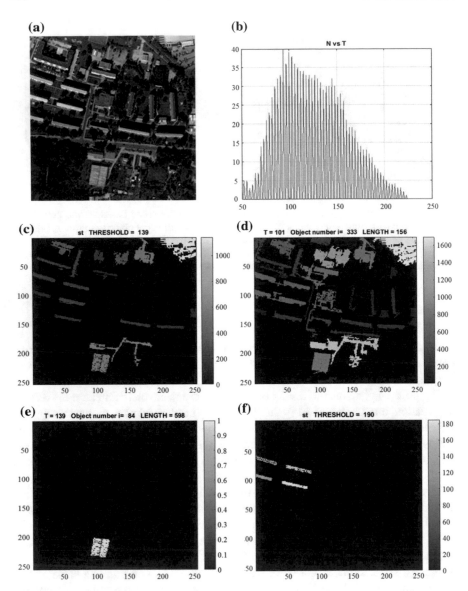

Fig. 6.14 Selection of objects on the area on the remote observation image: **a** raw image, **b** dependence of the number of connected objects on the threshold value, and the results of selection of objects by area: **c** merged objects at low threshold, **d** vanishing of objects at high thresholds, **e** selected objects, **f** false objects

remarkably, the natural background including land and water surface altogether form a base object that is characterized by P_c around 0.6 that is close to the theoretical expectation for the random grid. In contrast, various artificial objects are characterized by a variety of P_c values, with many of them well below the

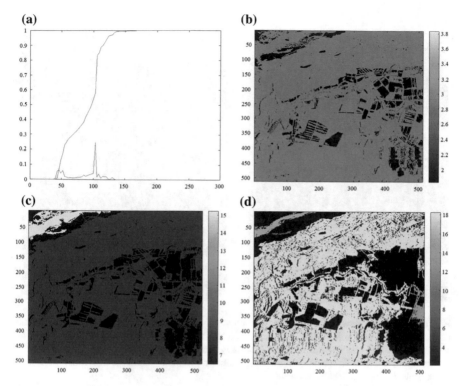

Fig. 6.15 Multi-threshold selection of largest objects for a series of thresholds: **a** function of threshold T, **b**–**d** multi-threshold selection for different levels. Only thresholds corresponding to significant changes in the maximum object size, i.e. the local peaks of the derivative of the percolation curve are being considered

theoretical expectation that may be treated as an indication of their non-random nature. Artifacts such as glimpse-like spots characterized by $P_c = 1$ in water areas may be attributed to excessive level quantization effects in the raw image.

Figure 6.17 shows the results of the combined analysis where the percolation based reconstruction have been followed by selection of particular threshold values according to the selection parameters P_L and P_C, respectively. Figure 6.17a shows that prolonged object can be easily distinguished from rather compact objects using the P_L parameter. Figure 6.17b shows that urban areas, as well as, other areas with indications of artificial modifications such as deforestation exhibiting sharp-edged objects can be successfully separated from natural environments by P_C parameter that easily distinguishes between rough-edges artificial planning and rather fractal-shaped natural landscapes, like in Fig. 6.16.

Figure 6.18a, d shows the raw microscopic images, while Fig. 6.18b, c show the results of their analysis using the local adaptive thresholding procedure. Here, the optimal threshold for each object is chosen independently of the others according to their individual P_S minima leading to nearly excellent selection of all objects.

(a)   **(b)**

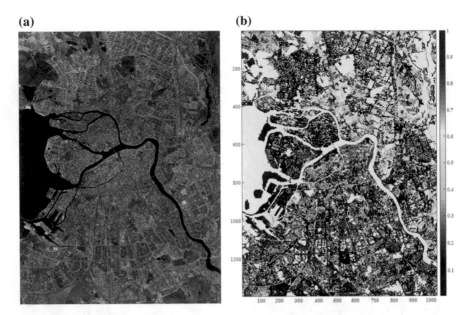

Fig. 6.16 An overview of the 3D representation of the entire scene based on the percolation concept: **a** $K_p = 0.6$, **b** $K_p = 0.8$. Individual objects (shown by different intensities) appear one above another, i.e., shades are selected separately from objects at different (higher) thresholds. Reconstruction depth is controlled by K_p parameter. Intensity denotes individual percolation coefficients P_c for each selected object

Notably, while objects with high intensity have $T_{opt} > 50$, those objects with low intensity have $T_{opt} < 30$. Besides, no background fragments have been selected during this procedure, although no dedicated preprocessing with intensity inhomogeneity correction have been performed. Figure 6.18c shows that the percolation based reconstruction also allows for a separate selection and visualization of both cells and cell nuclei as embedded objects.

As we have already noted above, the suggested percolation based framework can be easily generalized to the reconstructions of truly 3D objects. The following example represents a series of CLSM Z-stack layer images, where different layers represent the true vertical Z-coordinate. Figure 6.19a, b shows the 3D view of the biofilms depicted using standard visualization software containing both the intact *S. aureus* monoculture biofilms (Fig. 6.19a) and the biofilms treated with the plant protease ficin (Fig. 6.19b). Figure 6.19c, d shows the top view of the corresponding CLSM images percolation based analysis and object reconstruction, with 20–22 Z-stacks in each, where the intensity indicates P_C values for each of the reconstructed 3D objects. For the best selection of objects in each layer, all stacks have been preliminarily subjected to the local adaptive thresholding, thereby, leading to near optimal selection of cells and cell clumps in each layer according to the minimum of P_S parameter.

(a)   (b)

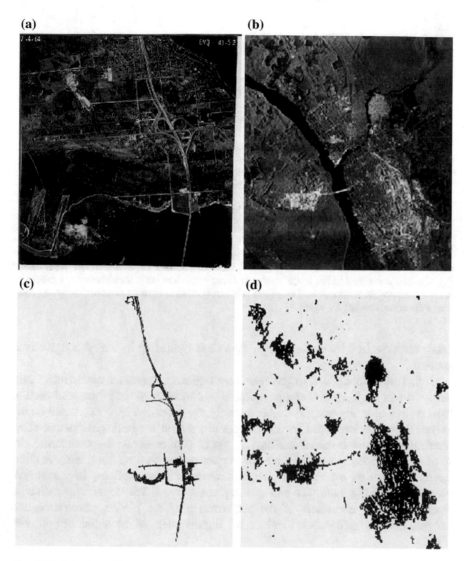

Fig. 6.17 Selection of individual objects from the percolation based 3D scene reconstruction by additional criteria based on P_L and P_C parameters, respectively: **a** highway image, **b** urban image, **c** and **d** the highway abd urban images from **a** and **b** with traces of artificial activity

Figure 6.19a, c shows that the *S. aureus* biofilm exhibits a typical mushroom-shaped structure characterized by continuous and rather smooth coating that can be clearly observed from the mainly uninterrupted coverage indicating small P_C values. By contrast, after Ficin treatment that leads to almost complete destruction of the biofilm structure [51], only isolated cell clumps characterized by high P_C values can be identified, confirming fragmentation of the biofilm into cell

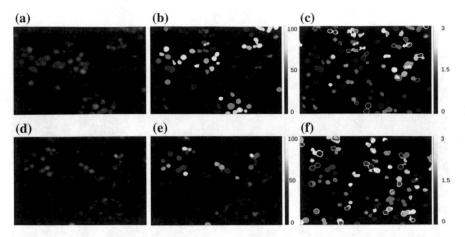

Fig. 6.18 Examples of monochrome microscopic images of Caco-2 cells culture: **a**, **d** raw monochrome microscopic images with significant intensity imbalance, **b**, **e** their best layer cuts at T_{opt} obtained according to the adaptive thresholding procedure suggested above, **c**, **f** results of percolation based reconstruction. The object color indicates the optimal threshold value T_{opt} (B), where the corresponding P_S value

debris characterized by typically high P_C values indicating its abrupt rough-edged structure.

CLSM reconstruction also indicates that regions with rather continuous uninterrupted cell agglomerates can be selected and their fractions in the total biofilm biomass can be easily quantified. Accordingly, the proposed approach based on the universal percolation based framework could be used for both quantitative characterization of the bacterial biofilms integrity (for example, by comparing the distributions of P_C values per layer or per image, as well as, their simpler characteristics such as average and standard deviation), as well as, its qualitative visualization using both 2D and 3D representations. The latter application is essential for the assessment of the performance of the biofilm penetrating and integrity destructing agents critical for the improvement of the antibiotics efficacy [51].

6.6 Conclusions

To summarize, we have considered several approaches to the multi-threshold analysis of monochromatic images and consequent interpretation of its results in computer vision systems. The key aspect of the proposed approach is that it is based on the complete scene reconstruction leading to the object based scene representation inspired by principles from percolation theory. As a generalization of the conventional image segmentation, we proposed a complete scene reconstruction

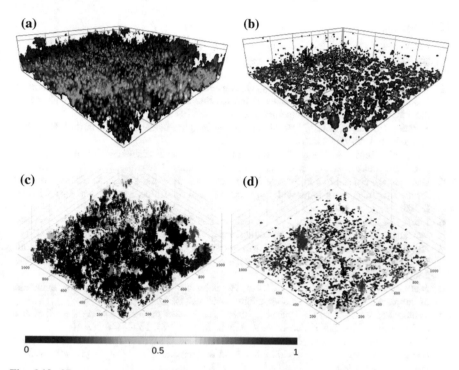

Fig. 6.19 3D reconstruction of CLSM images of *S. aureus* biofilm: **a** before treatment with ficin, **b** after treatment with ficin (the conventional representation of CLSM images), **c** before treatment with ficin, **d** after treatment with ficin (the percolation based on 3D reconstructions with color indicating the P_c values for each Z-stack)

framework providing a multi-scale hierarchy of objects thus allowing representation of embedded objects at different scales. Using the reconstruction, we next suggested a direct approach to the object selection as a subset of the reconstructed scene based on a posteriori information obtained by multi-thresholding at the cost of the algorithm performance. We consider several geometric invariants as selection quantities and validate our approach explicitly using prominent examples of synthetic images, remote sensing systems, and microscopic data of biological samples. Finally, we believe that the adaptive algorithms suggested above could be further supplemented by shape optimization analysis [52] resulting in more sophisticated selection of objects exhibiting arbitrary shapes.

Acknowledgements We like to acknowledge partial support of this research by the Ministry of Science and Higher Education of the Russian Federation in the framework of the basic state assignment of St. Petersburg Electrotechnical University (project No. 2.5475.2017/6.7)), as well as, by the Russian Science Foundation (project No. 16-19-00172).

References

1. Blaschke, T.: Object based image analyses for remote sensing. ISPRS J. Photogramm. Remote Sens. **65**, 2–16 (2010)
2. Lang, S., Baraldi, A., Tiede1, D., Hay, G., Blaschke, T.: Towards a (GE)OBIA 2.0 manifesto —achievements and open challenges in information & knowledge extraction from big Earth data. In: GEOBIA'2018, Montpellier, pp. 18–22 (2010)
3. Schlafer, S., Meyer, R.L.: Confocal microscopy imaging of the biofilm matrix. J. Microbiol. Methods **138**, 50–59 (2017)
4. Atale, N., Gupta, S., Yadav, U.C.S., Rani, V.: Cell-death assessment by fluorescent and nonfluorescent cytosolic and nuclear staining techniques. J. Microsc. **255**(1), 7–19 (2014)
5. Daemen, S., van Zandvoort, M.A.M.J., Parekh, S.H., Hesselink, M.K.C.: Microscopy tools for the investigation of intracellular lipid storage and dynamics. Mol. Metab. **5**(3), 153–163 (2016)
6. Liang, J.I., Piper, J., Tang, J.-Y.: Erosion and dilation of binary images by arbitrary structuring elements using interval coding. Pattern Recognit. Lett. **9**(3), 201–209 (1989)
7. Heydorn, A., Nielsen, A.T., Hentzer, M., Sternberg, C., Givskov, M., Ersbøll, B.K., Molin, S.: Quantification of biofilm structures by the novel computer program comstat. Microbiology **146**(10), 2395–2407 (2000)
8. Nattkemper, T.W., Twellmann, T., Ritter, H., Schubert, W.: Human vs. machine: evaluation of fluorescence micrographs. Comput. Biol. Med. **33**(1), 31–43 (2003)
9. Lempitsky, V., Rother, C., Roth, S., Blake, A.: Fusion moves for markov random field optimization. IEEE Trans. Pattern Anal. Mach. Intell. **32**(8), 1392–1405 (2010)
10. Klinger-Strobel, M., Suesse, H., Fischer, D., Pletz, M.W., Makarewicz, O.: A novel computerized cell count algorithm for biofilm analysis. PloS One **11**(5), e0154937.1– e0154937.22 (2016)
11. Bogachev, M.I., Volkov, VYu., Markelov, O.A., Trizna, EYu., Baydamshina, D.R., Melnikov, V., Murtazina, R.R., Zelenikhin, P.V., Sharafutdinov, I.S., Kayumov, A.R.: Fast and simple tool for the quantification of biofilm-embedded cells sub-populations from fluorescent microscopic images. PLoS ONE **13**(5), e0193267 (2018)
12. Gao, G.: Statistical modeling of SAR images: a survey. Sensors **10**(1), 775–795 (2010)
13. Vovk, U., Pernus, F., Likar, B.: A review of methods for correction of intensity inhomogeneity in MRI. IEEE Trans. Med. Imaging **26**(3), 405–421 (2007)
14. Carpenter, A.E., Jones, T.R., Lamprecht, M.R., Clarke, C., Kang, I.H., Friman, O., Guertin, D.A., Chang, J.H., Lindquist, R.A., Moffat, J., Golland, P., Sabatini, D.M.: Cellprofiler: image analysis software for identifying and quantifying cell phenotypes. Genome Biol. **7**(10), R100.1–R100.11 (2006)
15. Lamprecht, M.R., Sabatini, D.M., Carpenter, A.E.: Cellprofiler: free, versatile software for automated biological image analysis. Biotechniques **42**(1), 71–75 (2007)
16. Jones, T.R., Kang, I.H., Wheeler, D.B., Lindquist, R.A., Papallo, A., Sabatini, D.M., Golland, P., Carpenter, A.E.: Cellprofiler analyst: data exploration and analysis software for complex image-based screens. BMC Bioinformatics **9**(1), 482.1–482.17 (2008)
17. Kamentsky, L., Jones, T.R., Fraser, A., Bray, M.-A., Logan, D.J., Madden, K.L., Ljosa, V., Rueden, C., Eliceiri, K.W., Carpenter, A.E.: Improved structure, function and compatibility for cellprofiler: modular high-throughput image analysis software. Bioinformatics **27**(8), 1179–1180 (2011)
18. Schindelin, J., Arganda-Carreras, I., Frise, E., Kaynig, V., Longair, M., Pietzsch, T., Preibisch, S., Rueden, C., Saalfeld, S., Schmid, B., Tinevez, J.Y., White, D.J., Hartenstein, V., Eliceiri, K., Tomancak, P., Cardona, A.: Fiji: an open-source platform for biological-image analysis. Nat. Methods **9**(7), 676 (2012)
19. Zhou, W., Troy, A.: An object-oriented approach for analyzing and characterizing urban landscape at the parcel level. Int. J. Remote Sens. **29**(11), 3119–3135 (2008)

20. Gu, H., Han, Y., Yang, Y., Li, H., Liu, Z., Soergel, U., Blaschke, T., Cui, S.: An efficient parallel multi-scale segmentation method for remote sensing imagery. Remote Sens. **10**(4), 590.1–590.18 (2018)

21. Beyenal, H., Donovan, C., Lewandowski, Z., Harkin, G.: Three-dimensional biofilm structure quantification. J. Microbiol. Methods **59**(3), 395–413 (2004)

22. Sage, D., Donati, L., Soulez, F., Fortun, D., Schmit, G., Seitz, A., Guiet, R., Vonesch, C., Unser, M.: Deconvolutionlab2: an open-source software for deconvolution microscopy. Methods **115**, 28–41 (2017)

23. Naberukhin, Y.I., Voloshin, V., Medvedev, N.: Geometrical analysis of the structure of simple liquids: percolation approach. Mol. Phys. **73**, 917–936 (1991)

24. Dominik, K.G., Shandarin, S.F.: Percolation analysis of nonlinear structures in scale-free two-dimensional simulations. Astrophys. J. **393**, 450–463 (1992)

25. Xie, N., Shi, X., Feng, D., Kuang, B., Li, H.: Percolation backbone structure analysis in electrically conductive carbon fiber reinforced cement composites. Compos. Part B Eng. **43**, 3270–3275 (2012)

26. Chen, B., Guizar-Sicairos, M., Xiong, G., Shemilt, L., Diaz, A., Nutter, J., Burdet, N., Huo, S., Mancuso, M.A., Monteith, A., Vergeer, F., Burgess, A., Robinson, I.: Three-dimensional structure analysis and percolation properties of a barrier marine coating. Sci. Rep. **3**, 1177.1–1177.5 (2013)

27. Jarvis, N., Larsbo, M., Koestel, J.: Connectivity and percolation of structural pore networks in a cultivated silt loam soil quantified by X-ray tomography. Geoderma **287**, 71–79 (2017)

28. Langovoy, M., Wittich, O.: Detection of objects in noisy images and site percolation on square lattices. 1–14 (2009). arXiv:1102.4803v1

29. Langovoy, M., Habeck, M., Schölkopf, B.: Spatial statistics, image analysis and percolation theory. 1–12 (2011). arXiv:1310.8574v1

30. Langovoy, M., Wittich, O.: Randomized algorithms for statistical image analysis and site percolation on square lattices. Stat. Neerl. **67**, 337–353 (2013)

31. Arora, S., Acharya, J., Verma, A., Panigrahi, P.K.: Multilevel thresholding for image segmentation through a fast statistical recursive algorithm. Pattern Recognit. Lett. **29**(2), 119–125 (2008)

32. Yang, J., Yang, Y., Yu, W., Feng, J.: Multi-threshold image segmentation based on k-means and firefly algorithm. In: 3rd International Conference on Multimedia Technology. Atlantis Press, pp. 134–142 (2013)

33. Priyanka, P., Vasudevarao, K., Sunitha, Y., Sridhar, B.A.: Multi level fuzzy threshold image segmentation method for industrial applications. IOSR J. Electron. Commun. Eng. **12**(2), Ver. III, 6–17 (2017)

34. Fan, J., Meng, J., Saberi, A.A.: Percolation framework of the Earth's topography. Phys. Rev. E **99**, 022304.1–022304.6 (2019)

35. Cheng, J., Tsai, Y., Hung, W., Wang, S., Yang, M.: Fast and accurate online video object segmentation via tracking parts. In: 2018 IEEE/CVF Conference on Computer Vision and Pattern Recognition, Salt Lake City, UT USA, pp. 7415–7424 (2018)

36. Wang, M.A.: Multiresolution remotely sensed image segmentation method combining rainfalling watershed algorithm and fast region merging. Int. Arch. Photogramm. Remote Sens. Spatial Inf. Sci. XXXVII Part B 1213–1218. Beijing (2008)

37. Banimelhem, O., Yahya, Y.: Multi-thresholding image segmentation using genetic algorithm. 15th International Conference on Image Processing, Computer Vision, & Pattern Recognition, Las Vegas, Nevada, USA, pp. 1–6 (2012)

38. Cuevas, E., González, A., Fausto, F., Zaldívar, D., Pérez-Cisneros, M.: Multithreshold segmentation by using an algorithm based on the behavior of locust swarms. Math. Probl. Eng. **2015**, 805357.1–805357.25 (2015)

39. Volkov, V.: Extraction of extended small-scale objects in digital images. Int. Arch. Photogramm. Remote Sens. Spatial Inf. Sci. XL-5/W6, 87–93 (2015)

40. Bogachev, M., Volkov, V., Kolaev, G., Chernova, L., Vishnyakov, I., Kayumov, A.: Selection and quantification of objects in microscopic images: from multi-criteria to multi-threshold analysis. Bionanoscience **9**(1), 59–65 (2019)
41. Bunde, A., Havlin, S.: Fractals and Disordered Systems. Springer, Berlin, Heidelberg (1996)
42. Finn, H.M., Johnson, S.: Adaptive detection mode with threshold control as a function of spatially sampled clutter level estimators. RCA Rev. **29**(3), 414–465 (1968)
43. El-Mashade, M.B.: Performance improvement of adaptive detection of radar target in an interference saturated environment. Prog. Electromagn. Res. **2**, 57–92 (2008)
44. Rohling, H.: Radar CFAR thresholding in clutter and multiple target situations. IEEE Trans. AES-**19**(4), 608–621 (1983)
45. Volkov. V.Yu.: Adaptive and invariant algorithms for object detection in images and their modeling. Saint-Petersburg-Moscow-Krasnodar (in Russian), Lan (2014)
46. Gonzales, R.C., Woods, R.E.: Digital Image Processing, 4th edn. Pearson (2018)
47. Volkov, V.Yu.: Adaptive extraction of small objects in digital images. Izv. Vuzov Rossii. Radioelektronika. **1**, 17–28 (in Russian) (2017)
48. Levinshteln, M., Efros, L.: The relation between the critical exponents of percolation theory. Zh. Eksp. Teor. Fiz. **69**, 386–392 (1975)
49. Melnikov, V., Bogachev, M.I., Volkov, V.Y, Markelov, O.A.: Selection and analysis of objects in multi-threshold image processing. In: IEEE Conference on Russian Young Researchers in Electrical and Electronic Engineering (ElConRus), Saint Petersburg and Moscow, Russia, pp. 1202–1205 (2019)
50. Skolnik, M.: Radar Handbook, 3nd edn. McGraw-Hill (2008)
51. Baidamshina, D.R., Trizna, E.Y., Holyavka, M.G., Bogachev, M.I., Artyukhov, V.G., Akhatova, F.S., Rozhina, E.V., Fakhrullin, R.F., Kayumov, A.R.: Targeting microbial biofilms using Ficin, a nonspecific plant protease. Sci. Rep. **7**, 46068 (2017)
52. Krasichkov, A.S., Grigoriev, E.B., Bogachev, M.I, Nifontov, E.M.: Shape anomaly detection under strong measurement noise: an analytical approach to adaptive thresholding. Phys. Rev. E **92**(4), 042927.1–042927.9 (2015)

Part II
Medical Applications

Chapter 7
Vision-Based Assistive Systems for Deaf and Hearing Impaired People

Dmitry Ryumin, Denis Ivanko, Ildar Kagirov, Alexander Axyonov
and Alexey Karpov

Abstract Sign language is the way of communication among deaf and hearing impaired community and consist of a combination of hand movements and facial expressions. Successful efforts in computer vision-based research within the last years paved the path for first automatic sign language recognition systems. However, unresolved challenges, such as cultural differences in the sign languages of the world, lack of the representative databases for model training, relatively small size of the region-of-interest, issues due to occlusion, etc. keep automatic sign language recognition reliability still far from human-level performance, especially for the Russian sign language. To address this issue, we present a framework and an automatic system for one-handed gestures of Russian Sign Language (RSL) recognition. The developed system supports both online and offline modes and is able to recognize 44 classes of RSL one-handed gestures with almost 70% of accuracy. The system is based on color-depth Kinect v2 sensor and trained on TheRuSLan database using a combination of state-of-the-art deep learning approaches. The future research will focus on extracting additional features, expanding the data set, and increasing the amount of recognizable gestures with two-handed gestures. The developed vision-based RSL recognition system is meant as an auxiliary system for deaf and hearing impaired people.

D. Ryumin (✉) · D. Ivanko · I. Kagirov · A. Axyonov · A. Karpov
St. Petersburg Institute for Informatics and Automation of the Russian Academy of Sciences, SPIIRAS, 39, 14th Line, 199178 St. Petersburg, Russian Federation
e-mail: dl_03.03.1991@mail.ru

D. Ivanko
e-mail: denis.ivanko11@gmail.com

I. Kagirov
e-mail: kagirov@iias.spb.su

A. Axyonov
e-mail: a.aksenov95@mail.ru

A. Karpov
e-mail: karpov@iias.spb.su

© Springer Nature Switzerland AG 2020
M. N. Favorskaya and L. C. Jain (eds.), *Computer Vision in Advanced Control Systems-5*, Intelligent Systems Reference Library 175,
https://doi.org/10.1007/978-3-030-33795-7_7

Keywords Sign language · Gestures · Face detection · Computer vision · Machine learning · Artificial neural networks · CNN · LSTM

7.1 Introduction

This chapter focuses on sign language recognition methods and techniques of automatic lip reading. A description of methods that were developed by the authors to address these problems, is presented as well. Thus, the term vision-based systems for deaf and hearing-impaired people should be understood in this sense, i.e. systems that allow automatic lip reading and sign language recognition.

An improvement of the quality of life of deaf and hearing-impaired people is a topical issue today. According to the statistics provided by the World Federation of the Deaf, there are about 70 million deaf and hearing-impaired people around the world nowadays. Most of them use sign languages in everyday life, but have severe communicative problems when trying to interact with the hearing society. Sign languages as an independent communication system should be distinguished from spontaneous gesticulatory acts accompanying verbal communicative acts. Since the recognition system presented within this chapter deals with Russian sign language, we provide the reader with the information concerning RSL, as well: the Ministry of Health of the Russian Federation states, that there are about 122 thousand deaf and hearing-impaired citizens in Russia (according to 2010 census). In 2012, RSL was recognized as the language of communication in the Russian Federation, in the spheres of oral use of the state language of the Russian Federation. As a result, deaf and hearing-impaired people got the opportunity to apply to government institutions using Russian Sign Language. Nowadays, sign languages are the second state language in the United States, Finland, Spain, the Czech Republic, etc. In recent years, many investigations in the area of automatic sign language recognition have arisen. The main task of automatic recognition systems is to provide the deaf and hearing-impaired people with effective communicational tools and make it easier to interact with the hearing society.

Another powerful tool of communication used by the deaf is lip-reading. Automatic lip-reading systems are designed for the purpose of understanding the content of speech by interpreting the movements of the lips. Combined with a system of automatic sign language recognition, lip-reading systems can enhance the robustness level of vision-based interpreting systems. The common problem that vision-based systems share is that they are easily affected by noisy environments, illumination, etc. These challenges are treated in detail in this chapter.

The structure of this chapter is as follows. In Sect. 7.2, state-of-the-art methods and sources used for automatic lip-reading and sign language (i.e. hand gestures) recognition are presented. Methodological problems are discussed in Sect. 7.3. In Sect. 7.4, the database TheRuSLan collected for training an automatic hand gesture recognition system is described. Section 7.5 provides technical characteristics of the collected database. The framework proposed by the authors, as well as experimental results are in issue in Sect. 7.6. The conclusions are given in Sect. 7.7.

7.2 Related Work

This section investigates the recent approaches, techniques, features extraction methods, and databases being used for hand gestures recognition and lipreading in the context of sign language.

Generally, sign language is a form of communicative hand gestures and facial ex-pressions. Hand gestures are classified according to temporal relationship into two types: static and dynamic gestures. Techniques, commonly used to recognize static gestures are Support Vector Machine (SVM) [1–4], Hidden Markov Models (HMM) [5–8] or different types of Neural Networks (NN), such as Convolutional Neural Networks (CNN) [9–11], Long Short-Term Memory (LSTM) networks [12], NN with log-sigmoid activation function [1], NN with symmetric Elliot activation function [1], feed-forward back propagation NN [13–15], etc. A review focusing on sign language in a mobile context can be find in [16].

The work [1] explores the development of a mobile application for South African sign language recognition. The goal was to recognize the manual alphabet and manual numeric digits that have static gestures (31 gestures in total). The authors compared NN with a log-sigmoid activation function, NN with symmetric Elliot activation function and SVM with resulting reported accuracy of 94–99%. The mobile application was developed for Android.

In [2], the authors proposed a vision-based gesture recognition system, which can be applied in conditions with noisy background. The researchers developed a method to adaptively update the skin color model for different users and various lighting conditions. Three kinds of features were combined: principle component analysis, linear discriminant analysis, and SVM were integrated to construct a hierarchical classification scheme. The authors evaluated the proposed method on two datasets: the own collected CSL dataset and a public American Sign Language (ASL) dataset, in which images of the same gesture were captured in different lighting conditions. Method achieved the accuracies of 99.8 and 94%.

In paper [3], a novel framework was proposed to recognize images of several sign language gestures. Canny edge detection was implemented to segment the hand gesture from its background. Feature points then was extracted with speeded up robust features algorithm, whose features was derived through a bag of features method. SVM was subsequently applied to classify gesture image dataset. The proposed framework has been successfully implemented on smartphone platforms, and experimental results showed that it is able to recognize and translate 16 different gestures of ASL with an overall accuracy of 97.13%. The authors of the [8] introduced the first automatic Arabic sign language recognition system based on HMMs. A large set of samples has been used to recognize 30 isolated words from the standard Arabic sign language. For signer-dependent case, the system obtains a word recognition rate of 98.13%, for signer-independent case the system obtains a word recognition rate of 94.2%, respectively.

In the work [4], a feature learning approach based on Sparse Auto-Encoder (SAE) and principle component analysis was proposed for recognizing human actions, i.e. finger-spelling or sign language, for RGB-D inputs. The proposed model of feature learning consist of two components: First, features are learned respectively from RGB and depth channels, using sparse auto-encoder with convolutional neural networks. Second, the learned features from both channels is concatenated and fed into a multiple layer using Principle Component Analysis (PCA) to get the final feature. Experimental results on ASL dataset demonstrated the improvement of the recognition rate from 75 to 99%.

The goal of [9] was a real-time hand gesture-based human robot interface for mobile robots. The authors used Deep Neural Network (DNN) combining convolution and Max-Pooling CNN (MPCNN) for supervised feature learning and classification of hand gestures given by humans to mobile robots using colored gloves. The hand contour was retrieved by color segmentation, and then smoothened by morphological image processing. The developed deep MPCNN classifies 6 gesture classes with 96% accuracy.

The authors of [10] proposed a novel 3D CNN architecture, which extracts discriminative spatial-temporal features from raw video stream automatically without any prior knowledge, avoiding designing features. To boost the performance, multi-channels of video streams, including color information, depth clue, and body joint positions, were used as input to 3D CNN in order to integrate color, depth and trajectory information. The validation of the proposed model on a dataset collected with Microsoft Kinect demonstrated its effectiveness over the traditional approaches based on hand-crafted features. In the work [17], the authors proposed an approach to detect and recognize 3D one-handed gestures using deep CNNs for detecting hand shapes. The contribution of the work [11] was in building an Italian sign language recognition system using CNNs. The authors were able to recognize 20 Italian gestures with high accuracy. The predictive model was able to generalize on users and surroundings not occurring during training with a cross-validation accuracy of 91.7%.

Dynamic gestures are commonly recognized through Brute force comparison [18–22], Euclidian distance calculation [15], Mahalanobis distance calculation [23, 24], dynamic time wrapping [25], and Hierarchal temporal memory [26].

In [18], a new data glove approach for Malaysian sign language detection was presented. For the accuracy of 4 individuals who tested this device, total average accuracy for translating alphabets was 95%, numbers were 93.33% and gestures were 78.33%. The average accuracy of data glove for translating all types of gestures was 89%. The authors of [19] presented a bimanual communication aid, called AUTOSEM that instead of using the hand as a display surface for manual signs, uses combinations of different orientations of both hands to define a set of semaphores that can represent an alphabet. The contribution of work [15] was the development and implementation of an application that translates the alphabet and the numbers from 1 to 10 from sign language to text to help hearing impaired.

In the work [23], data from five-channel surface electromyogram and 3D accelerometer from the signer's dominant hand were analyzed using Intrinsic-Mode Entropy (IMEn) for the automated recognition of Greek Sign Language (GSL) isolated signs. Experimental results from the IMEn analysis applied to GSL signs corresponding to 60-word lexicon repeated ten times by three native signers have shown more than 93% mean classification accuracy using IMEn as the only source of the classification feature set. Researchers in [24] presented video based continuous Indian sign language recognition system. The performance of the proposed method with an average word matching score was around 85.58% and 90% for motion direction code and artificial neural network applications, respectively.

The authors of [25] proposed a weighted Dynamic Time Wrapping (DTW) method that weights joints by optimizing a discriminant ratio. Researchers demonstrated the recognition performance of proposed weighted DTW with respect to the conventional DTW and state-of-the-art. In the research [26], Hierarchical Temporal Memory-based (HTM) method for recognition of signed Polish words was proposed. The paper discusses the preparation issues of HTM and presents results of the recognition of 101 words used in everyday life at the doctors and in the post office.

In recent year's literature, there are a large numbers of different methods used for informative features extraction from sign speech. Among them we can distinguish several widely used techniques for feature extraction, such as: Viola-Jones robust real-time object detection [27, 28], Histogram of Oriented Gradients (HOG) [21], skin color segmentation [29–31], sensor-based features [32–34], wavelet features [13], discrete cosine transform [8], sparse auto-encoder [4], etc. In [27], researchers developed the mobile sign language translator that can directly interpret sign language submitted by deaf speech into written language. Feature extraction based on hand detection using Open source Computer Vision (OpenCV) Library implementation of Viola–Jones algorithm and translation of hand signals with K-NN classification. In [28], researchers developed system used to classify facial expressions for Arabic sign language translation and accomplished 92% recognition rate on 5 different people. The system employed already existing technical methods such as: Recursive PCA for feature extraction and multi-layer perceptron for classification.

Researchers in the work [21] developed HOG-based hand gesture recognition system on a mobile device. The proposed system was applied to ASL alphabet recognition problem. The experimental results demonstrated that the proposed recognition algorithm improves HOG's robustness under rotation change. In [29], a mobile vision-based sign language translation device for automatic translation of Indian sign language into speech was presented. The sign language gesture images are acquired using the inbuilt camera of the mobile phone. The developed system was able to recognize one handed sign representations of alphabets (A–Z) and numbers (0–9).

The work [30] introduces multimodal information fusion for human-robot interaction. This multimodal information consists of combining methods for sign language recognition and emotion recognition. The authors developed algorithms to determine the hands sign via a process called combinatorial approach recognizer equation. In the work [31] a methodology for feature extraction in Brazilian Sign Language (BSL) was presented. Methodology explores the phonological structure of the language and relies on RGB-D sensor, from which seven vision-based features are extracted. Each feature is related to one, two or three structural elements in BSL. Accuracy results above 80% on average using SVM-based classification.

Paper [32] presents a real-time portable sign language translation system. To discriminate between different hand gestures, the authors have sensors embedded into the glove so that the three most important parameters, such as the posture of fingers, orientation of the palm, and motion of the hand, can be recognized. Achieved accuracy for gesture recognition is up to 94%. The authors of [33] proved that activity recognition plays a key role in providing information for context-aware applications. The goal of this paper was to determine if hand posture can be used as a cue to determine the types of interactions a user has with objects in a desk/office environment. Experiments demonstrated that hand posture can be used to determine object interaction, with accuracy rates around 97%.

Researchers in [34] proposed a tangible interface using a customized glove equipped with flex sensors. The experiments conducted with the design prototype have shown significant interaction improvements among the society and deaf-mute people. In the paper [13], two new feature extraction techniques of combined orientation histogram and statistical features and wavelet features for recognition of static signs of numbers 0–9, of ASL was proposed. Wavelet features based system gave the best performance with maximum average recognition rate of 98.17%.

Along with the above methods, standard sign language databases are necessary for the reliable testing and comparison of hand gesture recognition algorithms. The availability of hand gesture and sign language databases was limited until the year 2007 and has been increased recently [35]. Existing databases can also be divided into static, dynamic, and both static and dynamic simultaneously. The most well known static gestures databases are the following: ASL finger spelling dataset with 24 classes, 9 subjects, and 65 K samples [36], dataset with 32 classes, 18 subjects, and gestures from Polish sign language and ASL [37], VPU hand gesture dataset with 12 classes and 11 subjects [38], NUS hand posture dataset I and II with 10 classes, 40 subjects, and 2,750 samples on complex background [39, 40], posture dataset with 10 classes, 10 subjects and 1,000 samples [41].

Dynamic gestures databases are: ChaLearn gesture challenge database with 62000 samples of dynamic hand gestures [42, 43], MSRC-12 Kinect gesture dataset with 12 classes, 30 signers, and 6,244 samples [44], ChaLearn multi-modal gesture data with 20 classes, 27 signers, and 13,858 samples [45], 6D motion gesture database with 20 classes, 28 signers, and 5,600 samples [46], Sheffield gesture dataset with 10 classes, 6 signers, and 2,160 samples recorded with Kinect cameras [47], Thesaurus of RSL (TheRuSLan) with 164 classes, 13 signers, and 10,660 samples of Kinect RGB-D video data [48].

Static and dynamic databases are: NATOPS aircraft handling signals database with 24 classes, 20 signers and 9,600 samples [49], gesture dataset with 10 classes, 15 signers; the database is meant for testing hand gesture and posture recognition algorithms that contains both movement patterns and specific hand shapes [50], Cambridge hand gesture data set with 9 classes and 2 signers recorded under different illumination conditions (the target task for this dataset is to classify hand shapes and motions at the same time) [51].

A comparative review of publicly available sign language and hand gesture datasets can be found in the work [35].

7.3 Methodology

In this section, three crucial issues are discussed. Detection of region of interest is presented in Sect. 7.3.1. Section 7.3.2 includes a description of feature extraction. Model training is described in Sect. 7.3.3.

7.3.1 Region of Interest Detection

Given a video of a sign speech to extract useful visual information, the first step is to locate Regions Of Interest (ROI) that contain the motion relevant to signs: hands and face regions. In general, we need to localize certain landmarks to correctly box wrists and mouth regions.

The release of color-depth (RGB-D) camera Kinect by Microsoft [52] created a revolution in gesture recognition by providing high quality depth images, addressing issues like complex backgrounds and illumination variation [35]. Kinect device provide features such as the coordinates of a skeletal model, which are utilized for gesture recognition (Fig. 7.1) [53].

The resulting skeletal model allows us to detect the approximate location of the areas of the hands and face. For a more accurate detection of the boundaries of these regions, we have implemented several different state-of-the-art ROI detectors: Haar cascade face detector, HOG for face detector, and deep learning based face detector.

Haar cascade detector was the state-of-the-art for many years since 2001, when it was introduced by Viola and Jones [54]. There have been many improvements in the recent years and OpenCV computer vision library [55, 56] contains many Haar based models implementations. This is a machine learning based approach where a cascade function is trained from a lot of positive and negative images. It is then used to detect objects in other images. Initially, the algorithm needs a lot of positive images (e.g., images of faces) and negative images (e.g. images without faces) to train the classifier. Then algorithm needs to extract features from it. For this, Haar features are used (Fig. 7.2).

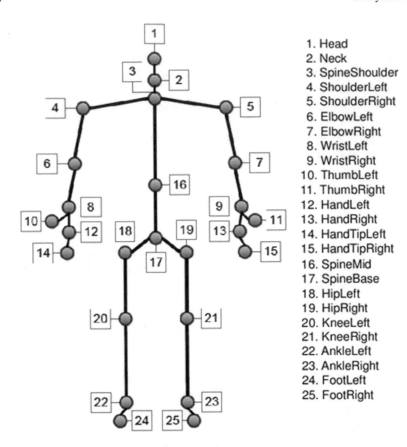

Fig. 7.1 Kinect 25 reference dots skeletal model

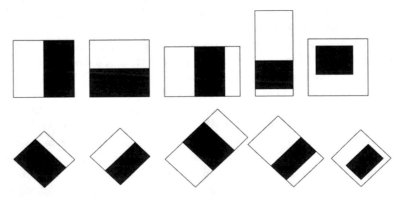

Fig. 7.2 Examples of Haar features used for face detection

Fig. 7.3 Face detection using
Haar cascades

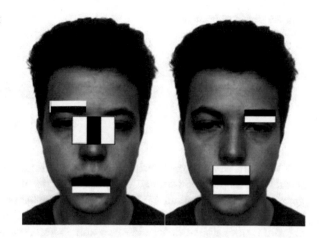

   The features are grouped into different stages of classifiers and applied one-by-one.
If a window fails the first stage, discard it. The algorithm doesn't consider the
remaining features on it. If it passes, apply the second stage of features and continue
the process. The window which passes all stages is a face region (Fig. 7.3).

   The main advantages of this method are the simple architecture, possibility of
detecting faces at different scales, and real-time working speed. However, the major
drawback of this method is that it gives a lot of false predictions, doesn't work on
non-frontal images, and doesn't work under occlusions.

   HOGs become widespread in 2005 [57]. The algorithm is also a features
extractor for the purpose of object detection. Instead of considering the pixel
intensities like Viola–Jones method, the technique counts the occurrences of gradient vectors represent the light direction to localize image segments. The method
uses overlapping local contrast normalization to improve accuracy. Dlib computer
vision library [58] provides pretrained models based on HOG features and SVM.
The model is built out of 5 HOG filters: front looking, left looking, right looking,
front looking but rotated left, and a front looking but rotated right. An example of
faces detected using this algorithm is shown in the Fig. 7.4.

Fig. 7.4 Examples of face
detection using the HOG
algorithm

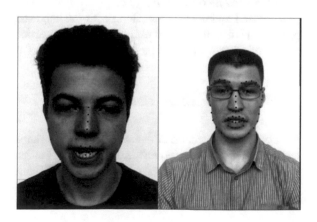

Fig. 7.5 Residual learning: a
building block

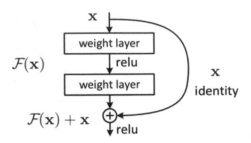

The main advantages of using this method are: algorithm works very well for frontal and slightly non-frontal faces, light-weight model is compared to the other approaches, and algorithm works under small occlusion. The disadvantages of the method are: the bounding box often excludes part of forehead and part of chin and algorithm does not work for side face and extreme non-frontal faces like looking down or up.

Together with HOG + SVM based face detector, CNN-based face detector also available in Dlib. This method uses Maximum-Margin Object Detector (MMOD) with CNN based features. The CNN based detector is capable of detecting faces almost in all angles. Unfortunately it is not suitable for real time video. It is meant to be executed on Graphics Processing Unit (GPU). The model is ResNet network with 29 conv layers. The architecture of such a network was initially proposed in [59] and contains a residual learning blocks (Fig. 7.5). Using this architecture allows to train a deeper network and address the degradation problem (with the network depth increasing, accuracy gets saturated and then degrades rapidly). About 3 million faces were trained to obtain the network. Since the models use deep learning to obtain face embeddings (face encodings), the process of obtaining face encodings takes most of the time.

The method has the following advantages. It works for different face orientations, robust to occlusions, and very easy to train. However, it works slowly on Central Processing Unit (CPU).

The detection of mouth region is the next step and is done according to the pipe-line presented in the Fig. 7.6.

The hand region detection, along with the hand shape classification, is performed using CNN described in the features extraction section. Comparison of the detection results of all the described methods is given in the experiments section.

7.3.2   Feature Extraction

For both tasks, detection of the hand region and shape of the hand classification, we trained CNN with inverted residuals and linear bottlenecks MobileNetV2 (Fig. 7.7), initially proposed for object detection in [60]. MobileNetV2 is a very effective feature extractor for object detection and segmentation. This module takes as an input a low-dimensional compressed representation, which is first expanded to high

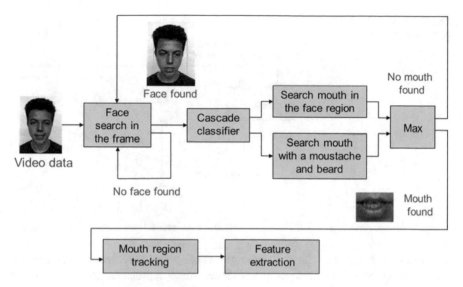

Fig. 7.6 Region-of-interest finding pipeline

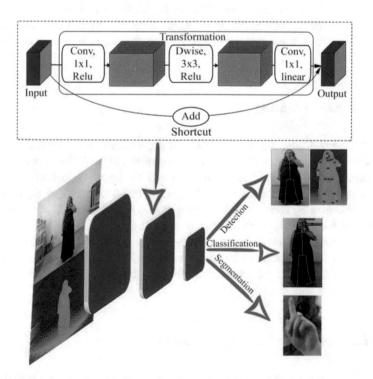

Fig. 7.7 Diagram of MobileNetV2 architecture

Fig. 7.8 Recurrent neural
network diagram

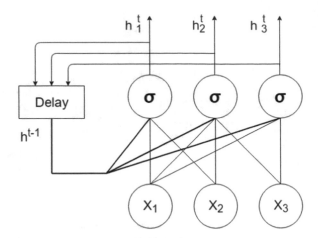

dimension and filtered with a lightweight depth-wise convolution. Features are subsequently projected back to a low-dimensional representation with a linear convolution.

The intuition is that the bottlenecks encode the model's intermediate inputs and outputs while the inner layer encapsulates the model's ability to transform from lower-level concepts such as pixels to higher level descriptors such as image categories. Finally, as with traditional residual connections, shortcuts enable better accuracy [60].

7.3.3   Model Training

In order to model the process of sign language, which is time-continuous, we use Long Short-Term Memory (LSTM) neural network [61]. LSTM network is a type of recurrent neural network. A recurrent neural network is a neural network that attempts to model time or sequence dependent behavior, such as sign language. This is performed by feeding back the output of a neural network layer at time t to the input of the same network layer at time $t + 1$ (Fig. 7.8).

LSTM network is a recurrent neural network that has LSTM cell blocks in place of the standard neural network layers. These cells have various components called the input gate, the forget gate, and the output gate [62]. The graphical representation of LSTM cell is shown in Fig. 7.9.

7.4   Russian Sign Language Database

The database used for the present research comprises 164 lexical units and clauses in RSL signed by 13 informants, in at least 5 iterations of each sign. The total size of the database is 3.8 TB, the total video duration is more than 8 h. Ten signers are

Fig. 7.9 LSTM cell diagram

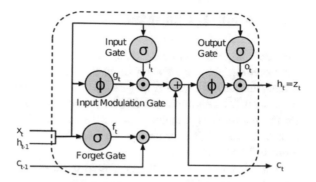

deaf students (male and female) in the age of 19–22 and the rest of the signers (female) are their hearing teachers in the age of 30–40. Since practically all the signers had come from different regions of Russia, all translations were pre-standardized by experts based on the dictionary [63]. The application of this publication seemed convenient to the researchers because the vocabulary included in that book is well-known to most of the educated speakers of RSL. All the gestures were accompanied by an overt lip articulation; however, no audio data were recorded.

The main subject area of the database vocabulary is supermarket products, classified into several categories: dairy, fish, bakery products, etc. Words and clauses related to such subject areas as "supermarket departments", "navigation in supermarket", and "requesting for products". The list does not include products that do not enjoy, according to the personal feelings of the authors of this work, a great popularity among the consumers (for example luxury or exotic products).

The list of commands includes lexemes related to orientation ("forward", "back", "right", "left" etc.), a number of common verbs of movement (let's go, let's go) and their modifiers (fast, slower), location and product requests (where?, show me and I need smth., I want smth.). The vocabulary of departments includes the names of the departments of the store; in fact, these are groups of goods united according to the type of raw materials, from which they are produced, or according to their designation. In practice, such groups of goods are often placed in separate locations. In addition, the same list includes tokens such as "cash", "exit", "toilet", etc. 85% of the entries belong to the category of "food and other goods", the remaining 15% of the entries are represented by user commands, department names, and questions and answers. Some lexemes are composites, i.e. consist of two (or more) components capable of functioning as independent lexemes, such as: BUCKWHEAT = PORRIDGE.BROWN, MANDARINE ORANGE = M.ORANGE, etc.

In total, the database consists of 48 different one-handed gestures and 116 two-handed gestures.

7.5   Technical Characteristics of the Collected Database

The recording was carried out using an automated system developed by the authors for recording 3D video stream using the Microsoft Kinect 2.0 rangefinder sensor. Translation stimuli were displayed on the computer screen and demonstrated to informants 5 times minimum for the purpose of the further training of the system for the automatic sign language recognition. The Kinect 2.0 camera was installed at a distance of 1.5–2 m from the speaker, while the speakers were recorded on a uniform light background.

A distinctive feature of the presented database is the recording of gestures in a three-dimensional (3D) format, which makes it a unique resource on RSL material. In fact, even the most extensive databases of those listed above represent a gesture in the form of a two-dimensional image. It allows us to record data in both manual and automatic modes. Lexical items to be signed by informants appear at the bottom of the screen, the minimal number of iterations was 5. The data obtained from the sensors (optical, depth sensor, and infrared camera) are stored in the database.

Distinctive feature of the presented database is that the gestures are recorded in 3D, which makes it a unique source of RSL vocabulary. Even the most extensive databases of sign languages represent a gesture as 2D signs. The use of the depth map introduces the third dimension, which makes it possible to deter-mine the position of articulators with a considerable precision. The distance between the active and passive articulators (i.e. hands), the distance of the hands from the body is a way to express various lexical meanings and connotations. 3D format was obtained due to the fact that the Kinect 2.0 camera has the ability to record not only video data in the optical Full HD format and the infrared range, but also in the "depth map" mode.

The color designations on the depth map correspond to the ranges of the spectrum of visible radiation, in other words, the groups of dots most distant from the camera are colored red, the closest ones are colored purple. Objects between these points are painted over with shades of green and yellow. The blue area corresponds to the "blind zone" of the camera: due to the peculiarities of the space and lighting (remote angle between the floor and the wall), the sensor could not determine the depth of this area. The infrared mode allows recording in dark or (too) bright places.

The obtained data have the following features:

- Binary video files containing the recording of a color camera in bin format without compression (with an optical resolution of 1920×1080 pixels at 30 frames per second, chromaticity—8 bits per pixel).
- Binary video files containing the recording of the depth map and the infrared camera in the bin format without compression (with an optical resolution of 512×424 pixels at 30 frames per second, color—16 bits per pixel).
- Text files in XML format with coordinates of the speaker's skeletal model, divided into 25 joints (elements); each specific point is the intersection of two

axes (x, y) on the coordinate plane and an additional double-precision coordinate value indicating the depth of the point, which is measured by the distance from the sensor to the object point in the range from 0 (1.2 m) to 1 (5.0 m).

The logical structure of the database can be represented as a directory tree containing information for each gesture shown by a separate announcer: (a) video recordings of the gesture shown in FullHD format in the infrared and in the depth map format; (b) data on the position of the skeleton on the video; (c) images in jpg format selected by frame from the video and necessary for markup.

7.6 Proposed Framework and Evaluation Experiments

A functional diagram of the method of one-handed RSL gesture recognition is presented in Fig. 7.10. In the case of the offline mode, the input data are color video data and a depth map in binary format (BIN), as well as text files in XML format with 3D and 2D coordinates of signer skeletal models from the collected multimedia database TheRuSLan [17]; in the online mode, the input data are color (RGB) video stream and the depth map received from the Kinect v2 sensor. The color depth for RGB is 8-bit with a video stream resolution of 1920 × 1080 (FullHD) pixels and a frequency of 30 fps, and for a 16-bit depth map with a video stream the resolution is 512 × 424 pixels, and the frame rate is analogous to that of a color video stream. If the Kinect sensor is not accessible or the required files are missing in TheRuSLan multimedia database [17], the method is automatically interrupted; otherwise cyclic processing of frames is performed with a check performed at each iteration to obtain a specific frame. At this stage, the abruption is possible in the following cases:

- An error occurred while receiving both RGB and Depth video frames.
- One of the aforementioned video streams is inaccessible.

In the online mode, a search for signers (up to 6 people) is performed on each 3D frame of the depth map with the use of a set of development tools (SDK, supplied with the Kinect v2 sensor) at the distance 1.2–3.5 m, as well as calculating 3D 25-point models of all human skeletons in the target area. As for the offline 3D mode, 25-point models of skeletons are built with the use of coordinates that are acquired from the depth map text files loaded in the previous step. After that, the closest skeleton model is found on the Z axis and tracked. Thereafter, if the method performed in the online mode, 3D coordinates are converted into 2D with the use of the SDK Kinect v2 or 2D coordinates are read from text XML files of the color video stream, as shown in Fig. 7.11.

A rectangular area containing the closest person on 2D frame of a color video stream is then formed up, as well as, his/her 2D 25-point skeleton model (offline mode), as can be seen in Fig. 7.12.

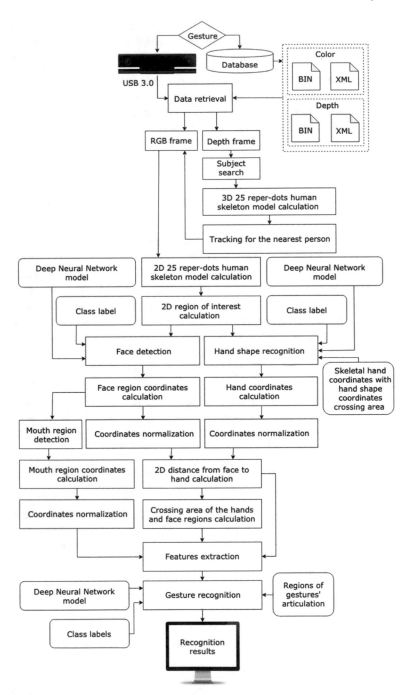

Fig. 7.10 Functional diagram of the method for one-handed RSL gesture recognition

Fig. 7.11 Examples of 25 reference dots skeleton models

Fig. 7.12 Examples of rectangular areas with signers

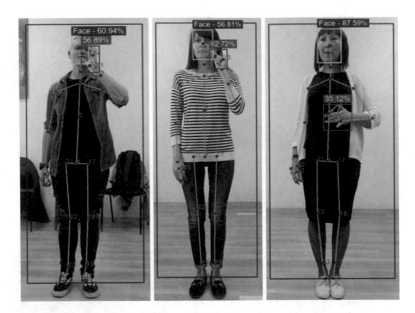

Fig. 7.13 Examples of face and hand graphical areas determination

At the next step, within the formed rectangular area containing a person, the graphical area of the face and form of the hand are determined with the help of two trained detectors (Fig. 7.13).

The following face detectors were tested with the help of TheRuSLan database in order to find the best result of the signer face detection [17]:

- Viola-Jones facial recognition method.
- Deep learning based face detector.
- HOG and SVM face detector.
- CNN-based face detection.

Viola–Jones algorithm, which was proposed by Jones and Viola [64, 65], makes it possible to find various objects within an image in real time mode. The main purpose of this algorithm is facial detection, but it can also be used in recognition of objects of other types. This algorithm is implemented in the library of computer vision OpenCV [55, 56] in compliance with the following basic principles:

- The image has an integral representation that makes it possible to find the required objects quickly.
- Search for desired objects with the use of Haar-like features [54].
- Use of boosting: selection of features that are the most appropriate for the desired object within the selected part of the image [66].
- The features are what are inputted. The classifier outputs "True" or "False".
- Cascades of features are used to quickly get rid of windows, in which no object was found.

Training this classifier takes a lot of time, but the search for objects (persons) is fast.

The deep-learning detector is based on Single Shot multibox Detector (SSD) [67] structure with a reduced ResNet-10 network architecture [59]. This detector is included in DNN module of OpenCV library [55, 56]. This detector is distinguished with the following features:

- The model was trained with the use images available on the Internet, but the source is not disclosed.
- There are 2 trained platforms: the floating-point Caffe platform [68, 69] and the TensorFlow platform [70, 71].

Face detector based on HOG [72] and SVM [73] functions is implemented in the Dlib computer vision and machine learning library [74]. This detector uses a classifier based on SVM and HOG methods to detect the area of interest. The detector model consists of 5 HOG filters: frontal view, left view, right view, frontal but turned to the left, and frontal but turned to the right.

Face detector based on CNN, analogously to the previous detector, is implemented in the Dlib library [74] and has the following features:

- The object detector MMOD [75] is used with functions based on CNN.
- ImageNet data sets [76, 77], PASCAL VOC [78], VGG [79], WIDER [80], Face Scrub [81] were used to train this model.

Table 7.1 presents a comparative analysis of facial detectors in question. Such metrics as Precision (Eq. 7.1), Recall (Eq. 7.2), and F-measure (Eq. 7.3), (which is the harmonic mean value between Precision and Recall), were used to indicate the quality.

$$Precision = \frac{TP}{TP + FP}, \tag{7.1}$$

$$Recall = \frac{TP}{TP + FN}, \tag{7.2}$$

$$F = 2\frac{Precision \times Recall}{Precision + Recall}, \tag{7.3}$$

Table 7.1 Comparative analysis of face detectors

Faial detectors	FPS	Precision	Recall	F-measure
Viola–Jones	≈26.7	0.8017	0.7229	0.7603
DNN face in OpenCV (Caffe)	≈8.6	0.8710	0.8732	0.8721
DNN face in OpenCV (TensorFlow)	≈8.7	0.8710	0.8732	0.8721
HOG face in Dlib	≈34.3	0.8913	0.6679	0.7636
CNN in Dlib	≈63.2	0.8976	0.8196	0.8568

Table 7.2 Experimental PC parameters

Central processor	RAM, Gb	Disc type	Video card
Intel Core i5-4210H 2.9 kHz	16	SSD	Nvidia GeForce GTX 860M

where TP is the true-positive rate, TN is the true-negative rate, FP is the false-positive rate, FN is the false-negative rate.

Thus, the last metric allows us to find a balance between the first two metrics.

In order to identify the optimal face detector, experiments were performed with PC. The parameters of the latter are presented in Table 7.2.

Thus, it was found out that CNN-based method implemented in Dlib library [74] is optimal for the task of defining the graphical area of the face. Compared to the other detectors, it works for different orientations of the face, is resistant to occlusions, and also can perform in real time both on the CPU and GPU.

As for the definition of the area containing human hand(s), CNN with the MobileNetV2 architecture [60] and default parameters was used. This architecture is included in the open source platform TensorFlow Object Detection API [82] and is based on an inverted residual structure. The neural network was trained on one-handed gestures annotated in respect for hand shapes, (44 classes) are presented in Fig. 7.14.

The entire annotation process was carried out using LabelImg tool [83]. Annotated areas are saved in a special PASCAL VOC format [78] in the form of XML text files. This format is widely used, for example, in ImageNet [76, 77].

For the training of the model, 4 iterations of 48 one-handed gestures of 12 signers were used. The iteration data were seen as model examples, and the rest was used as test data. Thus, a one-handed gesture dataset was split down into the training sample and the test sample with an approximate ratio of 75:25%.

Hand shape recognition is performed when:

- The trained neural network model identifies the handshape.
- The central coordinate of the arm obtained via Kinect v2 sensor is located within the recognized area containing the hand.

The following steps are aimed at calculating the coordinates of the face and arm areas with subsequent normalization. Then, 2D distance between the uppermost left coordinate of the face area and the same hand coordinate area is calculated. The plain area of the face and the arm intersection is calculated too.

The process of detecting the mouth area is carried out on the same step. The basic idea of the algorithm is to compare the statistical model of the shape and the appearance of the object with the use of a set of facial points of reference. Thus, it allows to find the area of the mouth in real time. This algorithm is implemented in Dlib library [74]. As in the case of the face and hands area coordinates, the mouth area undergoes normalization process.

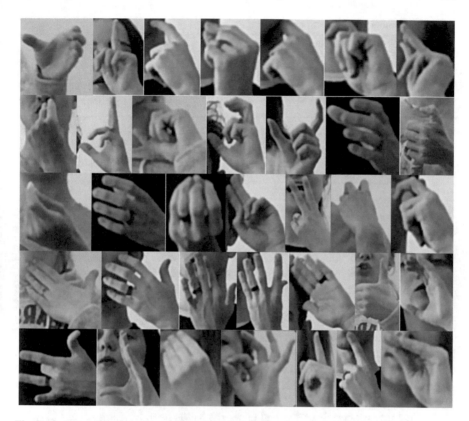

Fig. 7.14 Examples of the forms of the hands of one-hand gestures of RSL

Thus, the distinctive characteristics of a gesture at a certain point in time are:

- The normalized 2D distance from the face to the hand (articulation area).
- Normalized 2D area of intersection of the face and hand.
- Hand shape (represented by a numerical value).
- The result of the mouth area detection (represented by a numerical value).

The final stage is aimed at recognizing one-handed gestures of RSL using LSTM neural network. In general, LSTM network is a type of recurrent neural network. In turn, the recurrent neural network is a neural network that tries to model some behavior depending on time or sequence, e.g. gesture recognition. This is done by feedback of the output level of the neural network at time t with the input of the same network level at time $t + 1$. However, the recurrent neural network has the disadvantage of a vanishing gradient. This problem occurs, when the network tries to simulate a dependency within a long sequence of a training sample. This is due to the fact that small gradients or weights (values less than 1) multiply many times over several time steps, and that's why the gradients shrink to zero. This means that the weights of the earlier steps will not be significantly changed, and, therefore, the

Table 7.3 Results of the experiments

Experiment No.	Batch size	Number of epochs	Initial learning rate	Accuracy, %
1	16	1000	0.1	58.07
2	16	1200	0.1	53.62
3	32	500	0.1	67.59
4	32	700	0.1	60.98
5	**64**	**360**	**0.1**	**69.73**
6	64	600	0.1	61.19
7	128	300	0.1	67.43
8	128	400	0.1	62.95
9	256	200	0.1	64.13
10	256	300	0.1	61.53

network is not going not study long-term dependencies. LSTM network allows a solution of this problem, that's why we have preferred it to other types of networks.

The cores of this neural network were the functional cores of gestures, which consist of context-independent hand movements. In a broader understanding of LSTM, a neural network takes a sequence of N frames $\times$ 4 values from the characteristics of a gesture, in particular: the normalized 2D distance from face to hand is a floating point number, the normalized 2D area of the intersection between the face and the hand is also a floating point number, the shape of the hand is an integer in the range from 1 to 44, and the result of detecting the area of the mouth is represented by the numbers 0 (the area is not found) and 1 (the area is found).

The learning process is performed using the Keras deep learning library [84] and the open source library TensorFlow [70, 71]. The best results were achieved, when LSTM was trained with the package size of 64 at 360 epochs and an initial learning rate of 0.1. Table 7.3 presents some of the results of the experiments.

The worst recognition accuracy was shown by gestures consisting of similar hand shapes with the articulation area located the face area. In this case, it is possible to increase the accuracy by additional feature extraction, as well as, adding a new modality (a depth map).

7.7 Conclusions

The recognition of elements of sign languages is of great practical interest in the modern information world. Moreover, the problem of the effective automatic sign language recognition has not been resolved until now due to the number of reasons, such as cultural differences in the sign languages of the world, lack of the representative databases for training, different recording conditions, relatively small size of the region-of-interest, i.e. fingers area, etc. In this study, the great potential of using various architectures of DNNs for the task of automatic hand gestures recognition was demonstrated. Furthermore, the recognition results on the single-hand part of the collected TheRuSLan database was presented. It is already possible to assert about significant progress in the task of automatic recognition of RSL. Nevertheless, there is no doubt that with further development of the proposed framework and with collection of additional data, it will be possible to achieve even higher improvements in recognition results.

Future research will focus on extracting additional features expanding the data set and increasing the amount of recognizable gestures with the addition of two-handed gestures. The developing automatic RSL recognition system is meant as an auxiliary system for people with hearing impairments.

Acknowledgements This research was financially supported by the Ministry of Science and Higher Education of Russia, agreement No. 075-15-2019-1295 (reference RFMEFI61618X0095).

References

1. Seymour, M., Tsoeu, M.: A mobile application for South African sign language (SASL) recognition. In: Proceedings of the AFRICON, pp. 1–5 (2015)
2. Pan, T.Y., Lo, L.Y., Yeh, C.W., Li, J.W., Liu, H.T., Hu, M.C.: Real-time sign language recognition in complex background scene based on a hierarchical clustering classification method. In: IEEE 2nd International Conference Multimedia Big Data, pp. 64–67 (2016)
3. Jin, C.M., Omar, Z., Jaward, M.H.: A mobile application of American sign language translation via image processing algorithms. In: IEEE Region 10 Symposium, pp. 104–109 (2016)
4. Li, S.Z., Yu, B., Wu, W., Su, S.Z., Ji, R.R.: Feature learning based on SAE–PCA network for human gesture recognition in RGBD images. Neurocomputing **151**, 565–573 (2015)
5. Just, A., Marcel, S.: A comparative study of two state-of-the-art sequence processing techniques for hand gesture recognition. Comput. Vis. Image Underst. **113**(4), 532–543 (2009)
6. Chen, F.S., Fu, C.M., Huang, C.L.: Hand gesture recognition using a real-time tracking method and hidden Markov models. Image Vis. Comput. **21**, 745–758 (2003)
7. Marcel, S., Bernier, O., Viallet, J.E., Collobert, D.: Hand gesture recognition using input/output hidden Markov models. In: IEEE Automatic Face and Gesture Recognition, pp. 456–461 (2000)
8. Al-Rousan, M., Assaleh, K., Talaa, A.: Video-based signer-independent Arabic sign language recognition using hidden Markov models. Appl. Soft Comput. **9**(3), 990–999 (2009)
9. Nagi, J., Ducatelle, F., Di Caro, G.A., Cire, D., Meier, U., Giusti, A.: Max-pooling convolutional neural networks for vision-based hand gesture recognition. In: IEEE International Conference on Signal and Image Processing Applications, pp. 342–347 (2011)
10. Huang, J., Zhou, W., Li, H., Li, W.: Sign language recognition using 3D convolutional neural networks. In: IEEE International Conference on Multimedia and Expo, pp. 1–6 (2015)
11. Pigou, L., Dieleman, S., Kindermans, P.J., Schrauwen, B.: Sign language recognition using convolutional neural networks. In: Workshop at the European Conference on Computer Vision, pp. 572–578 (2015)
12. Liu, T., Zhou, W., Li, H.: Sign language recognition with long short-term memory. In: International Conference on Image Processing, pp. 2871–2875 (2016)
13. Thalange, A., Dixit, S.: Cohst and wavelet features based static ASL numbers recognition. Proc. Comput. Sci. **92**, 455–460 (2016)

14. Hartanto, R., Kartikasari, A.: Android based real-time static Indonesian sign language recognition system prototype. In: 8th International Conference on Technology and Electrical Engineering, pp. 1–6 (2016)
15. Vintimilla, M.G., Alulema, D., Morocho, D., Proano, M., Encalada, F., Granizo, E.: Development and implementation of an application that translates the alphabet and the numbers from 1 to 10 from sign language to text to help hearing impaired by android mobile devices. In: IEEE International Conference on Automatica, pp. 1–5 (2016)
16. Neiva, D.H., Zanchettin, C.: Gesture recognition: a review focusing on sign language in a mobile context. Expert Syst. Appl. **103**, 159–183 (2018)
17. Ryumin, D., Kagirov, I., Ivanko, D., Axyonov, A., Karpov, A.A.: Automatic detection and recognition of 3D manual gestures for human-machine interaction. Int. Arch. Photogramm. Remote Sens. Spatial Inf. Sci. **XLII-2/W12**, 179–183 (2019)
18. Shukor, A.Z., Mikon, M.F., Jamaluddin, M.H., bin Ali, F., Asyraf, M.F., Bahar, M.B.: A new data glove approach for Malaysian sign language detection. Proc. Comput. Sci. **76**, 60–67 (2015)
19. Khambadkar, V., Folmer, E.: A tactile-proprioceptive communication aid for users who are deafblind. In: IEEE Haptics Symposium, pp. 239–245 (2014)
20. Bajpai, D., Porov, U., Srivastav, G., Sachan, N.: Two way wireless data communication and american sign language translator glove for images text and speech display on mobile phone. In: 5th International Conference on Communication Systems and Network Technologies, pp. 578–585 (2015)
21. Prasuhn, L., Oyamada, Y., Mochizuki, Y., Ishikawa, H.: A HOG-based hand gesture recognition system on a mobile device. IEEE International Conference on Image Processing, pp. 3973–3977 (2014)
22. Sharma, R.P., Verma, G.K.: Human computer interaction using hand gesture. Proc. Comput. Sci. **54**, 721–727 (2015)
23. Kosmidou, V.E., Hadjileontiadis, L.J.: Sign language recognition using intrinsic-mode sample entropy on sEMG and accelerometer data. IEEE Trans. Biomed. Eng. **56**(12), 2879–2890 (2009)
24. Rao, G.A., Kishore, P.: Selfie video based continuous Indian sign language recognition system. Ain Shams Eng. J. **9**(4), 1929–1939 (2018)
25. Celebi, S., Aydin, A.S., Temiz, T.T., Arici, T.: Gesture recognition using skeleton data with weighted dynamic time warping. In: International Conference on Computer Vision Theory and Application, pp. 620–625 (2013)
26. Kapuscinski, T.: Using hierarchical temporal memory for vision-based hand shape recognition under large variations in hand's rotation. In: Rutkowski, L., Scherer, R., Tadeusiewicz, R., Zadeh, L.A., Zurada, J.M. (eds.) Artifical Intelligence and Soft Computing. LNCS, vol. 6114, pp. 272–279. Springer, Berlin, Heidelberg (2010)
27. Hakkun, R.Y., Baharuddin, A.: Sign language learning based on android for deaf and speech impaired people. In: International Electronics Symposium, pp. 114–117 (2015)
28. Elons, A., Ahmed, M., Shedid, H.: Facial expressions recognition for Arabic sign language translation. In: 9th International Conference on Computer Engineering & Systems, pp. 330–335 (2014)
29. Madhuri, Y., Anitha, G., Anburajan, M. Vision-based sign language translation device. In: International Conference Information Communication and Embedded Systems, pp. 565–568 (2013)
30. Luo, R.C., Wu, Y., Lin, P.: Multimodal information fusion for human-robot interaction. In: 10th Jubilee International Symposium on Applied Computational Intelligence and Informatics, pp. 535–540 (2015)
31. Almeida, S.G.M., Guimaraes, F.G., Ramírez, J.A.: Feature extraction in Brazilian sign language recognition based on phonological structure and using RGB-D sensors. Expert Syst. Appl. **41**(16), 7259–7271 (2014)
32. Kau, L.J., Su, W.L., Yu, P.J., Wei, S.J.: A real-time portable sign language translation system. In: 58th International Midwest Symposium on Circuits and Systems, pp. 1–4 (2015)

33. Paulson, B., Cummings, D., Hammond, T.: Object interaction detection using hand posture cues in an office setting. Int. J. Hum. Comput. Stud. **69**(1), 19–29 (2011)
34. Devi, S., Deb, S.: Low cost tangible glove for translating sign gestures to speech and text in Hindi language. In: 3rd International Conference on Computational Intelligence & Communication Technology, pp. 1–5 (2017)
35. Pisharady, P.K., Saerbeck, M.: Recent methods and databases in vision-based hand gesture recognition: a review. Comput. Vis. Image Underst. **141**, 152–165 (2015)
36. Pugeault, N., Bowden, R.: Spelling it out: real-time ASL finger-spelling recognition. In: 1st IEEE Workshop on Consumer Depth Cameras for Computer Vision, pp. 1114–1119 (2011)
37. Kawulok, M., Kawulok, J., Nalepa, J.: Spatial-based skin detection using discriminative skin-presence features. Pattern Recognit. **41**, 3–13 (2014)
38. Kollorz, E., Penne, J., Hornegger, J., Barke, A.: Gesture recognition with a time-of-flight camera. Int. J. Intell. Syst. Technol. Appl. **5**, 334–343 (2007)
39. Pisharady, P.K., Vadakkepat, P., Loh, A.P.: Attention based detection and recognition of hand postures against complex backgrounds. Int. J. Comput. Vis. **101**(3), 403–419 (2013)
40. Chuang, Y., Chen, L., Chen, G.: Saliency-guided improvement for hand posture detection and recognition. Neurocomputing **133**, 404–415 (2014)
41. Zhou, R., Junsong, Y., Zhengyou, Z.: Robust hand gesture recognition based on finger-earth movers distance with a commodity depth camera. In: ACM International Conference on Multimedia, pp. 1093–1096 (2011)
42. Malgireddy, M.R., Inwogu, I., Govindaraju, V.: A temporal Bayesian model for classifying, detecting and localizing activities in video sequences. In: IEEE Computer Vision and Pattern Recognition, Workshops, pp. 43–48 (2012)
43. Keskin, C., Kirac, F., Kara, Y., Akarun, L.: Randomized decision forests for static and dynamic hand shape classification. In: IEEE Computer Vision and Pattern Recognition, Workshops, pp. 31–46 (2012)
44. Simon, F., Helena, M.M., Pushmeet, K., Sebastian, N.: Instructing people for training gestural interactive systems. In: International Conference on Human Factors in Computing Systems, pp. 1737–1746 (2012)
45. Escalera, S., Gonzalez, J., Baro, X., Reyes, M., Lopes, O., Guyon, I., Athistos, V., Escalante, H.J.: Multi-modal gesture recognition change 2013: dataset and results. In: 15th ACM International Conference on Multimodal Interaction, pp. 445–452 (2013)
46. Chen, M., Al Regib, G., Juang, B, H.: 6DMG: a new 6D motion gesture database. In: IEEE Computer Vision and Pattern Recognition Workshops, pp. 83–88 (2011)
47. Liu, L., Shao, L.: Learning discriminative representations from RGB-D video data. Int. J. Artif. Intell. 1493–1500 (2013)
48. Ryumin, D., Ivanko, D., Axyonov, A., Kagirov, I., Karpov, A., Zelezny, M.: Human-robot interaction with smart shopping trolley using sign language: data collection. In: IEEE International Conference Pervasive Computing and Communications Workshops, pp. 1–6 (2019)
49. Yale, S., David, D., Randall, D.: Tracking body and hands for gesture recognition: NATOPS aircraft handling signals database. In: IEEE International Conference Automatic Face and Gesture Recognition, pp. 500–506 (2011)
50. Shen, X.H., Hua, G., Williams, L., Wu, Y.: Dynamic hand gesture recognition: an exemplar-based approach from motion divergence fields. Image Vis. Comput. **30**(3), 227–235 (2012)
51. Kim, T.K., Wong, S.F., Cipolla, R.: Tensor canonical correlation analysis for action classification. In: IEEE Computer Society Conference on Computer Vision and Pattern Recognition, pp. 1–8 (2007)
52. Zhang, Z.: Microsoft kinect sensor and its effect. IEEE Multimed. **19**(2), 4–10 (2012)
53. Rahman, M.W., Gavrilova, M.L.: Kinect gait skeletal joint feature-based person identification. In: IEEE 16th International Conference on Cognitive Informatics & Cognitive Computing, pp. 423–430 (2017)

54. Viola, P., Jones, M.: Rapid object detection using a boosted cascade of simple features. In: IEEE Conference on Computer Vision and Pattern Recognition, vol. 1, pp. 511–518 (2001)
55. Pulli, K., Baksheev, A., Kornyakov, K., Eruhimov, V.: Real-time computer vision with OpenCV. Commun. ACM **55**(6), 61–69 (2012)
56. OpenCV: OpenCV library. https://opencv.org. Accessed 7 Aug 2019
57. Dalal, N., Triggs, B.: Histograms of oriented gradients for human detection. In: International Conference on Computer Vision, pp. 886–893 (2005)
58. King, D.E.: Dlib-ml: a machine learning toolkit. J. Mach. Learn. Res. **10**, 1755–1758 (2009)
59. He, K., Zhang, X., Ren, S., Sun, J.: Deep residual learning for image recognition. In: IEEE Conference on Computer Vision and Pattern Recognition, pp. 770–778 (2016)
60. Sandler, M., Howard, A., Zhu, M., Zhmoginov, A., Chen, L.C.: MobileNetV2: inverted residuals and linear bottlenecks. In: IEEE Conference on Computer Vision and Pattern Recognition, pp. 4510–4520 (2018)
61. Hochreiter, S., Schmidhuber, L.: Long short-term memory. Neural Comput. **9**, 1–32 (1997)
62. Donahue, J., Hendricks, A.L., Guadarrama, S., Rohrbach. M., Venugopalan, S., Darrell, T., Saenko, K.: Long-term recurrent convolutional networks for visual recognition and description. In: IEEE Conference on Computer Vision and Pattern Recognition, pp. 2625–2634 (2015)
63. Geilman, I.: Russian sign language dictionary. In: vol. 2, St. Petersburg, Prana (2004)
64. Viola, P., Jones, M.: Robust real-time face detection. Int. J. Comput. Vis. **57**(2), 137–154 (2004)
65. Castrillyn, M., Deniz, O., Hernandez, D., Lorenzo, J.: A comparison of face and facial feature detectors based on the Viola-Jones general object detection framework. Mach. Vis. Appl. **22**(3), 481–494 (2011)
66. Freund, Y., Schapire, R.E.: Experiments with a new boosting algorithm. In: Thirteen International Conference on Machine Learning, vol. 96, pp. 148–156 (1996)
67. Liu, W., Anguelov, D., Erhan, D., Szegedy, C., Reed, S., Fu, C-Y., Berg, A.: SSD: single shot multibox detector. In: Leibe, B., Matas, J., Sebe, N., Welling, M. (eds) Computer Vision—ECCV 2016: European Conference on Computer Vision. LNCS, vol. 9905, pp. 21–37. Springer, Cham (2016)
68. Jia, Y., Shelhamer, E., Donahue, J., Karayev, S., Long, J., Girshick, R., Guadarrama, S., Darrell, T.: Caffe: convolutional architecture for fast feature embedding. In: 2nd ACM International Conference on Multimedia, pp. 675–678 (2014)
69. Caffe: a fast open framework for deep learning (2019). http://caffe.berkeleyvision.org. Accessed 7 Aug 2019
70. Abadi, M., Barham, P., Chen, J., Chen, Z., Davis, A., Dean, J., Devin, M., Ghemawat, S., Irving, G., Isard, M., Kudlur, M.: Tensorflow: a system for large-scale machine learning. In: 12th Symposium on Operating Systems Design and Implementation, pp. 265–283 (2016)
71. TensorFlow: an end-to-end open source machine learning platform. https://www.tensorflow.org. Accessed 7 Aug 2019
72. Déniz, O., Bueno, G., Salido, J., De la Torre, F.: Face recognition using histograms of oriented gradients. Pattern Recognit. Lett. **32**(12), 1598–1603 (2011)
73. Chang, C.C., Lin, C.J.: LIBSVM: a library for support vector machines. ACM Trans. Intell. Syst. Technol. **2**(3), 27 (2011)
74. Dlib: toolkit containing machine learning algorithms and tools for creating complex software. https://www.dlib.net. Accessed 7 Aug 2019
75. King, D.E.: Max-margin object detection (2015). arXiv:1502.00046
76. Deng, J., Dong, W., Socher, R., Li, L.J., Li, K., Fei-Fei, L.: Imagenet: a large-scale hierarchical image database. In: 2009 IEEE Conference on Computer Vision and Pattern Recognition, pp. 248–255 (2009)
77. Krizhevsky, A., Sutskever, I., Hinton, G.E.: Imagenet classification with deep convolutional neural networks. Advances in Neural Information Processing Systems, pp. 1097–1105 (2012)
78. Everingham, M., Van Gool, L., Williams, C.K., Winn, J., Zisserman, A.: The pascal visual object classes (VOC) challenge. Int. J. Comput. Vis. **88**(2), 303–338 (2010)

79. Parkhi, O.M., Vedaldi, A., Zisserman, A.: Deep face recognition. In: British Machine Vision Conference, pp. 41.1–41.12 (2015)
80. Yang, S., Luo, P., Loy, C.C., Tang, X.: Wider face: a face detection benchmark. IEEE Conference on Computer Vision and Pattern Recognition, pp. 5525–5533 (2016)
81. Ng, H.W., Winkler, S.: A data-driven approach to cleaning large face datasets. In: IEEE International Conference on Image Processing, pp. 343–347 (2014)
82. Huang, J., Rathod, V., Sun, Ch., Zhu, M., Korattikara, A., Fathi, A., Fischer, I., Wojna, Z., Song, Y., Guadarrama, S., Murphy, K.: Speed/accuracy trade-offs for modern convolutional object detectors. In: 30th IEEE Conference on Computer Vision and Pattern Recognition, pp. 3296–3297 (2017)
83. LabelImg: A graphical image annotation tool. https://github.com/tzutalin/labelImg. Accessed 7 Aug 2019
84. Chollet, F.: Keras. https://keras.io. Accessed 7 Aug 2019

Chapter 8
Methods of Endoscopic Images Enhancement and Analysis in CDSS

Nataliia Obukhova, Alexandr Motyko and Alexandr Pozdeev

Abstract In this chapter, the methods of medical image enhancement and analysis in Clinical Decision Support Systems (CDSS) are discussed. Three general groups of tasks in CDSS development are described: medical images quality improvement, synthesis of images with increased diagnostic value, and medical images automatic analysis for differential diagnostics. For the first group, the review and analysis of noise reduction methods are presented. The new state-of-the art algorithm for virtual chromoendoscopy is proposed as an illustration of the second group. Automatic images analysis concerning to the third group is shown on example of two algorithms: for the polyps' segmentation and bleeding detection. The algorithm for bleeding detection is based on two-stage strategy that gives the sensitivity and specificity scores of 0.85/0.97 for test set. The segmentation of polyps is based on deep learning technology and shows promising results.

Keywords Medical images processing · Clinical decision support system · Image enhancement · Virtual chromoendoscopy · Automatic images classification

8.1 Introduction

The ability to take pictures during endoscopy created a new and rapidly developing field of research: endoscopic image processing. Endoscopic images is a very wide class of medical images obtained by examining various organs and body cavities via the special optoelectronic device—endoscope. The types of endoscopy are:

N. Obukhova (✉) · A. Motyko · A. Pozdeev
Saint Petersburg State Electro Technical University "LETI", Professora Popova str. 5, Saint-Petersburg 197022, Russian Federation
e-mail: natalia172419@yandex.ru

A. Motyko
e-mail: motyko.alexandr@yandex.ru

A. Pozdeev
e-mail: puches4@gmail.com

© Springer Nature Switzerland AG 2020
M. N. Favorskaya and L. C. Jain (eds.), *Computer Vision in Advanced Control Systems-5*, Intelligent Systems Reference Library 175,
https://doi.org/10.1007/978-3-030-33795-7_8

gastroscopy for examination of stomach, colposcopy for examination of cervix, laparoscopy for examination of abdominal cavity, etc. There are more than 30 types of endoscopes. During endoscopy, a physician performs visual examination of organ surface and makes decision about diagnosis, possible biopsy; also he/her can perform surgical interventions (endosurgery).

The effectiveness of endoscopy is strongly connected with physician qualification and experience, because video data obtained during endoscopy have high level of variability connected with personal features of each patient, and also images may have strong degradation connected with difficult conditions of their obtaining. As a result, endoscopic images are very difficult for correct interpretation.

The modern trend aimed to increase the effectiveness of endoscopy investigation is CDSS development. Such systems are designed to detect and/or classify abnormalities and, thus, assist medical expert in improving the accuracy of medical diagnosis. Clinical decision support systems integrate automatic image analysis results with the results obtained by the physician, as well as use the information from the system database. This integration allows for higher sensitivity and specificity of the diagnosis compared to cases of diagnoses being made independently by a physician or the system.

The main goal of this chapter is to describe new and effective methods of endoscopic image processing providing image enhancement, comfortable visualization, and image segmentation and analysis for CDSS in gastroscopy.

CDSS development requires to solve the following groups of tasks for image processing and analysis:

- First group improves the quality of medical images providing new level of their ergonomics.
- Second group synthesizes the medical images with new properties and with increased diagnostic value for effective visual analysis by physician.
- Third group realizes automatic analysis of medical images for differential diagnostics with high values of sensitivity and specificity.

Solving of the first and second groups of tasks are aimed to realize more convenient and ergonomic conditionals for physician analysis, and as a result must give more high level of physician diagnostics effectiveness. Solving of the third group of tasks ought to provide a possibility of effective automatic analysis. The methods from all three groups are the base of CDSS. The main requirements for image processing methods of the first group are:

- High effectiveness of the methods taking in account the features of endoscopic images. The main features of endoscopic images, which make popular algorithms of digital processing low effective for application in endoscopic images pipeline are: uneven contrast (there are very dark and very bright areas in the same image) and large areas of image with very low amount of detail.
- The process of endoscopic images enhancement ought to be organized as a pipeline. So, each algorithm must be designed as a step of the endoscopic pipeline, which implies high processing speed. Computational cost of

algorithms ought to provide a possibility of real-time pipeline realization, also the result of one step must be improved by the next step.

Additionally, the designed algorithms and software have to be adapted to real-time processing of endoscopic video data with high resolution (up to FHD, 4K) and frame rate (up to 30–60 fps).

The main pipeline steps can be divided in two groups. The first group ought to provide the obtainment of quality images from a sensor. The algorithms of the second group are aimed at removing artifacts caused by difficult observation conditions. Such conditions include the lack of light, complex shape of objects under observation, organ muscular breathing, movement of the sensor, which leads to the following artifacts: high presence of noise, nonlinear brightness and contrast characteristics, blur and low sharpness. The critical procedures of this group are: noise reduction and contrast enhancement. These tasks are especially difficult in endoscopy because of the fact the endoscopic images suffer from various types of degradation. Moreover, these algorithms are the most computationally expensive. Thus, our research includes a development of real-time algorithms of the second group, such as noise removing, brightness and contrast enhancement, taking into account the main features of endoscopic images.

The following new real-time algorithms for the endoscopic pipeline are proposed:

- Algorithm of noise removal, which is based on obtained experimental results of modern noise reduction algorithms assessment. The algorithm determines the level of high frequency in the image fragment (level of detail in the fragment). The median/bilateral filtering and Non-Local Mean (NLM)/Block Matching and 3D (BM3D) filtering are used for low detailed fragments and high detailed fragments respectively. The algorithm enables one to obtain high quality images with reasonable computational cost.
- Algorithm of image enhancement. For image sharpening and contrast enhancement, it is proposed to consider the local features of the image on the basis of the found functional relationship between the correction strength and the normalized brightness variance in the image fragment. The algorithm carries out contrast enhancement without significant stressing of the noise component, which is one of main disadvantages of modern non-linear enhancement algorithms, especially in low detail images.

The second group of methods is required for CDSS realization. They are methods of visualization. The goal of these methods to increase diagnostic value of images for physician analysis. Methods of this group must provide easier assessment and specificity of the morphology of a lesion by highlighting the mucosal microstructure and microvasculature features that are needed for adequate and accurate diagnosis and treatment. The effective way to increase diagnostic value of gastroscopy images is ChromoEndoscopy (CE).

CE can be classified into the dye-based CE and electronic CE methods. Dye-based CE is an endoscopy technology that consists of spray application of dyes, which are harmless to the human body, onto the mucosal surface of interest. The application of dye improves visualization of the microstructure and vascular patterns of lesions under investigation. CE is equipped with only a spray catheter providing a relatively simple and cost-effective method of dye application. Despite these advantages, the application of CE to screening programs remains limited due to absence of standardized methods and analysis, resulting in uncertainty in lesion identification.

Electronic endoscopes obtain images in a form of electronic signals that can be analyzed using various image-processing techniques. Electronic CE, therefore, eliminate the need for time-consuming methods of CE including spraying and suctioning of dyes and the disadvantage of solution pooling in depressed-type lesions that obstruct visual inspection. There are two types of electronic endoscopes that may be used to enhance certain mucosal or vascular characteristics: in-chip and off-line, so called virtual chromoendoscopy. Two new in-chip technologies providing the enhancement with in-chip processor are known as narrow band imaging and auto-florescence imaging.

Virtual chromoendoscopy is one of the modern and effective approach to enhance features of endoscopic images. It is commonly used to improve diagnostic accuracy by converting color and structure-based diagnostic information into more objective and quantitative indicators. Among modern technologies of virtual chromoendoscopy the most famous examples are Flexible spectral Imaging Color Enhancement (FICE) [1] and i-SCAN [2, 3]. Studies showed that i-SCAN and FICE technology could improve the diagnostic accuracy.

Thus, it is important to realize function of virtual chromoendoscopy in CDSS. The stressing of different tissue features and vessels structure with color gives possibility for more easy and accurate interpretation and diagnostics by physician. We proposed new method of virtual chromoendoscopy based on digital processing of images obtained in the white light. The main feature of proposed method of virtual chromoendoscopy is local nonlinear processing each color plane.

Both of described above groups (methods of enhancement and visualization) give a possibility to make visual analysis by physician more convenient and effective. Thus, they are aimed to realize first feature of CDSS—physician diagnostic with high sensitivity and specificity.

Another important group of tasks for CDSS development is connected with automatic image analysis with the aim to realize differential diagnostics. Today the modern trend of image classification is deep learning technology application. Deep learning technologies are very strong and effective especially if input images has high level of variability. So deep learning must be effective in the task of medical images classification, because one of the principle feature of these images is variability. However, deep learning technologies demand very large (or even extremely large) datasets for classifier training. In real condition, the obtainment of large dataset of medical images verified by physician is a difficult problem. Usually, there are no large datasets with ground true made by physician. It makes application of deep learning technologies low effective. The main task of our research is to propose a method that gives a possibility to obtain the satisfactory results of Convolution Neural Networks (CNN) application in conditions of small dataset.

To solve this task, we propose two-stage classification. Method is realized and tested for polyp detection and segmentation. Polyp is a factor of high oncology risk. It is important to realize the polyps' early detection and removing. The polyp detection takes central part in stomach and colon endoscopy. The main idea of the proposed method is to preliminarily determine the presence of polyps by analyzing global signs of an image, and then if they are confirmed, use CNN to localize them.

Another method of this group is aimed on bleeding detection and segmentation. The bleeding detection is also very important, because it is sign of different pathologies [3–5]. Automatic detection of bleeding is especially very important for Wireless Capsule Endoscopy (WC) [6, 7]. To perform WCE, a patient swallows a small capsule containing a light source, lens, camera, radio transmitter, and batteries. Propelled by peristalsis, the capsule then travels through the digestive system for about eight hours and automatically takes more than 50,000 images. The large amount of images for analysis made high actual task bleeding automatic detection. However, the problem of dataset with large amount of images is also very hard, because the frames from one video have high correlation. Thus, the main idea of bleeding segmentation is based on local features application in classification task and on splitting an initial image in blocks [8]. Block-based approach solves both task bleeding detection and segmentation in the same time. All proposed methods of endoscopic images enhancement, visualization and analysis were tested using real endoscopic images obtained in clinic conditions and on free KVASIR dataset. The experimental results proved their correctness and effectiveness.

The chapter is organized as follows. Related work is presented in Sect. 8.2. The methods of endoscopic images enhancement are given in Sect. 8.3. The proposed method of virtual chromoendoscopy is described in Sect. 8.4. Section 8.5 is devoted to endoscopic image analysis. Section 8.6 is concluded the chapter.

8.2 Related Work

There is a significant number of works devoted to endoscopy images processing, for example, an extremely comprehensive review with more than 300 works in the field of endoscopy image analysis and processing was made in work [9]. Upon the analysis of such works in the field of endoscopy image processing, subsequent conclusion could be reached. Algorithms proposed in the works devoted to particular task of enhancement, for example, contrast enhancement [10, 11] or distortion correction [12, 13]. It is important to take note that the algorithms described in these works were implemented to solve separate tasks, without taking into account the pipeline steps.

The crucial point for quality representation and processing of endoscopic images is a noise reduction due to the fact that the noise is one of the most common problems in medical image processing. Even a high-resolution medical video is bound to have some noise in it. As for the endoscopic video, it is a constant problem, because the lack of light forces one to use high gain which leads to high level of noise. In [14], the researchers made a comparison of noise reduction algorithms for medical images and concluded that NLM produces the best results, but requires high computational costs. Therefore, there appears to be no algorithm, which satisfies such requirements as high quality and low computational cost.

Contrast is one of the key characteristics of an image. Images with low contrast usually display non-significant visual information, whereas high-contrast images maximize the available visual information. The conditions, under which an image is captured, affect greatly on its contrast. In endoscopic images, a contrast improvement is crucial for a correct diagnosis, but, on the other hand, the conditions for obtaining medical images are very difficult. Therefore, because of the fact that medical images suffer from degradation, it is necessary to eliminate such degradation before images are demonstrated to a physician.

Nowadays traditional image enhancement is based on global linear transformation [15]. In [16], authors proposed to acquire a set of grayscale images using FICE technology [17], then selected the image with the maximum entropy and restored its color to increase the contrast. In other works [18], the adaptive sigmoid function was utilized to convert the intensity component values of the image. Such enhancement is not effective for medical images due to the fact that overall conditions of medical image obtainment have nonlinear contrast and brightness changes. Therefore, there are very dark and very bright areas in one image, which demands for special correction techniques. In [19], it was attempted to solve this problem by splitting the image into the base and detailed layer, as well as, using different methods of processing for fragments with good illumination, low illumination, and extremely low illumination. This algorithm is parallelizable; however, it is computationally costly, which makes it difficult to utilize in real-time systems in conjunction with other algorithms. The other problem of modern contrast enhancement techniques is noise stressing, especially in the part of image with low amount of detail. Therefore, it is required to design nonlinear local brightness and contrast enhancement algorithms, as well as, algorithms for medicine image adaptation.

The most famous example of virtual chromoendoscopy are FICE [1] and I-SCAN [2, 3]. FICE is often used that can simulate an infinite number of wavelengths in real time. Here, the software applies an algorithm to the real-time endoscopic image, which is reconstructed to determine a wavelength for each of the three colors (red, green, blue). The image is instantaneously reconstructed after changing the wavelengths with virtual electronic filters. I-scan by PENTAX modifies pixel sharpness, hue and contrast in real time.

It is necessary to stress that FICE and I-scan technologies are rather closed technologies. Moreover, FICE method is most closely to hardware realization because it has rather complete calibration procedure, which gives possibility to estimate images at three wavelengths, or spectral images using the following 3×3 matrix. The aim of calibration procedure is to find matrix, which is defined by a correlation matrix for the spectral radiance and camera output, and an auto-correlation matrix for the camera output. In calibration procedure the special devices such as spectrometer is used.

On the other hand, i-SCAN and FICE exploit the color tone enhancement to extract unique features from the different spectral response. Similarly, image processing algorithms that leverage the unique characteristics of different spectral response in the endoscopic image can provide better enhancement in terms of better visualization of the anomaly in the image. So, there are additional methods of virtual chromoendoscopy with the aim to make endoscopy images easier for visual analysis and interpretation by means image processing algorithms, the most famous example is TRI–SCAN [20].

TRI–SCAN has three stages: Tissue and Surface Enhancement (TSE), Mucosal Layer Enhancement (MLE), and Color Tone Enhancement (CTE). TSE is employed using modified linear unsharp masking. MLE is performed in Red (R) plane of the sharpened color image using adaptive sigmoid function. Finally, in CTE the pixels of three color planes are uniformly distributed to increase the contrast level and to create an enhanced color tone.

MLE and CTE steps of TRI–SCAN method are based on global contrast enhancement methods. These methods apply the same transformation to all image pixels. Although global contrast enhancement methods are simple, they cannot be used successfully to all images, since they tend to exhibit degraded appearance amplified noise or other annoying artifacts. The main reason for this is that the global methods cannot adapt to local features because they use the same transformation over the whole image. Thus, TRI–SCAN approach has some problems endoscopic images with very dark and very bright areas in the same endoscopic image, and for such images this method is not effective.

Thus, the aim of our investigation is to develop a method based on digital processing of images obtained in the white light. Method ought to realize visual effect corresponding visual effect of images obtaining with modern technologies of virtual chromoendoscopy (I-SCAN and FICE) and ought to provide a possibility to get visual effect superior than modern methods of tone enhancement, for example TRI-SCAN.

There are numerous approaches for polyp's detection and segmentation in video endoscopic systems. The common problem in medical Artificial Intelligent (AI) systems design is a lack of training data. Thus, the most of solutions are connected with the so called conventional Machine Learning (ML) techniques (Support Vector Machine (SVM), Random Decision Forests (RDF), and logistic-regression) for they are able to solve the task with relative small amount of training data. Deep learning methods and especially CNN is now the trend in modern AI, and researchers more and more often try to use them in medical video systems and the endoscopic systems aren't an exception.

In [21], authors found out that the global-feature approach outperforms the deep neural network approach and gives better results than a convolution neural network for almost all classes, except for a few cases of tumors. In the article [22], the authors propose to apply Otsu threshold and select the largest connected component among all candidate areas to reduce the number of false positives.

Methods of automatic detection and segmentation of polyps use different characteristics, such as geometric primitives, color spaces, texture descriptors. Polyps have pronounced geometric features such as a well-defined shape or protrusion from the mucosal surfaces [1]. Methods based on color traits are ineffective in the task of polyp detection, since often their color does not differ from the adjacent healthy surface. At the same time, textural features have a wide variety. A strong influence on the texture of polyps has a movement of the gastric mucosa, appearance of bubbles, as well as, distortion and blurring caused by sharp movement of the camera. Effective detection of polyps can provide a combination of numerous features having different types: color, texture, and geometry. For this, various descriptors could be applied. Another approach to segmentation of polyps is the use of CNN. CNN is a standard approach for solving the problem of finding objects in images. It allows to capture a variety of features and can be used for the segmentation of polyps as well.

The aim of our research is to design a method as combination traditional ML technologies and deep learning technologies to achieve a reasonable quality of classification in conditions of small data set for training.

The existing approaches for automatic bleeding detection realize different methods and algorithms of digital image processing and machine learning. They could be classified by the image features they rely as the base for analysis and by the decision strategy they use for detection. Suspected blood indicator is a technique to detect bleeding from endoscopic images. However, in accordance with research and meta reports, its sensitivity and specificity are not sufficiently well. Many approaches use color or texture features as the base for analysis [23]. The others are based on contour analysis or various unsupervised segmentation techniques [24]. As for the decision-making mechanism, it is sometimes simple rule-based algorithm or some type of Naive Bayesian approach, but often it is based on powerful techniques as SVM or artificial neural networks. Therefore, recent advancement in the area of endoscopic bleeding detection algorithms resulted in establishing a great number of image processing techniques. However, it is obvious that there is still a strong need for a detailed study of the particular visual features, which are considered by the physicians during the examination of the images, and which were not sufficiently analyzed and described yet. Moreover, every solution should be adapted for specific video system, sensor, data, and environment. Thus, the task of automatic bleeding detection cannot be considered as solved today. It requires more researches, new algorithms, and solving strategies.

8.3 Methods of Endoscopic Images Enhancement

This section includes the discussion of methods and algorithms for noise reduction on medical endoscopic images (Sect. 8.3.1) and endoscopic images contrast and brightness enhancement (Sect. 8.3.2).

8.3.1 Methods and Algorithms for Noise Reduction in Medical Endoscopic Images

Modern methods of noise reduction can be divided into two groups. The methods of the first group work with a signal of image. These methods include smoothing filters, morphological filters, median and rank filters, the adaptive median filter, bilateral filter, and NLM filter. The methods of the second group are based on the decomposition of the image signal into a basis followed by decomposition transform processing.

As a transform, Discrete Cosine Transform (DCT), discrete Fourier transform, and wavelet-based processing are used. The actual trends in the development of this group methods are aimed at eliminating detail suppression and artifacts caused by the basic decomposition. New methods of this group (such as BM3D method) use three-dimensional DCT and a sliding window for processing of static images.

In the first group, the adaptive median filter, bilateral filter, and NLM filter are the most effective in terms of noise reduction and the original image preservation. In the second group, the BM3D method based on effective filtering in 3D transform domain by combining sliding-window transform processing with block-matching is one of the most promising.

Thus, to study the methods of noise reduction for their using in endoscopic pipeline were chosen next methods: adaptive median filter, bilateral filter, NLM filter, and 3D block matching method. The bilateral and NLM filters are well known improvement of the Gaussian filter in order to reduce image blurring after filtering [15]. Bilateral filter replaces the intensity of each pixel with a weighted average of intensity values from nearby pixels. This weight can be based on a Gaussian distribution.

NLM is the modern algorithm for image denoising. Unlike "local mean" filters, which take the mean value of a group of pixels surrounding a target pixel to smooth the image, non-local means filtering takes a mean of all pixels in the image weighted by how similar these pixels are to the target pixel. These results in much greater post-filtering clarity and less loss of detail in the image compared with local mean algorithms [25].

An adaptive median filter is the improving of median filter aimed to reduce impulse high density noise while preserving details. Unlike the usual median filter, the size of the analyzed area (the size of filter structural element) is variable. The basis of the filter is the analysis of median value L_{med} found in the processed area. If found median value L_{med} does not satisfy the condition ($L_{min} < L_{med} < L_{max})$, where L_{min} and L_{max} are the maximum and minimum brightness values of the analyzed area, respectively), then the size of the processed area is increased. If the condition is true, then the filter response is formed. If the size of processed area has reached the maximum possible, then the filter response is the brightness value L_c of pixel in the center of processed area.

BM3D method [26] is a new approach to noise reduction developed for static images. It is based on the using three-dimensional DCT. The most often digital image processing uses a two-dimensional DCT applied to a square block (for example, an 8 by 8 pixel block is used in JPEG standard). To ensure the possibility of applying a three-dimensional DCT in BM3D method, each processing block is associated with an array of blocks. The array includes image blocks that have a high level of correlation with the block being processed. For the formed array, three-dimensional DCT is performed followed by threshold processing of decomposition transformants. Threshold processing of decomposition transform provides a noise reduction. The described procedure is repeated for all blocks of image, and the blocks processing is performed in accordance with the principle of the sliding window. The resulting brightness estimation for each pixel of the processed image is formed as a weighted average founding over all blocks that overlap in a given pixel.

To find the image blocks Z_{xi}, the most similar to the block being analyzed Z_{xR} the procedure of block matching is used. The procedure of block matching is well developed in the video compression standards for the motion compensation. The original image has noise component, which makes it difficult to realize the block matching according to correlation function. To reduce the effect of noise, the correlation function estimating the measure of blocks similarity is formed on the basis of two-dimensional DCT transformants with threshold processing. In accordance with this, the correlation function has the next form:

$$d(Z_{xR}, Z_{xi}) = N_1^{-1} \left\| \begin{matrix} \Psi\left(\mathbf{T}_{2D}(Z_{xR}), \lambda_{thr2D}\sigma\sqrt{2\log(N^2)}\right) \\ -\Psi\left(\mathbf{T}_{2D}(Z_{xi}), \lambda_{thr2D}\sigma\sqrt{2\log(N^2)}\right) \end{matrix} \right\|_2$$

where Z_{xR} and Z_{xi} are the square blocks of the image with the coordinates of the upper left corner at the points xR and xi, respectively, $\mathbf{T}_{2D}$ is the operator of two-dimensional DCT, $\| \|_2$ is the L_2 norm. The operator Ψ determines the threshold processing of the two-dimensional DCT transform:

$$\Psi_{2D}(\lambda, \lambda_{thr}) = \begin{cases} \lambda & \text{if } |\lambda| < \lambda_{thr} \\ 0 & \text{otherwise} \end{cases}.$$

As a result of correlation matching, a set of blocks S_x with different value of the similarity measure $d(Z_{xR}, Z_{xi})$ with the processed block is formed. For further analysis, select the set of blocks for which the similarity measure value is less than the specified threshold value:

$$S_x = \{x \in X, \ d(Z_x, Z_x) < \tau_{match}\}.$$

The blocks included in the three-dimensional array for processing are ordered according to similarity measure value $d(Z_{xR}, Z_{xi})$. Three-dimensional DCT is applied to the obtained three-dimensional array. To reduce the noise level, the resulting transformants of three-dimensional are subjected to threshold processing. Then the inverse DCT transform is performed:

$$\hat{\mathbf{Y}}_{\mathbf{x}R} = \mathbf{T}_{3D}^{-1}\left(\Psi\left(\mathbf{T}_{3D}(\mathbf{Z}_{\mathbf{x}R}), \lambda_{thr3D}\sigma\sqrt{2\log(N^2)}\right)\right),$$

where $\hat{\mathbf{Y}}_{\mathbf{x}R}$ is the square block of the image after the inverse DCT transformation, $\mathbf{T}_{3D}$ and $\mathbf{T}_{3D}^{-1}$ are the operators of the three-dimensional DCT and inverse DCT, respectively, Ψ is the operator determines the threshold processing of DCT transform, λ_{thr3D} is the fixed threshold parameter.

For each block obtained after applying the inverse three-dimensional DCT, the weighting coefficients w_{xR} are calculated. Its value is determined by the number of non-zero transforms N_{NzT} after the threshold processing:

$$w_{xR} = \begin{cases} \frac{1}{N_{NzT}} & \text{if } N_{NzT} \geq 1 \\ 1 & \text{otherwise} \end{cases}.$$

After performing the described procedures for all image blocks, each pixel of the initial image will be associated with a set of brightness estimations. This is ensured by overlapping blocks (principle of sliding window). To calculate the result pixel brightness value on the restored image, the obtained estimations are averaged taking into account the found weight coefficients w_{xR}.

The basic approach can have extension with Wiener filtering. The linear Wiener filter replaces the nonlinear hard-thresholding operator on the step of denoising in three-dimensional transform domain.

The methods discussed above were implemented in special software and tested on real endoscopic images. The images were divided in three groups:

- Initial images are whole endoscopic pictures obtained by sensors in various medical devices. They have various resolutions according to the specific sensor. In Fig. 8.1, the examples of images with relative high resolution (1280 × 1024)

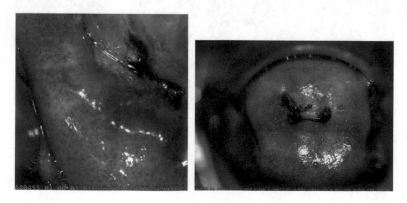

Fig. 8.1 The examples of colposcopic images with high resolution

are shown. They were obtained with colposcopic device. The images from video gastroenteroscopy are presented in Fig. 8.2. They have relative low resolution (320 × 320).

- Set of images with rich number of details obtained by cropping initial images (Fig. 8.3).
- Set of images with low number of details obtained by cropping initial images (Fig. 8.4).

The reason of this dividing was to check the performance of the denoising algorithms in various circumstances. The low detailed images (or large areas of images) are typical case in endoscopic observations of the organs with smooth tissues. That's why it is very important for denoising algorithms to cope with this type of flat surfaces.

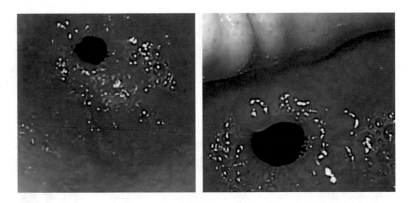

Fig. 8.2 The images from gastroenteroscopy with low resolution

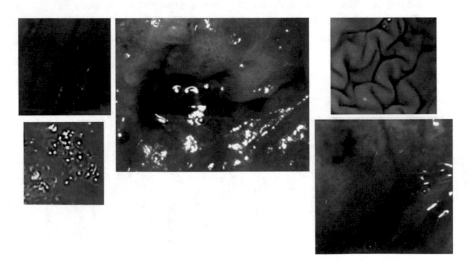

Fig. 8.3 The fragments with rich number of details

Fig. 8.4 The fragments with low number of details

However, the vascular patterns, different wrinkles and other high detailed parts of images represent often the most interesting part of the picture for physician. Of course, the algorithm for noise reduction should do its best on these fragments.

In our research we use the described three groups of images with added Gaussian noise with three different standard deviations ($\sigma = 10$, 20, 30 for 24 bits RGB images). In Fig. 8.5 the image with different level of noise is shown.

Tables 8.1, 8.2 and 8.3 represent the results. As a measure, we used the standard Peak Signal to Noise Ratio (PSNR) metric. In Table 8.4 and the corresponded Fig. 8.16, the overall data for all noise levels from previous tables is shown.

It should be mentioned that the PSNR metrics doesn't provide the full information about the quality of the noise reduction for medical images. According to the charts and tables the BM3D algorithm has great advantage for all other methods. In fact, this isn't absolutely correct. In Fig. 8.6, the result of application of BM3D and NLM algorithm are shown. It can be noticed that BM3D impose some distortion in the image. The picture has some rectangular artifacts and looks cartoonish.

The obtained value of PSNR (Tables 8.1, 8.2, 8.3 and 8.4) and visual analysis of images after processing with different noise removing filters give a possibility to do the intermediate conclusions:

1. On low detailed fragments, which dominate on endoscopic images, the computationally costly algorithms do not provide well expressed effects in comparison with relatively simple approaches. The difference is approximately 2 dB. It is necessary to stress that endoscopic images have rather big area of fragments with low level of detail about 70–80% of the full image square. So, for general video processing pipeline for endoscopic CDSS system it is not useful to implement some complex algorithms with large processing time. The well-known relative simple and fast adaptive median filter provides very close results for the most cases of endoscopic images.

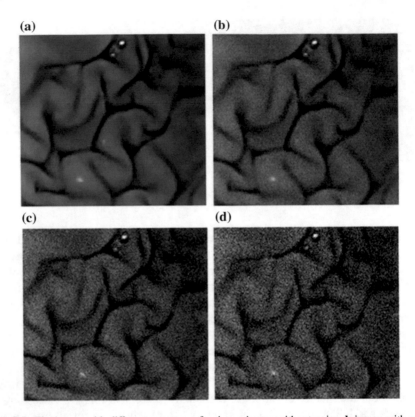

Fig. 8.5 The image with different presence of noise: **a** image without noise, **b** image with noise $\sigma = 10$, **c** image with noise $\sigma = 20$, **d** image with noise $\sigma = 30$

Table 8.1 Results of noise reduction algorithms for initial images (dB)

σ	Noised	Adaptive median	Bilateral	NLM	BM3D
10	28.23	33.69	38.16	34.84	41.47
20	22.31	29.15	29.54	31.6	38.09
30	18.94	26.03	23.01	27.79	35.95

Table 8.2 Results of noise reduction algorithms for fragments with high details (dB)

σ	Noised	Adaptive median	Bilateral	NLM	BM3D
10	28.17	32.32	35.87	32.36	39.11
20	22.23	28.33	28.77	30.13	35.58
30	18.86	25.44	22.65	26.93	33.52

Table 8.3 Results of noise reduction algorithms for fragments with low details (dB)

σ	Noised	Adaptive median	Bilateral	NLM	BM3D
10	28.16	34.86	41.02	35.51	43.56
20	22.20	29.40	29.97	32.06	40.94
30	18.80	25.97	22.94	28.12	39.21

Table 8.4 Overall results of noise reduction algorithms (dB)

Images	Noised	Adaptive median	Bilateral	NLM	BM3D
Initial	23.05	30.07	31.31	31.90	41.24
High detailed	23.16	29.62	30.24	31.41	38.50
Low detailed	23.09	28.70	29.10	29.81	36.07

(a)   (b)

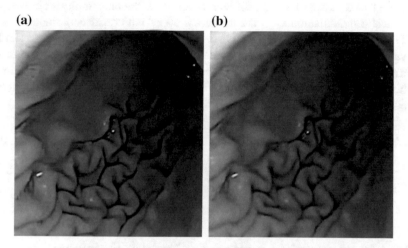

Fig. 8.6 The result of image processing using: **a** BM3D algorithm, **b** NLM algorithm

2. For the case, when the presence of noise is very high the implementation of the modern, BM3D algorithm can make sense. But the developer should keep in mind the specific of BM3D algorithm. It is relative slow, has many parameters and can provide some artifacts on the image. The implementation of this algorithm in video endoscopic system should be guided by physicians, who will be able to estimate the allowable level of imposed distortions.
3. The approach of complex technique is to use resource-consuming BM3D on some important (high-detailed) fragments and relative quick adaptive median filtering. For general video processing, a pipeline for endoscopic CDSS system it is not useful to implement some complex algorithms with large processing time. The well-known relative simple and fast adaptive median filter provides very close results for the most cases of endoscopic images.

Based on this position and taking into account both requirements (quality and low computational cost) we propose the following algorithm:

1. Image processing by the operator of a multiscale morphological gradient in order to emphasize the high-frequency component.
2. Splitting the image into blocks with $N \times M$ size and determining for each block the detail level estimation $D_M(k, l)$ according to Eq. 8.1.

$$D_M(k,l) = D(k,l)/D_{\min},$$
$$D(k,l) = \sum_{j=1}^{M} \sum_{i=1}^{N} \Lambda(x_k + i, y_l + j), \qquad (8.1)$$
$$D_{\min} = \text{moda}\{D(k,l)\}\, k = 1\ldots KK, l = 1\ldots LL,$$

where $\Lambda(x,y)$ is the pixel brightness in the image obtained from the initial one $L(x,y)$ as a result of spatial differentiation with operator multi-scale morphological differentiation, x_k, y_l are the block upper left corner coordinates.

3. Threshold processing. The detail estimations obtained for all blocks is random value. Analysis and study of its histograms suggests that it has Rayleigh distribution. Let $\hat{\sigma}^2$ be the variance of $D_M(k,l)$ values calculated for all image blocks, p_0 be the quantile of Rayleigh distribution. In this case, one is able to calculate the threshold value according to Eq. 8.2.

$$\lambda = \sqrt{(-2\hat{\sigma}^2 \ln(p_0))} \qquad (8.2)$$

4. If $D_M(k,l) < \lambda$, then median or bilateral filter is used for block processing else NLM (BM3D) filter is applied.

8.3.2 Methods and Algorithms for Endoscopic Images Contrast and Brightness Enhancement

Numerous contrast and brightness enhancement techniques exist in literature. Among them there are two main groups: linear enhancement and non-linear enhancement. Non-linear algorithms can be categorized to global and local. Global contrast enhancement algorithms apply the same transformation to all image pixels. This category comprises the logarithm transformation, power-law transformation and histogram equalization. Although global contrast enhancement methods are simple, they cannot be used successfully to all images, since they tend to exhibit degraded appearance, amplified noise or other annoying artifacts. The main reason for this is that the global methods cannot adapt to local features because they use the same transformation over the whole image.

Local enhancement techniques take into account the local features of the image and apply different transformations to every pixel or region. In our investigation, we used only non-linear enhancement techniques. It is connected with main feature of images: there are very dark and very bright areas in the same image. The main methods for non-linear contrast enhancement under investigation were the following:

- Multi Scale Image Contrast Enhancement (MSICE) [27].
- Adaptive and Integrated Neighborhood Dependent Approach for Nonlinear Enhancement (AINDANE) [28].

- Locally Tuned Nonlinear Technique for Color Image Enhancement (LTSNE) [29].

The key feature of the first method is three spatial scales by employing different region sizes. As a result, contrast enhancement is performed in different spatial frequencies of the original image. The key features of last two methods are intensity enhancement, contrast enhancement, and color enhancement. The difference between AINDANE and LTSNE methods takes place on the step of intensity enhancement. In AINDANE method, this step is based on global information about image, in LTSNE—on local information, so it makes adaptive intensity enhancement. On steps of contrast and color enhancement both methods use local information and provide a possibility to apply different transforms to every region.

For effectiveness investigation we realized all three algorithms. According to our tests the results of AINDANE and LTSNE methods are quite similar, but in the most cases the quality of images after LSTNE method is slightly higher than after AINDANE enhancement. The reason of this situation is clear, because LTSNE method is advanced modification of AINDANE, so for further investigation we selected LTSNE. MSICE method is not effective for difficult cases on images with big very dark fields, but in the case of image with moderate degradations it is very useful. The quality of correction for images with moderate degradations is very high and comfortable for visual analysis. The illustration of this conclusion is given in Fig. 8.7. Here, the enhancement was achieved only with LTSNE method. MSICE method is not effective for difficult cases (with big very dark fields). So, for difficult cases we can advise LTSNE method, otherwise MSICE method.

(a)   **(b)**   **(c)**

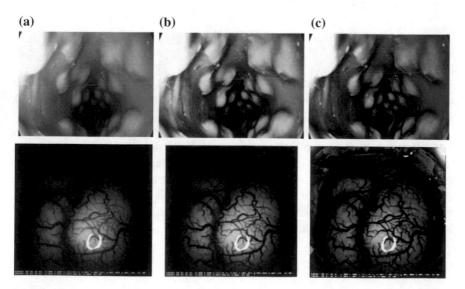

Fig. 8.7 Visual results of various enhancement algorithms: **a** initial images, **b** images processed by MSICE, **c** images processed by LSTNE method

Table 8.5 Results of calculating focus values for various contrast enhancement algorithms

	Initial	MSICE	LSTNE
Focus value	0.021	0.034	0.059

To evaluate the difference between LSTNE and MSICE methods, the focus value was measured for images corrected with various approaches. The test set included 20 images. The averaged results are presented in Table 8.5.

On the stage of contrast enhancement, LSTNE applies the transform function to every pixel:

$$S(x, y) = 255 \cdot I_{E_n}(x, y)^{E(x,y)},$$

$$E(x, y) = r(x, y)^P = \left(\frac{I_M(x, y)}{I(x, y)} \right)^p,$$

where $S(x, y)$ is the pixel brightness after correction, $I_{En}(x, y)$ is the normalized brightness, $I(x, y)$ is the initial pixel brightness, $I_M(x, y)$ is the brightness of a local fragment of an image after Gaussian filtering, p is the parameter for correction "strength".

The described methods have common drawback. Simultaneously with the increase of the contrast, the methods significantly emphasize the noise, especially in areas of the image with low detail. To solve this problem, we propose to consider the local features of the image on the base of the found functional relationship between the correction "strength" p and the normalized variance of brightness in the image fragment [30]:

$$p = \log \left(\frac{k\sigma^2}{\sigma_{\min}^2} \right),$$

where σ is the standard deviation of brightness in a fragment of the image, $\sigma_{\min}$ is the minimum standard deviation of brightness among all fragments of the image, k is the parameter of the regression model.

Our investigation show that the signal-to-noise ratio of the image processed by LTSNE and AINDANE methods falls from 40 to 25 dB. Figures are obtained on low-detail fragments of the image. The developed method provides signal-to-noise ratio of 35–38 dB on the same fragments of processed images. It is important to stress that the method uses information about contrast in local part of image and on this base realize non-linear enhancement. Thus, additionally to contrast and brightness enhancement it realizes edge sharpening.

8.4   Method of Endoscopic Images Visualization (Virtual Chromoendoscopy)

RGB planes carry different spectral responses of the surface. These spectral responses are mainly dependent on the camera sensor and its spectral sensitivity. From the observation of spectral sensitivities ranging from 300 to 700 nm of CMOS and CCD cameras and their effect on endoscopic images, it can be noticed that R plane dominates higher wavelength, G plane dominates mid-wavelength, and B plane dominates shorter wavelength regions. As a result, these individual spectral responses of R, G, and B planes carry different spatial characteristics.

For example, the subtle blood vessels and microvessels located deeper in the mucosal layer are better visible in R plane, primarily carrying information related to deep mucosa layer than the other plane. It is possible to highlight these subtle features by enhancing these different spatial characteristics separately. The proposed enhancement method works on each plane separately to highlight these subtle features to help the gastroenterologists to inspect the tissue characteristics, mucosa structures, and abnormal growths better than the original image.

The main features of proposed method are:

1. Procedures for processing each channel of the original image obtained in white light. Their principal feature is the use of adaptive (local) nonlinear contrasting technology; in all existing solutions global transforms are used. Global transform means a transform, the type and parameters of which are constant for all image elements. In the proposed local transforms, the parameters change for each region of the image depending on its features. The use of local algorithms for endoscopic imaging is more efficient than global. This is determined by the important property of endoscopic images: the simultaneous presence of dark and light areas of considerable area. This property is due to the difficult conditions for obtaining endoscopic images.

2. The absence of a calibration procedure for obtaining the effect of virtual chromoendoscopy: the proposed method is completely based on the technology of digital image processing.

Proposed algorithm is based on idea to divide the initial images obtained using normal white light into its RGB components, and subsequently converts and re-synthesizes them into a new image to produce images, in which subtle hue differences highlight vascular patterns or subtle changes in the mucosa. For the better blood vessels visualization, we used enhancement only in R channel, and then resynthesizes new images. For R channel processing, we proposed to use nonlinear local method of contrast enhancement—MSICE method [26].

MSICE method employs the following adjustable non-linear transformation functions:

$$G(x) = \frac{(B+A) \cdot x}{A+x} \quad \forall x \in [0, B], \quad A, B \in R, \tag{8.3}$$

$$H(x) = \frac{A \cdot x}{A+B-x} \quad \forall x \in [0, B], \quad A, B \in R, \tag{8.4}$$

where x is the input data, B is the maximum value of x, A is the factor that regulates the degree of the non-linear transformation. B is defined according to the range of input data varying A can result to different nonlinear curves, controlling the transformation between the input x and output $G(x)$ or $H(x)$. In this method, since the input data will be intensity values, the input range is [0, 255] and thus $B = 255$.

For every pixel (i, j) of the original image, the difference between the pixel Y_{ij} and its mean surrounding intensity S_{ij} should increase. If $Y_{ij} > S_{ij}$, then Eq. 8.3 is employed in order to calculate a new pixel value $G(Y_{ij}) \geq Y_{ij}$ and, thus, increase the intensity difference with its surround S_{ij}. Similarly, if $Y_{ij} < S_{ij}$, then Eq. 8.4 is employed in order to calculate a new pixel value $H(Y_{ij}) \leq Y_{ij}$ and, thus, increase the intensity difference with its surround S_{ij}.

The mean surrounding intensity S_{ij}^K for scale K is calculated as follows:

$$S_{ij}^K = \frac{1}{(2d_k+1)^2} \sum_{y=i-d_K}^{i+d_K} \sum_{x=j-d_K}^{j+d_K} Y_{yx}, \tag{8.5}$$

where d_K is the size (in pixels) of the surrounding region for scale K.

Factor A, which determines the degree according to which the original pixel value Y_{ij}, is either increased or decreased, should be adjusted by the initial difference between Y_{ij} and S_{ij}^K. Small differences should result to greater changes in the pixel values in order to increase the local contrast. On the contrary, large differences should result to smaller changes, since the contrast in these cases is already satisfactory. The regulation of the non-linearity factor A, as well as, the combination of Eqs. 8.3–8.5 is described by the following equations, which are the contrast enhancement function of this method:

$$A(x) = \begin{cases} \frac{M}{x} & \forall x \in [1, B] \\ M & \text{if } x = 0 \end{cases},$$

$$Out_{ij}^K(Y, S) = \begin{cases} \frac{[B+A(Y_{ij}-S_{ij}^K)] \cdot Y_{ij}}{A(Y_{ij}-S_{ij}^K)+Y_{ij}} & \text{if } Y_{ij} \geq S_{ij}^K \\ \frac{A(Y_{ij}-S_{ij}^K)Y_{ij}}{A(S_{ij}^K-Y_{ij})+B-Y_{ij}} & \text{if } Y_{ij} < S_{ij}^K \end{cases},$$

where $Out_{ij}^K(\cdot)$ is the new intensity value of pixel (i, j) for scale K, M is a constant value that determines the degree of contrast enhancement.

Small M values result to strong enhancement, while high values result to moderate enhancement. Since image contrast enhancement can be a rather subjective application, M value should be determined by the user according to the enhancement degree that he/she wants to apply. The recommended value for medicine images for M is 5,000 (obtained from experiments). The proposed method is applied independently to three different spatial scales and the final result is the average between their results.

The second part of proposed method is a tone enhancement [31]. It includes separate enhancement in each channel. We propose to use Contrast–Limiting Adaptive Histogram Equalization (CLAHE) algorithm for separate processing of R, G, B channels [32]. CLAHE based on the classical contrast enhancement technique Histogram Equalization (HE) and it advanced variant Adaptive Histogram Equalization (AHE). HE is global method, so it has good performance in ordinary images. This method increases the contrast of an image globally by spreading out the most frequent intensity values. HE has been generalized to a local histogram equalization—AHE. AHE formulates each histogram of sub-image to redistribute the brightness values of the images.

CLAHE differs from ordinary AHE in contrast limiting. CLAHE introduced clipping limit to overcome the noise amplification problem. CLAHE limits the amplification by clipping the histogram at a predefined value before computing the Cumulative Distribution Function (CDF). In CLAHE technique, an input original image is divided into non-overlapping contextual regions called sub-images, tiles or blocks.

CLAHE has two key parameters: Block Size (BS) and Clip Limit (CL). CLAHE method applies histogram equalization to each contextual region. The original histogram is clipped pixels are redistributed to each gray level. The redistributed histogram is different with ordinary histogram, because each pixel is limited to a selected maximum. But the enhanced images and the original image have the same minimum and maximum gray levels. Thus, the proposed method has next main steps presented in the Fig. 8.8. The endoscopic image is divided into three primary color planes: R, G, and B. For R-plane we apply MSICE method. Then we realize separate processing of R, G, B channel using CLAHE algorithm.

The experiments were carried out with open KVASIR dataset of endoscopic images [33]. The dataset consists of 4,000 images with different resolutions from 720×576 up to 1920×1072 pixels. The images are separated on eight types and represent different endoscopic cases. The rich variability of the data allows to comprehensively investigate the proposed method in various conditions. Figure 8.9 shows the initial images of different cases and the results of TRI-Scan and proposed method application.

In Fig. 8.10, there are the comparative results with the pictures from FICE atlas.

In our research we used Focus Value (FV) as a quantitative measure of image enhancement. FV represents the ratio of AC and DC energy of DCT image [34]. The results of tone enhancement obtained using KVASIR dataset for different types of pathologies are presented in the Table 8.6.

Experiments show that the proposed method implements:

Fig. 8.8 Block diagram of proposed method

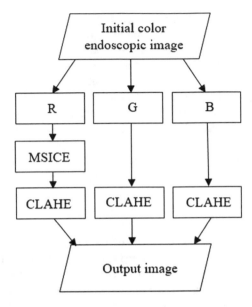

- Correction of image brightness providing the ability to obtain the necessary visual information from both very dark and overexposed fragments.
- Sharpening the image emphasizing small details and vessels.

An expert evaluation of the images obtained shows that the visual effect of the proposed method is superior to the TRI-Scan tone correction algorithm and also corresponds to or, in some cases, exceeds the visual effect of the closed technologies of virtual endoscopy I-scan and FICE.

8.5 Methods of Endoscopic Images Automatic Analysis

In this section, two-stage method for polyp detection and segmentation and block-based algorithm for bleeding detection are considered in Sects. 8.5.1 and 8.5.2, respectively.

8.5.1 Two-Stage Method for Polyp Detection and Segmentation

Central part of CDSS is an automatic images analysis. We propose two-stage classification method that gives a possibility to obtain satisfactory results of CNN application in conditions of small dataset.

(a)   (b)   (c)

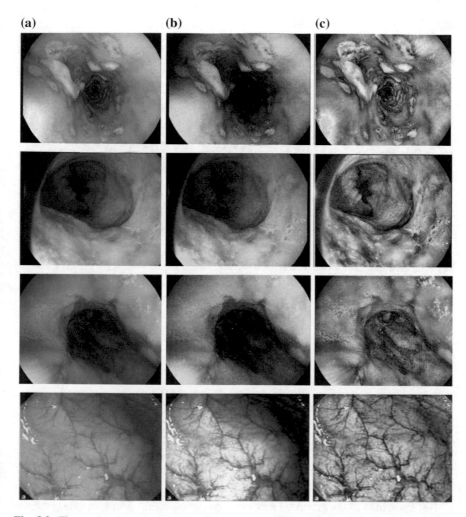

Fig. 8.9 The results of image processing from open KVASIR dataset: **a** initial images, **b** images processed by TRI-Scan, **c** images processed by the proposed method

Method was realized and tested for polyp detection and segmentation. The reasons to select this task is the following. Polyps are high cancer risk factor and it is important in medicine practice to realize early polyps' detection and removing. Polyps has high variability of its shape, size and appearance. The probability of polyps' correct detection is strongly connected with physician experience. So, polyp detection must be realized in CDSS. From the other hand, polyps in different organs are similar in visual representation, so it gives a possibility to use different datasets in our experimental research. Thus, we used in our research the expanded KVASIR dataset contains 8,000 endoscopic images of the gastrointestinal tract and CVC-ClinicDB dataset of frames extracted from colonoscopy videos.

(a)   (b)   (c)

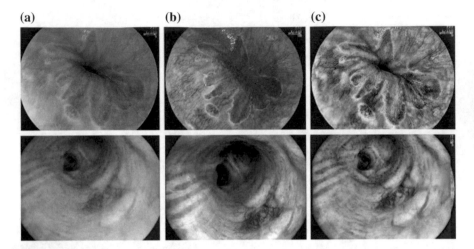

Fig. 8.10 The results of image processing from FICE atlas: **a** initial images, **b** images processed by TRI-Scan, **c** images processed by the proposed method

Table 8.6 The results of contrast enhancement

Type of pathology	Initial image	Processed image
Esophagitis	0.015	0.043
Dyed lifted polyps	0.031	0.069
Dyed resection margins	0.033	0.072
Normal seccum	0.025	0.067
Normal pylorus	0.014	0.043
Normal z-line	0.015	0.042
Polyps	0.025	0.059
Ulcerative colitis	0.031	0.075

The main idea of the proposed method is to preliminarily determine the presence of polyps by analyzing global signs of the image and then, if they are confirmed, use CNNs to localize them. Thus, method based on two-stage classification (Fig. 8.11), including:

Stage 1. Binary classification—preliminary classification on the basis of global signs and traditional machine learning technologies. The result of the preliminary classification is the decision on the presence of a polyp in the image.
Stage 2. Polyp segmentation—classification based on convolution neural networks in order to localize pathology.

The basic principles for implementing each of the stages are described below.
The basic purpose of the first stage is to analyze the input image and decide whether this frame contains a polyp or not. The common developing process of the

Fig. 8.11 Block scheme of proposed algorithm

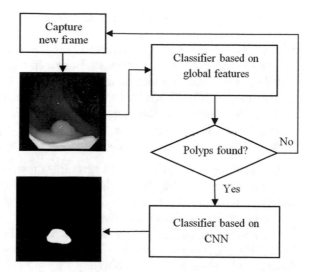

algorithm based on traditional machine learning approach consists of the following stages.

- Data acquisition and preprocessing.
- Features set definition, extraction, analysis and filtering.
- Choosing the best suitable machine learning algorithms for current task and— finding the best solution.
- The results post-processing.

The first and very important part is to choose proper features for the automatic analysis.

Images can be represented using various features, which basically can be divided into two categories: global and local. Global features describe the whole image, for example, a group of MPEG-7 descriptors [35]. While local features describe local fragments of images, for example for color and brightness estimation R, G, B, r, g, b, B/R, G/R, X, Y, Z, L, a, b, Y, Cr, Cb, H, S, V and for texture feature estimation based on Rosenfeld-Troy measure or Histogram of Oriented Gradients (HOG) [36].

The extracted global features should take into account the main distinctive characteristics of polyps. In our investigation we used the following global features:

1. Joint Composite Descriptor (JCD) [37]: It contains color and texture information: a color and edge directive handle plus a fuzzy color histogram.
2. Tamura [38]: correspond to human visual perception. They identified six textural features that include coarseness, contrast, directionality, line-likeness, regularity and roughness and compared them with psychological measurements for people.
3. ColorLayout: Features describe the spatial distribution of colors on the grid with DCT.

4. EdgeHistogram: determines vertical, horizontal, 45° diagonal, 135° diagonal and non-directional edges.
5. AutoColorCorrelogram [39] determines the dependence of the spatial correlation of color pairs on the distance. A color histogram captures only color distribution and does not include information about texture and edges.
6. Pyramid PHOG (PHOG) [36]: consists of a histogram of orientation gradients for each sub-region of the image at each resolution. The distance between the two PHOG image descriptors then shows the degree to which the images contain the same shapes and correspond to their spatial layout.

Extracted features can be used in two different ways. In the first case, they are concatenated into one vector and then sent to the classifier (early fusion). In the second case, its own classifier is used for each feature vector, and class membership is determined by a majority vote (late fusion).

In our study, we decided on early fusion and formed a feature vector of dimension 1185 (168 JCD + 18 Tamura + 33 ColorLayout + 80 EdgeHistogram + 256 AutoColorCorrelogram + 630 PHOG).

The next important step in algorithm developing is to choose the best suitable machine learning algorithms for current task and finding the best solution.

In our investigation we used the following ML methods [40]:

- Linear discriminant analysis—as the base level evaluation of the performance.
- SVM with radial-based function kernels—as the most suitable algorithm in case of relatively small datasets.
- RDF.
- AdaBoost.

The last two methods are very good also in case of nonlinear separability of classes. That's why it was the reasonable strategy to include them in investigation.

The basic purpose of the second stage is to segment one or several polyps in case their presence on the image was confirmed at the first stage. Our approach is based on deep neural network with Unet-like architecture, one of the most popular for segmentation tasks and medical images analysis. The choice of this architecture is due to the following reasons:

- Network can be trained end-to-end from relative few images and outperforms the most popular and general method (a sliding-window convolutional network).
- Network input/output resolution (512 × 512–388 × 388) and receptive fields fit well the native resolution of endoscopic images.
- The architecture consists of a contracting path to capture context and a symmetric expanding path that enables precise localization.

Neural network training requires a huge amount of input data, so insufficient size of training set is a common problem. Dataset with a marked ground truth includes a total of 650 endoscopic images. Pictures represents series of frames from several video samples. Thus, many members of dataset have strong cross-correlation. Actually, the dataset has only about 80 unique video fragments. Based on the

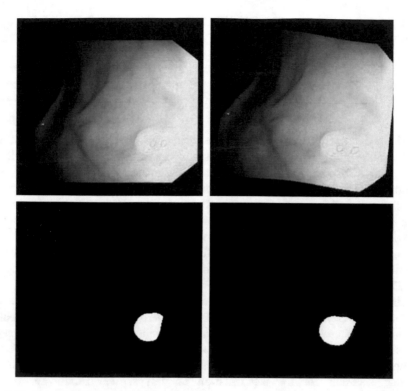

Fig. 8.12 Example of sinusoidal image transform

assumption that the endoscopic context could be represented as the slightly mor-phable object (unlike, for example, bones or some rigid organs) we used special data augmentation (along with the classic approaches)—the sinusoidal image transform (Fig. 8.12). Each frame was randomly transformed before feeding into the network, therefore the network never saw same pictures during training, which allowed to solve the problem of insufficiently large endoscopic images dataset.

In our investigation we used two open KVASIR dataset and CVC-ClinicDB [41]. The expanded KVASIR dataset contains 8,000 endoscopic images of the gastrointestinal tract—8 different classes (about 1,000 images with polyps), checked and marked by experienced endoscopists. The resolution of images varies from 720×576 up to 1920×1072. The images were obtained by different devices in various conditions and examinations.

CVC-ClinicDB is a dataset of frames extracted from colonoscopy videos. These frames contain several examples of polyps. Each frame has the ground truth for the polyps. This ground truth consists of a mask corresponding to the region covered by the polyp in the image. CVC-ClinicDB contains 612 images with associated polyp and background segmentation masks obtained from 31 polyp video sequences acquired from 23 patients. Resolution: from 720×576 up to 384×288.

Table 8.7 Performance
comparison of classifiers

Classifier	Accuracy	Sensitivity	Specificity
Linear	0.83	0.84	0.83
SVM	0.88	0.93	0.82
RDF	0.87	0.92	0.82
AdaBoost	0.86	0.88	0.84

The main task of investigation for the stage of binary classification was to find the most effective classification strategy among linear discriminant analysis, SVM, RDF, and AdaBoost. Table 8.7 represent classification results. The results were obtained by using 10-fold cross-validation on KVASIR dataset with 1000 positive samples and 1,015 negative samples. SVM classifier with radial basis function kernel, as well as, RDF gave the same results. Given the higher speed performance of SVM, we prefer it as the main decision.

Thus, as classification strategy for the first stage we used RDF with next vector of global features JCD, Tamura, ColorLayout, EdgeHistogram, AutoColorCorrelogram, and PHOG.

The main task of investigation for the stage of polyp segmentation was to estimate effectiveness of CNN application for extremely small dataset, but in conditions of presence the preliminary binary classification results. For this part of investigation, KVASIR dataset cannot be used, because it does not contain any ground truth information about the presence of polyps on the images. Thus, the training classifier for the segmentation of polyps, we carried out on the CVC-ClinicDB dataset [41]. To evaluate our neural network, we used the Dice [42] coefficient, which is determined by the ratio between doubled ground truth and net result intersection and their sum:

$$DCS = \frac{2|A \cap B|}{|A| + |B|},$$

where $|A|$ and $|B|$ are the cardinalities of the ground truth and CNN result sets, respectively.

After 15,000 iterations of generating new images, the distribution of Dice-score has reached saturation, which means that the network does not receive new information from the dataset. In Fig. 8.13 we can see the Dice curve, it is in stagnation. Stagnation of test Dice shows that the network has learned all from the given data. And, although the average mean of Dice is not high and is not enough for a final decision, but it shows that even in the case of an extremely small dataset (about 600 correlated images from 20 videos), the approach is promising and workable.

We realized additional test in our investigation of CNN effectiveness for second stage. We used CNN trained on CVC-ClinicDB dataset for further polyp segmentation in KVASIR dataset. This dataset doesn't have ground truth for polyp segmentation, so we cannot calculate Dice. We estimate only the quality of polyp

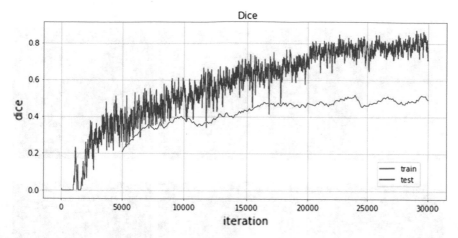

Fig. 8.13 Learning curve

segmentation. Our results in this test show that the neural network prediction is rather good! KVASIR pictures were never seen by the network, they were achieved with different devices and conditions, but the quality of segmentation is comparable with results obtained for testing set from CVC-ClinicDB. Figures 8.14 and 8.15 illustrate this position.

Two stages approach was realized for polyp segmentation: binary classification based on classical machine learning strategy with global features and segmentation based on deep learning. The following numeric results were achieved for test set on the stage of binary classification: accuracy equals 0.88, sensitivity equals 0.93, and specificity equals 0.82. The estimation of segmentation stage based on CNN shows that network obtained the suitable features from data. The learning curves show that the task can be solved successfully with the enhancement of the dataset. Thus, we can make conclusion that the designed approach works for given task and can be strong base for further investigation in the direction of AI endoscopic system design.

8.5.2   Block-Based Algorithm for Bleeding Detection

For bleeding detection, we proposed block-based approach for features extraction. Block size may vary depending on the application and characteristics of a particular sensor. Block diagram of proposed algorithm is shown in Fig. 8.16. The main advantages of this approach are following:

(a)

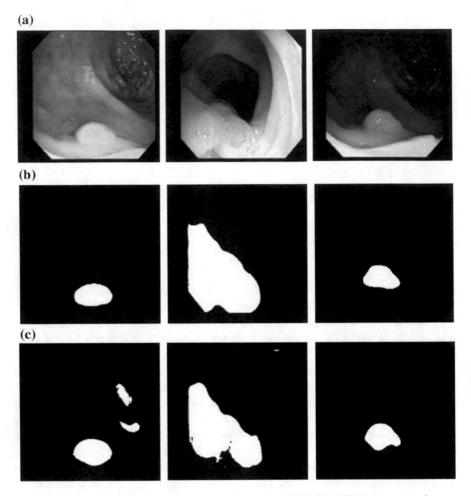

(b)

(c)

Fig. 8.14 Examples of polyp segmentation on the marked CVC-ClinicDB dataset: **a** test images, **b** ground-truth, **c** prediction (first column: Dice-score equals 0.62, second column: Dice-score equals 0.87, third column: Dice-score equals 0.86)

- Small number of images with bleeding does not allow to train global classifier that processes the entirely image as one sample. But the extraction the small samples with and without bleeding significantly enlarge the dataset and enhance the diversity of samples, represented in each class.
- The block-based method allows not only classify the image on correspondence to some class, but provides the more valuable result of segmentation.

In our investigation for bleeding detection we used also free-access KVASIR dataset [21, 32]. However, this dataset does not contain any ground truth information about the presence of bleeding on the images. Thus, in our research we had to prepare the dataset by ourselves. We searched through all the images and

(a)

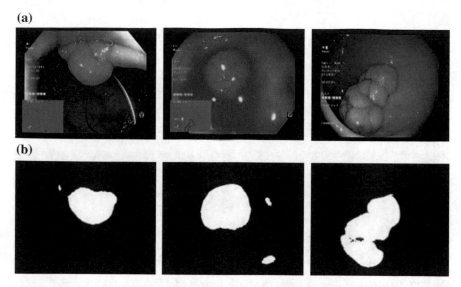

(b)

Fig. 8.15 Examples of polyp segmentation on the unmarked KVASIR dataset: **a** test images, **b** prediction

Fig. 8.16 Block diagram of proposed algorithm

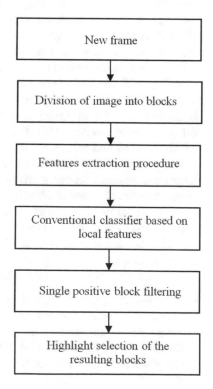

(a)

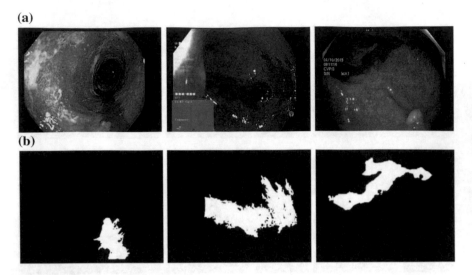

(b)

Fig. 8.17 Examples of manual segmentation: **a** images with bleeding from KVASIR dataset, **b** ground truth information

manually collect about 90 images with bleeding from the entire dataset. The examples of those pictures are represented in Fig. 8.17a. Next the ground truth information had to be obtained. By using image editor, we created manually the masks for selected images. The examples are shown in Fig. 8.17b.

For features automatic extraction, the special software was designed. It gives a possibility to divide the images on blocks (16×16 pixels is default size that can be tuned) and for each block there is possible to calculate different characteristics (brightness or color information, the amount of details etc.). According to the created masks of bleeding, the software is able to separate the blocks by two classes —with bleeding and without it.

The process of data acquisition is very flexible. It can be tuned, from which images the negative samples (without bleeding) should be extracted (from the images with the presence of bleeding samples, from the images without bleeding at all etc.), the software provides different data augmentation procedures (random crop, scaling). The screenshots of feature extraction procedure is shown in Fig. 8.18. The image is divided into blocks. If more than half of the block area falls into the bleeding area on the ground image, this block is attributed as positive class.

Thus, the algorithm of features mining consists of the following steps:

- To collect the pictures with bleeding.
- To make the ground truth masks.
- To divide the images to blocks.
- To attribute all the blocks with bleeding from corresponded images as positive samples, the blocks without bleeding from the images randomly selected from the whole DB as negative samples.

(a)   **(b)**

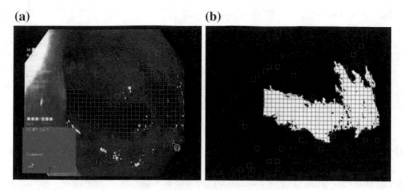

Fig. 8.18 The screenshots of the software for feature extraction: **a** the marked image, **b** the mask

- As the amount of pictures with bleeding is small, the data augmentation procedure for positive samples has to be processed to balance the number of samples in two classes.

It should be mentioned that the final datasets contain about 80,000 negative samples and about 40,000 positive samples.

The following features (calculated as the mean values for extracted blocks) were extracted and investigated: color and brightness: R, G, B, r, g, b (normalized), B/R, G/R, X, Y, Z, L, a, b, Y, Cr, Cb, H, S, V and texture: Rosenfeld-Troy measure, HOG. The features from different color spaces allows to highlight various characteristic of the samples from both classes. The normalized R, G, B values and their ratios show the presence and weight of different channels in samples.

At the next stage, the obtained features were investigated. The main goals are:

- To eliminate the features with high mutual linear dependence—as they only enlarge the features space dimensionality but don't provide any additional information useful for classification.
- To filter the features, which do not differ from different classes—as they don't provide separability.

The covariation matrix was computed, and the features with high covariance were rejected as they may worsen the learning process. The rest of features were investigated for their separability potential. The final features set (after features filtering) included: H, S, a, b, and Rosenfeld-Troy measure. The covariation matrix of the final features set is displayed in Table 8.8.

It is important to mention, that the final features set seems very reasonable. There are not features connected with brightness, because brightness isn't the real feature of bleeding. The strongest features are connected with hue, saturation, color distribution, and the amount of details. That corresponds to prior expectations.

In this investigation, we used the same set of ML methods as in the task of binary classification for polyps detection: linear discriminant analysis, SVM with radial-based function kernels, RDF, and AdaBoost.

Table 8.8 The covariation matrix of the final features set

	H	S	a	b	Ros
H	1	−0.045	−0.750	−0.425	−0.026
S	−0.045	1	0.199	0.303	−0.145
a	−0.750	0.199	1	0.750	0.089
b	−0.425	0.303	0.750	1	0.007
Ros	−0.026	−0.145	0.089	0.007	1

The last two methods are very good also in case of nonlinear separability of classes. That is why it was the reasonable strategy to include them in investigation.

For classifier effectiveness estimation, we calculate the following metrics for the so-called confusion matrix: True Positives (TP), False Positives (FP), True Negatives (TN), and False Negatives (FN).

The numeric results were obtained by 10-fold cross validation method [34]. In 10-fold cross-validation, the original sample is randomly partitioned into 10 equal sized subsamples. Of the 10 subsamples, a single subsample is retained as the validation data for testing the model, and the remaining 9 subsamples are used as training data. The cross-validation process was then repeated 10 times, with each of the 10 subsamples used exactly once as the validation data. The results were then averaged to produce a single estimation. The obtained results are represented in the Table 8.9.

The achieved metrics can be considered as successfully result of classifier building. Moreover, the additional inference test was provided.

The dataset with 90 images containing bleeding was divided on the training set (70 images) and the test set of 20 images. Additional 20 images without bleeding were added to the test set for correct measurement of the false positive triggering. After building the classifier on the base of training set (70 images with bleeding) the following metrics were achieved for the test set: accuracy equals 0.95, sensitivity equals 0.85, and specificity equals 0.97. The results for reserved test set confirm the success of classifier building and of solving bleeding detection task. Thus, the second task of current research—to investigate different classification strategy for automatic bleeding detection and estimate their effectiveness was successfully solved.

The following inference results (Fig. 8.19) represent the analysis of reserved test set. The main things to mention here:

Table 8.9 The 10-fold cross validation result

Classifier	Accuracy	Sensitivity	Specificity
Linear	0.928	0.742	0.972
SVM	0.943	0.855	0.964
RDF	0.944	0.847	0.968
AdaBoost	0.929	0.788	0.964

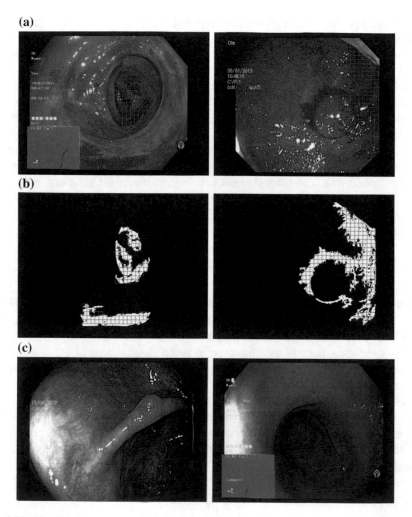

Fig. 8.19 The inference result of bleeding detection: **a** the marked images, **b** their masks, **c** images without bleeding

- The bleeding is successfully detected on the images obtained by different devices in various conditions, shapes, light, and contrast.
- The images from negative class (without bleeding) did not trigger much false positives.

To improve the result, we used single positive block filtering as post-processing procedure. The idea is based on assumption that the bleedings area in overwhelming most cases is manifested as several spatially connected blocks. Thus, the single positive blocks are related to false positives.

The magnitude of the segmentation error in allocating an object of interest is an important characteristic of the segmentation method using blocks. The segmentation error in this case is an incomplete or redundant allocation of the object of interest. In the first case, only the part of the corresponding image is assigned to the object: the selected part is significantly smaller than the area of the object of interest in the image plane. Redundant segmentation is the assignment of image parts to an object that are not relevant to it: the selected part is much larger than the real area of the object in the image plane.

The image of the object of interest resulting from the segmentation is a set of blocks. Even if the bleeding was successful, the segmented image will not exactly match the real one. The latter describes an object with pixel accuracy, segmented with block accuracy. There is a systematic error due to non-compliance of the discretization levels. There is a systematic error caused by non-compliance of the discretization levels:

$$E_{syst} = \left| S_e - N_p^0 * s_b \right|,$$

where N_p^0 is the number of blocks assigned to the object as a result of segmentation, S_e is the area of the object of interest in the image (the number of pixels related to the object), s_b is the block area in pixels.

The systematic error is the difference between the real area of the object in the image and the area of its block approximation, which most closely corresponds to the real image of the object.

Random error is the accuracy of determining the number of blocks related to the object of interest N_p. Blocks that do not belong to it can be assigned to an object of interest or, conversely, the corresponding blocks are not included. Random error in the ith image is the difference between the numbers of blocks assigned to the objects N_p^i and N_p^0:

$$E_r^i = \left| N_p^0 - N_p^i \right|.$$

Figure 8.20 shows an example of an image fragment with an ideal block segmentation and segmentation using a trained classifier. The number of blocks for the ideal case is 25, while the classifier has found 24 blocks. Thus, the random error for a current image fragment is 4%.

Estimates of random and systematic errors across the entire dataset with bleeding for block size of 16 are following:

$$m\left(E_r^i\right) = 11.2\%, \quad m\left(E_{syst}^i\right) = 12.2\%.$$

The random and systematic errors are permissible, which makes it possible to use the block approach in the problem of bleeding segmentation.

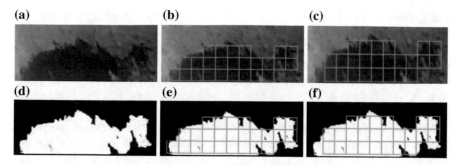

Fig. 8.20 Example of block segmentation: **a** image fragment, **b** ideal segmentation, **c** real segmentation, **d** ground truth segmentation, number of blocks equals 0, **e** ideal block segmentation, number of blocks equals 25, **f** segmentation using a trained classifier, number of block equals 24

The block-based image analysis gives us the possibility to solve at the same time two tasks: sample classification and image segmentation. Here, the general advantages are the following: fast processing and training combined with precision results. The results show high quality of segmentation for images obtained in different condition with various sensors, different light sources, and different diagnosis.

8.6 Conclusions

The main result is the set of new methods for endoscopic images processing and analysis, which can be used as base for CDSS construction. We propose methods aimed as on permitting high effective analysis by physician and on permitting high effective automatic analysis. So, we tried to solve both kinds of tasks actual in CDSS developing.

The important issue for high effective physician analysis is a high quality of images. For solving of this task, we propose methods of noise reduction and image enhancement taking in account main features of endoscopic images and with computational cost permitting real-time realization. The software realization of the proposed algorithms allows to process the endoscopic video with 4K resolution in real time. The achieved speed is about 30 frames per second. The proposed set of new algorithms gives possibility to obtain in real-time the high contrast sharp image of organ under supervision with a high signal/noise ratio.

For diagnostic value image increasing, we propose new method of virtual chromoendoscopy. Method consists of two stages: the first stage is a visualization of tissues and surfaces of mucous membranes including vessels structure stressing and the second stage is a tone enhancement. The method realized possibility to create image with contrast enhancement and color manipulation the subtle image features to be voluminous. The visual effect corresponds or, in the significant set of cases, exceeds visual effect of closed virtual endoscopy technologies I-Scan and FICE.

The experimental test of proposed method was carried out with open KVASIR dataset of endoscopic images. The emphasizing the structural differences in the areas of the tissues are examined. In all 8 classes of images with different diagnosis from KVASIR dataset, FV obtained after method applying was increased in two times in comparing with initial image.

For differential diagnostic realization, we proposed methods for polyp and bleeding detection and segmentation in conditional of small dataset for training. The method of polyp detection is based on combination traditional ML technique (RDF) and CNN. It includes two stage. The stage of polyp detection provides the following characteristics: accuracy equals 0.88, sensitivity equals 0.93, and specificity equals 0.82. The Dice coefficient on the stage of segmentation is about 0.6 for training dataset about 900 images. The obtained value of Dice coefficient is 10% high then Dice coefficient obtained by separate CNN application with the same value dataset. The block-based image analysis proposed for bleeding detection gives a possibility to solve at the same time two tasks: sample classification and image segmentation.

The experimental test on open KVASIR dataset gave next estimation of main characteristics: accuracy equals 0.95, sensitivity equals 0.85, and specificity equals 0.97. Estimates of random and systematic errors in segmentation across the entire dataset with bleeding for a block size of 16 are the following: 11 and 12%. The results show high quality of segmentation for images obtained in different condition with various sensors, different light sources, and different diagnosis.

Thus, our investigation shows that for endoscopic images enhancement and visualization (chromoendoscopy) the effective trend is nonlinear local enhancement application. Also profitable decision is to use different methods of processing for image fragments, which have different level high frequency energy. Such combination gives a possibility to union effective and computational high cost algorithms in real-time. For automatic analysis, main conclusion is in that for medical image processing a problem of verified dataset is strong actual, and one of possible ways here is to combine the traditional ML technology with deep learning technology. It permits to obtain rather good characteristics of images classification for images with high level variability and use in the same time small dataset in training.

References

1. FICE atlas of spectral endoscopic images. https://en.fujifilmla.com/products/endoscopy/catalogs/pdf/index/fice-atlas-esp.pdf. Accessed 8 Aug 2019
2. PENTAX medical I-scan mini-atlas for gastroenterology. https://www.i-scanimaging.com/fileadmin/user_upload/PENTAX_i-scan_Mini-Atlas.pdf. Accessed 8 Aug 2019
3. Han, S., Fahed, J., Cave, D.R.: Suspected blood indicator to identify active gastrointestinal bleeding: a prospective validation. Gastroenterology Res. 11(2), 106–111 (2018)

4. Liao, Z., Gao, R., Xu, C., Li, Z.S.: Indications and detection, completion, and retention rates of small-bowel capsule endoscopy: a systematic review. Gastrointest. Endosc. **71**(2), 280–286 (2010)
5. Jung, Y.S., Kim, Y.H., Lee, D.H., Kim, J.H.: Active blood detection in a high resolution capsule endoscopy using color spectrum transformation. In: International Conference on BioMedical Engineering and Informatics, vol. 1, pp. 859–862 (2008)
6. Pan, G., Xu, F., Chen, J.: A novel algorithm for color similarity measurement and the application for bleeding detection in WCE. Int. J. Image, Graphics and Signal Process. **3**(5), 1–7 (2011)
7. Xiong, Y., Zhu, Y., Pang, Z., Ma, Y., Chen, D., Wang, X.: Bleeding detection in wireless capsule endoscopy based on MST clustering and SVM. IEEE Work. Signal Process. Syst. **35**, 1–4 (2015)
8. Brzeski, A., Blokus, A., Cychnerski, J.: An overview of image analysis techniques in endoscopic bleeding detection. Int. J. Innov. Res. Comput. Commun. Eng. **1**(6), 1350–1357 (2013)
9. Nishimura, J., Nishikawa, J., Nakamura, M., Goto, A., Hamabe, K., Hashimoto, S., Okamoto, T., Suenaga, M., Fujita, Y., Hamamoto, Y., Sakaida, I.: Efficacy of i-scan imaging for the detection and diagnosis of early gastric carcinomas. Gastroenterol. Res. Pract. 1–6 (2014)
10. Münzer, B., Schoeffmann, K., Böszörmenyi, L.: Content-based processing and analysis of endoscopic images and videos: a survey. Multimed. Tools Appl. **77**(1), 1323–1362 (2018)
11. Sheraizin, S., Sheraizin, V.: Endoscopy imaging intelligent contrast improvement. In: 27th Annual International Conference on IEEE Engineering in Medicine and Biology Society, pp. 6551–6554 (2006)
12. Asari, K., Kumar, S., Radhakrishnan, D.: A new approach for nonlinear distortion correction in endoscopic images based on least squares estimation. IEEE Trans. Medical Imaging **18**(4), 345–354 (1999)
13. Barreto, J., Swaminathan, R., Roquette, J.: Non-parametric distortion correction in endoscopic medical images. In: 2007 3DTV Conference, pp. 1–4 (2007)
14. Oulhaj, H., Amine A., Rziza M., Aboutajdine, D.: Noise reduction in medical images—comparison of noise removal algorithms. In: International Conference on Multimedia Computing and Systems, pp. 1–6 (2012)
15. Damiani, E., Dipanda, A., Yetongnon, K., Legrand, L., Schelkens, P., Chbeir, R. (eds.): Signal Processing for Image Enhancement and Multimedia Processing. Springer US, New York, PA (2007)
16. Imtiaz, M.S., Khan, T.H., Wahid, K.A.: New color image enhancement method for endoscopic images. In: 2nd International Conference on Advances in Electrical Engineering, pp. 263–266 (2013)
17. Miyake, Y., Kouzu, T., Takeuchi, S., Yamataka, S., Nakaguchi, T., Tsumura, N.: Development of new electronic endoscopes using the spectral images of an internal organ. In: 13th Color and Imaging Conference, pp. 261–263 (2005)
18. Imtiaz, M.S., Wahid, K.A.: Image enhancement and space-variant color reproduction method for endoscopic images using adaptive sigmoid function. In: Computational and Mathematical Methods in Medicine, pp. 607407.1–607407.19 (2015)
19. Xia, W., Chen, E., Peters, T.: Endoscopic image enhancement with noise suppression. Healthc. Technol. Lett. **5**(5), 154–157 (2018)
20. Imtiaz, M.S., Mohammed, S.K., Deeba, F., Wahid, K.A.: Tri-Scan: a three stage color enhancement tool for endoscopic images. J. Med. Syst. **41**(6), 1–16 (2017)
21. The Kvasir dataset. https://datasets.simula.no/kvasir/. Accessed 8 Aug 2019
22. Akbari, M., Mohrekesh, M., Nasr-Esfahani, E., Soroushmehr, S.M.R., Karimi, N., Samavi, S., Najarian, K.: Polyp segmentation in colonoscopy images using fully convolutional network. arXiv:1802.00368, pp. 1–5 (2018)
23. Khan, T.H., Mohammed, S.K., Imtiaz, M.S., Wahid, K.A.: Color reproduction and processing algorithm based on real-time mapping for endoscopic image. Springerplus **5**(17), 1–16 (2016)

24. Siddharth, V., Bhateja, A.: Modified unsharp masking algorithm based on region segmentation for digital mammography. In: 4th International Conference on Electronics Computer Technology, pp. 63–67 (2012)
25. Wenchao, J., Qi, J.: An improved approximate K-nearest neighbors nonlocal-means denoising method with GPU acceleration. In: Yang, J., Fang, F., Sun, C. (eds.) Intelligent Science and Intelligent Data Engineering, LNCS, vol. 7751, pp. 425–432. Springer, PA (2012)
26. Dabov, K., Foi, A., Katkovnik, V., Egiazarian, K.: Image denoising by sparse 3D transform-domain collaborative filtering. IEEE Trans. Image Process. **16**(8), 2080–2095 (2007)
27. Vonikakis, V., Andreadis, I.: Multi-scale image contrast enhancement. In: 10th International Conference on Control, Automation, Robotics and Vision, pp. 17–20 (2008)
28. Tao, L., Asari, V.K.: Adaptive and integrated neighborhood dependent approach for nonlinear enhancement of color images. SPIE J. Electron. Imaging **14**(4), 1–14 (2005)
29. Arigela, S., Asari, V.K.: A locally tuned nonlinear technique for color image enhancement. WSEAS Transl. Signal Process. **4**(8), 514–519 (2008)
30. Obukhova, N., Motyko, A., Alexandr Pozdeev, A.: Review of noise reduction methods and estimation of their effectiveness for medical endoscopic images processing. In: 22nd Conference on FRUCT Association, pp. 204–210 (2018)
31. Obukhova, N., Motyko, A.: Image analysis in clinical decision support system. In: Favorskaya, M.N., Jain, L.C. (eds.) Computer Vision in Control Systems-4, ISRL, vol. 136, pp. 261–298. Springer International Publishing, Switzerland (2018)
32. Yadav, G., Maheshwari, S., Agarwal, A.: Contrast limited adaptive histogram equalization based enhancement for real time video system. In: 2014 International Conference on Advances in Computing, Communications and Informatics, pp. 2392–2397 (2014)
33. Pogorelov, K., Randel, K.R., Griwodz, C., Eskeland, S.L., de Lange, T., Johansen, D., Spampinato, C., Dang-Nguyen, D.-T., Lux, M., Schmidt, P.T., Riegler, M., Halvorsen, P.: Kvasir: a multi-class image dataset for computer aided gastrointestinal disease detection. In: 8th ACM on Multimedia Systems Conference, pp. 164–169 (2017)
34. Shen, C.H., Chen, H.H.: Robust focus measure for low-contrast images, consumer electronics. In: 2006 Digest of Technical Papers International Conference on Consumer Electronics, pp. 69–70 (2006)
35. Manjunath, B.S., Salembier, P., Sikora, T. (eds.): Introduction to MPEG-7: Multimedia Content Description Interface. Wiley (2002)
36. Bosch, A., Zisserman, A., Munoz, X.: Representing shape with a spatial pyramid kernel. In: 6th ACM International Conference on Image and Video Retrieval, pp. 401–408 (2007)
37. Chatzichristofis, S.A., Boutalis, Y.S., Lux, M.: Selection of the proper compact composite descriptor for improving content based image retrieval. In: Signal Processing, Pattern Recognition and Applications, pp. 134–140 (2009)
38. Tamura, H., Mori, S., Yamawaki, T.: Textural features corresponding to visual perception. IEEE Trans. Syst., Man, Cybern. **8**(6), 460–473 (1978)
39. Huang, J., Kumar, S.R., Mitra, M., Zhu, W.-J., Zabih, R.: Image indexing using color correlograms. In: IEEE Computer Society Conference on Computer Vision and Pattern Recognition, pp. 762–768 (1997)
40. Flach, P.: Machine Learning: The Art and Science of Algorithms That Make Sense of Data. Cambridge University Press, New York, NY, USA (2012)
41. CVC Colon DB. http://mv.cvc.uab.es/projects/colon-qa/cvccolondb. Accessed 8 Aug 2019
42. Dice, L.R.: Measures of the amount of ecologic association between species. Ecology **26**(3), 297–302 (1945)

Chapter 9
Tissue Germination Evaluation on Implants Based on Shearlet Transform and Color Coding

**Aleksandr Zotin, Konstantin Simonov, Fedor Kapsargin,
Tatyana Cherepanova and Alexey Kruglyakov**

Abstract The chapter is devoted to computational methods for evaluation the indicators of the tissue regeneration process using as an example the medical data of mesh nickelide titanium implants obtained during clinical experiment. Processing and analysis of scanning electron microscopy and classical histological data are performed using a set of algorithms and their modifications, which allows simplify the data analysis procedure and improve the accuracy of estimates (15–20%). The proposed technique as a computational toolkit for analyzing the dynamics of the process under study, as well as. For highlighting the internal geometric features of the experimental images of objects of interest contains algorithms of shearlet and wavelet transforms and the algorithms for elastic maps generation with color coding, which allows to obtain more representative visualization of spatial data. An important aspect of the proposed methodology is a use of brightness correction by algorithm based on Retinex technology. It allows to obtain unified average brightness of analyzed images and, in some cases, increase local contrast, as a result

A. Zotin (✉)
Reshetnev Siberian State University of Science and Technology,
31 Krasnoyarsky Rabochy pr., 660037 Krasnoyarsk, Russian Federation
e-mail: zotin.sibsau@gmail.com

K. Simonov
Institute of Computational Modelling of the Siberian Branch of the Russian
Academy of Sciences, 50/44 Akademgorodok, 660036 Krasnoyarsk, Russian Federation
e-mail: simonovkv50@gmail.com

F. Kapsargin · T. Cherepanova
V.F. Voino-Yasenetsky Krasnoyarsk State Medical University, 1 Partizana Geleznyaka st.,
660022 Krasnoyarsk, Russian Federation
e-mail: kapsargin@mail.ru

T. Cherepanova
e-mail: grakova@list.ru

A. Kruglyakov
Siberian Federal University, 79 Svobodny pr., 660041 Krasnoyarsk, Russian Federation
e-mail: piggsyy@gmail.com

© Springer Nature Switzerland AG 2020
M. N. Favorskaya and L. C. Jain (eds.), *Computer Vision in Advanced Control
Systems-5*, Intelligent Systems Reference Library 175,
https://doi.org/10.1007/978-3-030-33795-7_9

it affects the quality of application of the computer-based evaluation tools offered in the work. Thus, the estimation errors are reduced by 1–5% in compared to processing without brightness correction.

Keywords Medical image · Image processing · Noise reduction · Retinex · Wavelet transform · Shearlet transform · Median filter · Elastic maps

9.1 Introduction

Nowadays, the use of mesh implants is widely used in various fields of surgery [1–5]. Studies for the histological evaluation of the suitability of various materials as implants are conducted in various countries [6–12]. One of the tasks of modern hernioplasty solved by the use of implants is the reconstruction of the defects of the anterior abdominal wall, i.e. the surgical treatment of hernial defects [13, 14]. As part of this task, experiments using different materials, as well as various methods of histological tissue analysis are carried out [15, 16]. Currently, polypropylene and polytetrafluoroethylene implants are widely used all over the world. Despite the fact that synthetic materials presented on the market are chemically inert, they do not possess super elastic properties inherent to living organisms. Polymers implanted into tissues cause a chronic inflammatory reaction of the periprosthetic tissues with the possible formation of "fibrous tissues" [17, 18].

In recent years, research has been conducted on the use for implantation of a class of super-elastic alloys based on titanium nickelide at Kuznetsov Tomsk State University, Siberian Physical-Technical Institute under the leadership of Gunter [1, 2]. These implants were created solely for medical purposes. The works [15, 16] show the ability of nickel-plated titanium meshes after implantation to integrate the complex "living tissue-implant" as a single reinforced area with high mechanical strength. Modern technology of reconstruction of the defects of the anterior abdominal wall using biocompatible superelastic mesh implants from nickel titanium is an urgent task [13, 14]. Previously, numerous studies have been conducted on the reaction of tissue to the implantation of various forms (monolithic, porous, reticular) titanium nickelide when placed in organs and tissues [4, 5].

Currently studies are being conducted in the area of implantation of mesh titanium nickelide in the body tissue [16, 19]. The studies carried out on rabbits of the Chinchilla breed. The obtained experimental material is subjected to traditional histological study, as well as, analysis under an electron scanning microscope. Quantitative histological analysis along with scanning electron microscopy is used for tissue analysis by various researchers [20–23]. Herewith, the evaluation is made taking into account a specific set of morphological and morphometric parameters [20].

One of the most common solutions as a computational tool, for the analysis of graphic data, is the wavelet and shearlet transformation [24–28]. Thus, in the works [29, 30] the methods of complex processing of experimental clinical data are described. Shearlet-transformation is used as basis of multidimensional analysis of spatial data and medical images for solving histological tasks.

Taking into account the research experience in this subject area of quantitative histology and algorithmic developments in the field of spectral decomposition and visualization of experimental clinical data, a new algorithmic solution to the problem of quantitative morphological (geometric) representation of visual data is proposed. It is based on the use of shearlet transformations and color-coding [31]. This solution is aimed at reducing the complexity and time-consuming of natural experiment, while analyzing the dynamics of tissue growth on implants, as well as, formation of more clearly distinguish the internal geometric features of objects of interest in experimental images.

The remainder of this chapter is organized as follows. In Sect. 9.2, the related work is analyzed. Section 9.3 describes medical experiment and analyzing data. Section 9.4 presents the proposed method of medical images analysis. Section 9.5 includes description of experimental research and analysis of obtained results. The conclusions are given in Sect. 9.6.

9.2   Related Work

Modern studies in the field of quantitative histology to assess the survival of the implant are carried out using classical histological analysis and advanced approaches using scanning electron microscopy. In this case, mathematical methods are widely used to solve the problem of interpretation of measurements made on slices [20–23]. Quantitative methods in histology are aimed at increasing the objectivity of the evaluation of tissue characteristics, the development of new criteria for assessing tissue function, as well as, markers of pathological processes [22, 32–34]. The approaches of quantitative histological analysis allows to find the relationship between the structure and function of tissues, while increasing the accuracy of the estimates and reducing the influence of the subjective factor on the results of the analysis [35–39]. This creates the conditions for the algorithmization and automation of the procedure of clinical research and diagnostics.

The material obtained during the experimental histological research must be presented in a convenient form so that a medical specialist can give a correct assessment. Usually microscopic digital photographs of stained tissue are used. The material (images) obtained by scanning electron microscopy and microscopic digital photograph depend on the quality of technical equipment. Quite often defects appear on digital images caused by the presence of noise, irregular contrast, and low brightness, which obviously reduces the quality of the analysis and the accuracy of the obtained estimates.

In this regard, the first step in the analysis of medical images is a preliminary processing aimed at improving an image's quality. Within the framework of pre-processing, two main tasks are solved: noise reduction and brightness adjustment (leveling) with contrast correction. Problem of noise reduction can be solved using various filters. Thus, many different studies in the field off noise reduction of medical images are being conducted.

In [40], different medical images like MRI, cancer, X-ray, and brain images were studied. The authors tried out various median base filtering techniques to suppress salt and pepper noise in MRI and X-ray images. Sudha et al. [41] recommends a thresholding algorithm for denoising speckle noise in ultrasound images with wavelets. An improved adaptive median filtering method for denoising impulse noise was used by Mamta et al. [42].

Arin et al. [43] studied different filters, such as average (mean), Gaussian, and median, for the noise reduction in medical images affected by different types of undesired noises. However authors have not been found a universal solution. Based on the data, Kanmani et al. [44] believed that median filter was best in cancer histology noise removal. To improve contrast and brightness distribution researchers have widely used histogram equalization techniques as a powerful preprocessing tool [45–47]. Apart from histogram equalization, Retinex-based algorithms are also used to improve quality of medical images [48–50]. Biswajit et al. [48] demonstrated good suitability of Retinex-based algorithm for microscopic image enhancement.

As part of image analysis, it is required to evaluate the experimentally determined features of germinated tissue on the implant and its surroundings. To solve such problems, methods of geometric (vector) image analysis are used. Thus, for allocation of linear and non-linear features (edges, borders, and contours of the objects) the methods such as Sobel, Prewitt, Roberts, Canny, Laplacian of Gaussian (LoG) [51–55] and shearlet transform [24–31] can be used.

The papers and reviews [21–23, 56] describe computational schemes for performing specific studies using particular specialized algorithms as solution. The described algorithms for performing quantitative histological analysis make it possible to effectively find the dependencies between the structure and function of tissues, increase the accuracy of the estimates, reduce the influence of the subjective factor on the analysis results, and automate the research and diagnostic procedures.

9.3 Medical Experiment Description

Experimental studies were conducted on 39 rabbits of the Chinchilla breed, of both sexes. The average weight of rabbits was 2,700–3,000 g, between the ages of 6 months and 1.5 years. The choice of rabbits as experimental animals was due to the best fit of the biological response of animal tissues to the implantation of synthetic material and the response of human tissues, and, at the same time, providing an opportunity to obtain a sufficient amount of biopsy material of a relatively large size.

For the study, animals without external signs of disease were selected, quarantined under vivarium conditions. The experiment involved 3 comparison groups of 11 animals each and one control group including 6 animals.

The following materials were used in the laboratory study:

- PDS plus thread is a monofilament synthetic absorbable suture material, which is a polydioxanone thread containing polydioxanone and triclosan.
- The standard "Esfil", is a classic mesh endoprosthesis made from monofilament polypropylene.
- Mesh implant based on titanium nickelide (brand TN-10), which has bio-chemical and biomechanical compatibility with mobile anatomical structures of the body.

The greatest interest to medical specialists was a result of experiment on a group using an implant woven by textile technology from superelastic nickel-titanium yarn (brand TH-10) 5 × 4 cm in size, 60 μm thick, with cell sizes of 120–240 μm. The filament is a composite material comprising a core of nanostructured mono-lithic titanium nickelide and a porous surface layer (5–7 μm) of titanium oxide.

As a part of the histological study, a section of the anterior abdominal wall of the animal in the implant zone of the graft in the comparison groups was subjected to morphological analysis. In each case, a 1.5 × 1.5 cm full-thickness anterior abdominal wall flap with an implant and surrounding tissue/es was excised. The sections were stained with hematoxylin eosin, silver, according to the method of Van-Gizona and Pikro-Mallori, and then subjected to Axioskop 40 microscopy and a digital camera.

Classical counting was performed on Bimam microscope at a magnification of 10 × 1.5 × 20 and 10 × 15 × 40 using an ocular micrometer and a grid with equidistant points. To standardize the data, morphometry of tissue structures was performed according to the guidelines of Avtandilov [20].

The histomorphometric indicators of tissue structures in the zone of former defects were obtained in the marginal, middle, and central zones. They were sum-marized, and the average value was calculated. During the histological study, the connective tissue cells (fibroblasts, fibrocytes), cell elements of the inflammatory series, amorphous component, vessel size, thickness of fibrous elements, and elastic and collagen fibers in the studied group and comparison groups were evaluated. As a separate parameter, signs of chronic inflammation—the presence of giant cells were evaluated. An example of histological images obtained on the 7th day of the experiment is shown in Fig. 9.1. The images demonstrate changes in the tissues after suturing PDS thread. The tissues were stained with hematoxilin and eosin.

Additional medical study was conducted with scanning electron microscope. It has certain advantages over transmission electron microscopes. One of them is a simpler sample preparation method. The preparation of the samples did not require the manufacture of ultrathin sections and staining in the three experimental groups. In Group III, the nickelide titanium filament was not removed from the samples, thus the microscopic structure was not damaged, and the orientation of collagen fibers formed during the growth and maturation in the porous structure of the nickelide titanium implant was preserved.

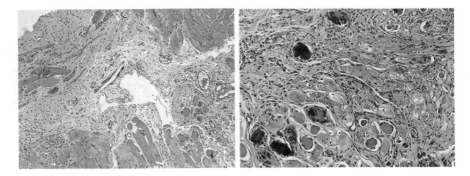

Fig. 9.1 Example of histological images

As a part of data evaluation using scanning electron microscopy, a study of implant fragments of all three groups was performed using Hitachi TH-3500 scanning electron microscope at the Institute of Physics named after academician Kirensky, Center for collective use of the Krasnoyarsk Scientific Center of the Siberian Branch of the RAS.

Determination of the degree of tissue germination in the mesh structures of an implant based on titanium nickelide was carried out on a high resolution microscope TM 3500 electron microscope up to × 30,000 using QUANTAX 70 EDS software. Examples of images obtained using electron scanning microscopy are shown in Fig. 9.2. The figure shows the germination of tissue on the implant on the 7th and 14th day of implantation. On the images can be seen that the titanium nickelide mesh implant is surrounded by granulation tissue. The arrows indicate: 1 —bundles of collagen fibers around the implant, 2—red blood cells.

In the case of each period (day) of sampling, 10–15 shots were taken. This in turn significantly increased the amount of work for researchers. The most time consuming evaluation of tissue germination on an implant is the calculation of the area of collagen fibers, the number of red blood cells, etc. indicators demonstrating the dynamics of the studied process. To speed up and improve the accuracy of calculations, it was decided to use a processing based on shearlet transform and color coding [31].

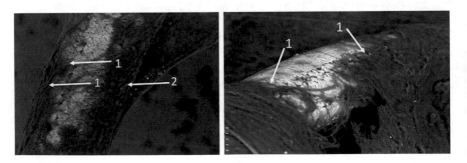

Fig. 9.2 Example of electron scanning microscopy images

9.4   The Proposed Method of Medical Images Analysis

Purpose of the proposed computational methodology for the quantitative assessment of the main histological indicators is to improve the accuracy of the analysis of experimental visual data. This is achieved by modifying and adapting the algorithmic support of the solution, which allows performing quantitative morphological analysis based on the shearlet transform. In addition, this technique is proposed to be used to assess the indicators from images obtained by electron scanning microscopy.

Thus, the method [31] was taken as the basis, where for the approximation and spectral decomposition of shearlet transform known as Fast Finite Shearlet Transform (FFST) is used, since the results of the analysis may be distorted due to the presence of noise arising in the digital matrix or the nature of the lighting area of the test sample. We propose to use the following generalized phases for medical image analysis:

1. Definition of parts of the object-structure under study, within which the structural and textural features of the reference objects are determined.
2. Application of pre-processing of the image in order to suppress the noise resulting from the formation of the image, as well as, to correct the brightness and local contrast of the analyzed images.
3. Transformation of the image to the form, which is convenient for visual interpretation and further analysis. At this stage, the parameters of the calculation algorithm of the shearlet transform and the color coding map are determined.
4. Simulation of computed images based on the shearlet transform and defined color coding scale for obtaining optimal segmentation and color visualization of specific objects of interest.
5. Quantitative evaluation of the geometric, textural and morphological characteristics of the selected objects of interest to solve the stated substantive problem.

Taking into account described phases for medical image analysis, the following steps to computer processing of the medical images are suggested (Fig. 9.3):

Step 1   Applying a noise reduction filter.
Step 2   Conducting brightness correction and contrast enhancement.
Step 3   Forming a contour representation.
Step 4   Color coding of the objects in the contour representation and formation of edge and region maps.
Step 5   Conducting a data analysis.

The generalized form of processing procedure to process medical images is shown in Fig. 9.3. In view of the proposed method the data analysis step depends on the data (images) and tasks.

The analysis of the medical images for clinical experiments on implants consists in revealing the geometric features of the texture, extracting fibers, and vessels of

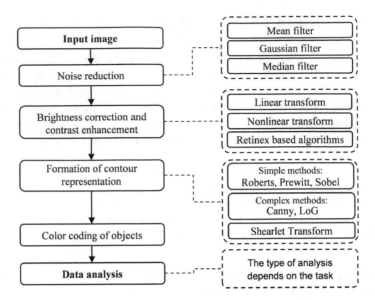

Fig. 9.3 Block diagram of the proposed method of medical images analysis

different types and their quantifying, as well as, segmentation and contouring amorphous formations with the estimation of their sizes and forms. When assessing the germination of tissues, the area of coverage of the corresponding implant is also taken into account.

Sections 9.4.1 and 9.4.2 provide information about preprocessing steps of the proposed method. Section 9.4.1 briefly describes filter selected for noise reduction, while Sect. 9.4.2 contains information of brightness correction algorithm based on Retinex technique. The next Sect. 9.4.3 briefly describes step of forming a contour representation using conventional methods and discrete shearlet transform. Section 9.4.4 gives a description of final step before data analysis i.e. color coding of objects.

9.4.1 Noise Reduction

At the preprocessing stage, as the first stage, noise suppression arising due to equipment imperfections is performed. Various filters are used for image preprocessing. The primary purpose of these filters is a noise reduction, but filter can also be used to emphasize certain features of an image or remove other features. Almost all contemporary image processing involves discrete or sampled signal processing. Most of image processing filters can be divided into two main categories: linear filters and nonlinear filters. Nonlinear filters include order statistic filters and adaptive filters. The choice of filter is often determined by the nature of the task and the type and behavior of the data. Noise, dynamic range, color accuracy, optical artifacts, and many more details affect the outcome of filter in image processing.

The median filter was chosen as the main filter for noise suppression due to the fact that it shows the best performance in suppressing impulse noise and performs processing, while maintaining the sharpness of the borders of contrasting objects. The median filter, as its name implies, replaces the value of the pixel spectrum intensity by the median of the spectrum intensity values in the neighborhood of this pixel.

Mathematically, the median filter can be described by Eq. 9.1 where I_{new} and I_{old}, are the new and old values of the image pixels spectrum, respectively, K_{xy} is the kernel window with the dimensions $K_{Hs} \times K_{Ws}$ centered at (x, y). The original pixel value is included in the computation of a median.

$$I_{new}(x, y) = \underset{(kx,ky) \in K_{xy}}{med} \{I_{old}(kx, ky)\} \tag{9.1}$$

As an auxiliary filter for suppressing small additive noise, it was decided to use a performance optimized Gaussian filter. Gaussian filter is a convolution filter, in which computations are conducted according to Eq. 9.2, where G is the Gaussian function, RH and RW are the constants defining the rank of the filter vertically and horizontally, respectively.

$$I_{new}(x, y) = \sum_{dy=-RH}^{RH} \sum_{dx=-RW}^{RW} G(dx, dy) \times I_{old}(x + dx, y + dy) \tag{9.2}$$

The equation of Gaussian function in one dimension is expressed by Eq. 9.3. For two dimensions, it is the product of two such Gaussians, one in each dimension (Eq. 9.4), where x is the distance from the origin in the horizontal axis, y is the distance from the origin in the vertical axis, σ is the standard deviation of the Gaussian distribution.

$$G(x) = \frac{1}{\sqrt{2\pi} \cdot \sigma} e^{-\frac{x^2}{2\sigma^2}} \tag{9.3}$$

$$G(x, y) = \frac{1}{2\pi \cdot \sigma^2} e^{-\frac{x^2 + y^2}{2\sigma^2}} \tag{9.4}$$

The convolution can be performed fairly quickly because the equation for 2D isotropic Gaussian is separable into y and x components [57]. Thus, instead of one two-dimensional convolution, two one-dimensional convolutions can be applied (in rows and columns).

9.4.2   Brightness Correction with Contrast Enhancement

Medical images may have different brightness characteristics depending on the shooting conditions. For example, a series of images may have low or high brightness and contrast. In the proposed method to improve the image quality, it was decided to use a modification of Multi-Scale Retinex (MSR) in Hue, Saturation, Value (HSV) color space, in which wavelet transform was used to accelerate the calculation [58].

As a result of the application of the wavelet transform to the image, four areas of LL, HL, LH, and HH are formed by alternating one-dimensional wavelet transform of rows and columns. To calculate the corrected brightness values by MSR algorithm, LL region is used and then the inverse wavelet transformation is performed. During the inverse transform, high-frequency components (detail) can additionally increase the level of local contrast.

Contrast control (decrease or increase) is possible due to the usage of two intensity correction coefficients of high-frequency components. The first coefficient (k_{div}) is responsible for uniform correction of values, and the second (k_h) allows to conduct linear correction in case of exceeding the local contrast threshold (T_H). To reduce the computational complexity in the software implementation of wavelet transform, it was decided to use Look-Up Table (LUT), which is formed according to Eq. 9.5.

In this case, the *detail* (*H*) component, when processing a one-dimensional discrete signal $S = \{s_j\}_{j \in Z}$, will be formed depending on the difference of neighboring pixels ($S_{2j}-S_{2j+1}$) using Eq. 9.6. It should be noted that the inverse wavelet transform remains unchanged.

$$H_{LUT}(i) = \begin{cases} \frac{i}{k_{div}} & if \quad i \leq T_H \\ \frac{(i-T_H) \cdot k_h + T_H}{k_{div}} & if \quad i > T_H \end{cases} \tag{9.5}$$

$$H_j = sgn(S_{2j} - S_{2j+1}) \cdot H_{LUT}\left(\left|S_{2j} - S_{2j+1}\right|\right) \tag{9.6}$$

The response of MSR function (R_{MSR}) typically gives both negative and positive values, and the obtained range limits will be arbitrary. Taking this into consideration, there is a need to convert the obtained values into the display domain i.e. intensity range [0, 255]. Depending on the source image in the distribution of the output values of MSR function, the average value can be shifted relative to zero. Therefore, for the formation of the output image, it was decided to use the range stretching taking into account the adaptive adjustment of the stretching degree. The adjustment is based on the boundary thresholds taking into account the desired range size (*Pr*) and the shift of the distribution center (alignment with zero). Thus, the calculation of output image brightness value (I_{MSR}) of a pixel with the desired range of visualization (I_{TR}) is conducted using Eq. 9.7, where k_{offset} is the brightness

(a)   **(b)**

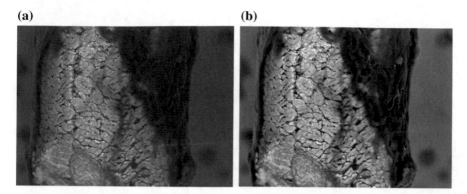

Fig. 9.4 Example of Retinex algorithm image processing: **a** original image, **b** image processed by modified Retinex algorithm

offset factor (by default, $k_{offset} = 127$), $Cl(\cdot)$ is the cut-off function for values outside the desired range.

$$I_{MSR}(x, y, \boldsymbol{\sigma}) = Cl\left(\frac{R_{MSR}(x, y, \boldsymbol{\sigma}) - \text{avg}(R_{MSR}(\boldsymbol{\sigma}))}{(\max(R_{MSR}(\boldsymbol{\sigma})) - \min(R_{MSR}(\boldsymbol{\sigma}))) \cdot Pr} \cdot I_{TR} + k_{offset}\right) \quad (9.7)$$

The example of quality improvement for one image from raster microscopy image set using modified MSR algorithm is shown on Fig. 9.4. A clearer visual separation of the titanium implant and tissues (collagen fibers, erythrocytes and etc.) can be observed in the processed image (Fig. 9.4b).

9.4.3 Formation of Contour Representation

Contour representation can be generated using different algorithms. We can use classic simple methods such as Roberts, Prewitt, Sobel, and more complex, for example Canny and LoG [51–55].

Roberts cross operator provides a simple approximation to the gradient magnitude. It uses a 2×2 kernel. Method, which uses Sobel or Prewitt operators, computes an approximation of gradients along the horizontal (x) and the vertical (y) directions (2D spatial) of the image intensity function at each pixel and highlights regions corresponding to edges. It computes the approximations of the derivatives using two 3×3 kernels (masks) in order to find the localized orientation of each pixel in an image. The difference between Prewitt and Sobel kernels are kernel coefficients. Typically, formation of LoG contour representation consists of complex calculation. However, LoG may be implemented by precalculated kernel, which requires far fewer arithmetic operations.

Thus, 2D LoG function centered at zero and with Gaussian standard deviation σ has a form of Eq. 9.8.

$$LoG(x, y) = -\frac{1}{\pi \cdot \sigma^2}\left[1 - \frac{x^2 + y^2}{2\sigma^2}\right]e^{-\frac{x^2+y^2}{2\sigma^2}} \tag{9.8}$$

Canny algorithm for the contour detection uses the following steps: image smoothing, searching gradients, suppression of "false" peaks, double threshold filtering (potential contours defined by the thresholds), and tracing the area of ambiguity.

The use of these methods depending on the source image can give different results, and in most cases additional processing of contour representation will be required for a more successful analysis. Previous research shows that shearlet transform provides more effective way to describe the characteristics of the directivity of image elements based on affine transformation. Shearlets possess a uniform construction for both the continuous and the discrete setting. They further stand out since they stem from a square-integrable group representation and have the corresponding useful mathematical properties.

The shearlet ($\psi_{a,s,t}$) is generated by the dilation, shearing, and translation of a function $\psi \in L^2(R^2)$ called the mother shearlet in the following way (Eq. 9.9):

$$\psi_{a,s,t}(x) = a^{-\frac{3}{4}}\psi\left(A_a^{-1}S_s^{-1}(x - t)\right), \tag{9.9}$$

where $t \in R^2$ is the translation, A_a is the scaling (or dilation) matrix, S_s is the shearing matrix defined according to Eq. 9.10.

$$A_a = \begin{pmatrix} a & 0 \\ 0 & \sqrt{a} \end{pmatrix} \quad a \in R^+ \quad S_s = \begin{pmatrix} 1 & s \\ 0 & 1 \end{pmatrix} \quad s \in R \tag{9.10}$$

The anisotropic dilation A_a controls the scale of the shearlets by applying a different dilation factor along the two axes. The shearing matrix S_s, not expansive, determines the orientation of the shearlets. The normalization factor $a^{-3/4}$ ensures that $\|\psi_{a,s,t}\| = \|\psi\|$, where $\|\psi\|$ is the norm in $L^2(R^2)$. In the classical setting, the mother shearlet ψ is assumed to factorize in Fourier domain as Eq. 9.11:

$$\hat{\psi}(\omega_1, \omega_2) = \hat{\psi}_1(\omega_1)\hat{\psi}_2\left(\frac{\omega_2}{\omega_1}\right), \tag{9.11}$$

where $\hat{\psi}$ is the Fourier transform of ψ, $\hat{\psi}_1$ is one dimensional wavelet, $\hat{\psi}_2$ is any non-zero square-integrable function.

The shearlet transform $SH(f)$ of a signal $f \in L^2(R^2)$ is defined by Eq. 9.12, where $<f,\psi_{a,s,t}>$ is the scalar product in $L^2(R^2)$.

$$SH(f)(a, s, t) = \langle f, \psi_{a,s,t} \rangle \tag{9.12}$$

For the calculations s of the shearlet transform, algorithm referred to as Fast Finite Shearlet Transform (FFST) and presented in [24] is used. It is based on the discrete fast direct and inverse Fourier transformation. Meyer wavelet is used for the present implementation as the mother wavelet.

For finite discrete shearlets, digital images in $R^{M \times N}$ are considered as functions sampled on the grid. The discrete shearlet transform not only discretize the involved parameters a, s, and t but also consider only a finite number of discrete translations t. Additionally, they discretizes the translation parameter t on a rectangular grid and independent of the dilation and shear parameter. To obtain a discrete shearlet transform, let us note that a number of considered scales equals $j_0 = [\frac{1}{2}\log_2 \max \{M, N\}]$. Then, the dilation, shear, and translation parameters are discretized by Eqs. 9.13–9.16.

$$a_j = 2^{-2j} = \frac{1}{4^j}, \quad j = 0, \ldots j_0 - 1 \tag{9.13}$$

$$s_{j,k} = k2^{-j} = \frac{1}{4^j}, \quad -2^j \leq k \leq 2^j \tag{9.14}$$

$$t_m = \left(\frac{m_1}{M}, \frac{m_2}{N}\right), \quad m \in G \tag{9.15}$$

$$G = \{(m_1, m_2) : m_1 = 0, \ldots, M - 1, m_2 = 0, \ldots, N - 1\} \tag{9.16}$$

With these notations, shearlets become as Eq. 9.17.

$$\psi_{j,k,m}(x) := \psi_{a_j, s_{j,k}, t_m}(x) = \psi\left(A_{a_j, \frac{1}{2}}^{-1} S_{s_{j,k}}^{-1}(x - t)\right) \tag{9.17}$$

Now the discrete shearlet transform is defined as:

$$SH(f)(j, k, m) = \langle f, \psi_{j,k,m} \rangle. \tag{9.18}$$

Studying FFST algorithm shows that the object contours can be obtained as a sum of the coefficients of the shearlet transform for the fixed scale and all the possible values of the shift and bias parameters. In this connection, using this peculiarity in solving the given approximation problem is suggested.

The contours of the objects can be obtained as a sum of the coefficients of the shearlet transform fixed parameter values for the last shift of the scale and various parameter values as shown in Eq. 9.19, where SH_ψ is the shearlet coefficient for the scale j^*, cornering (orientation) k and displacement m, k_{max} is the maximum number of turns, m_{max} is the maximum amount of displacement.

$$f_{cont} = \sum_{k=0}^{k_{max}} \sum_{m=0}^{m_{max}} SH_\psi(f(j^*, k, m)) \tag{9.19}$$

An alternative approach for calculating the shearlet transform referred as ShearLab is described in [28].

9.4.4 Color Coding of Objects

In order to better understand represented medical data and simplify further analysis, it is proposed to use the method of contrasting medical images based on color coding. This method allows to attract the attention of a medical specialists to certain areas of the analyzed image. At the same time, it conducts a preparation of further analysis, during which the tasks needed for quantitative histology are solved.

For color coding (selection of colors and distribution of contour density in the image), an algorithmic procedure is used with the application of the method of constructing an elastic map from spatial data [31, 59]. It is assumed that the analyzed object is a limited two-dimensional variety embedded in a set of studied data. In this case, the feature map of object is a part of the curved surface located inside the data cloud of the analyzed image.

Consider a two-dimensional rectangular grid of nodes, where there are p nodes horizontally and q vertically. The nodes of the grid will be enumerated using two indices $y_{i,j}$, $i = 1,...,p$ $j = 1,...,q$. The constructed grid is located in the set of the data points that each point should be associated with the nearest node of the grid. The data points have a form $t_{i,j}(x, y, M)$, where (x, y) are their relative coordinates, M is the intensity (color) characterizing the peculiarities of the object under study.

Such a method allows partitioning a set of data into pq subsets of $K_{i,j}$ with the subset points being closer to $y_{i,j}$ than to any other node. Let $K_{i,j}$ be as Eq. 9.20.

$$K_{i,j} = \left\{ t | t \in P_k, \left\| y_{i,j} - t \right\|^2 \leq \varepsilon \right\} \tag{9.20}$$

In the case of Euclidean metric, the functional D (quality functional) is quadratic in the position of the nodes $y_{i,j}$. This means that with the given set of data points partitioned into taxa, it is necessary to solve the system of linear equations $pq \times pq$ for its minimization.

It should be noted that implementation of the computational scheme takes into account a flat image character, so the functional D as a measure of the grid proximity to the data point in simplified form can be represented by Eq. 9.21, where the $|P_k|$ is the number of points in cell.

$$D = \frac{\sum\limits_{i,j} \sum\limits_{t_n \in K_{ij}} \left\| t_n - y_{i,j} \right\|^2}{|P_k|} \rightarrow \min \tag{9.21}$$

In generalized form the algorithm can be represented as follows:

Step 0. The grid nodes are anyway located in the data space.
Step 1. Given the positions of the grid nodes the data set is partitioned into taxa—subsets K_{ij}.
Step 2. With the given partition of the set of data points into taxa the functional D is minimized.

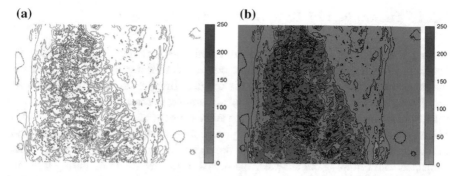

Fig. 9.5 Example of the color coding of image from Fig. 9.4b: **a** edge map, **b** regions' map

Step 3. Steps 1 and 2 are repeated until the functional D stops sensibly changing.

In the framework of the implementation of color coding, an object of interest is represented as the initial (central) part of the grid construction, with respect to which essential isolines are formed. As a result, a map is formed, which is further colored according to the physical attribute for the studied data.

An example of visual data generated during the application of the proposed technique using color coding according to the "Cool" scheme is shown in Fig. 9.5.

Figure 9.5 shows two representations of the medical image processing results. They are presented in the form of colored contours and filled regions of the visual data under study. The definitive result of image processing is a clear selection of both the threads of the implant itself and characteristics of tissue germination on it.

9.5   Experimental Research

Evaluation of the proposed methodology was carried out using data obtained during medical experiments on rabbits of the Chinchilla breed, which were conducted at Voino-Yasenetsky Krasnoyarsk State Medical University, Department of Surgery. The purpose of the medical experiment was to identify the features of the tissue reaction depending on the material used with the same method of implantation.

Section 9.5.1 contains a description of experimental medical data, which were used to evaluate proposed method. The evaluations of brightness characteristics of medical images data set are presented in Sect. 9.5.2, also in the section showed examples of brightness correction and obtained results for further analysis. The estimations of medical experts, as well as, the experimental results based on the proposed techniques are presented in Sect. 9.5.3. The efficiency evaluation of the proposed method from a viewpoint of medical experts is presented in Sect. 9.5.4.

9.5.1 Description of Experimental Medical Data

During the medical experimental research, histological data were obtained for analysis, as well as images of scanning electron microscopy. For the evaluation of the proposed technique, more than 40 histological images allowing to evaluate the morphological changes of tissues in the implantation zone were taken. Obtained images have different resolution, which varies from 1292 × 968 to 3264 × 2448 pixels. As a part of the histological study, a section of the anterior abdominal wall of the animal in the implant zone of the graft in the comparison groups was subjected to morphological analysis. Examples of images obtained during different days of medical experiment are shown in Fig. 9.6. They demonstrate a histological data characterizing the change in tissues after the implantation of an endoprosthesis based on titanium nickelide at 7, 14, 21, and 45 days, respectively.

It can be seen that the chromatic, as well as, the brightness and contrast characteristics of the images obtained during medical experiments are very different. Such instabilities characteristic of original images lead to the need for careful adjustment of the algorithms for the qualitative analysis of images.

(a)   **(b)**

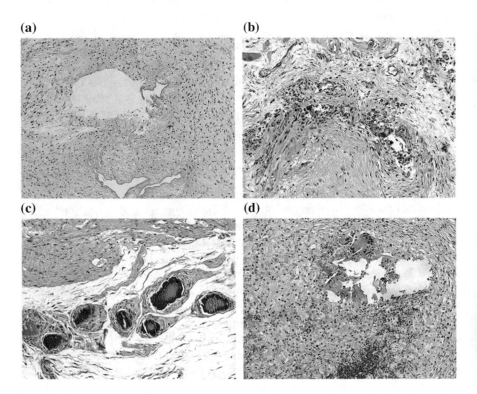

(c)   **(d)**

Fig. 9.6 Histological data characterizing the change in tissues after the implantation of an endoprosthesis based on titanium nickelide at: **a** 7 days, **b** 14 days, **c** 21 days, **d** 45 days

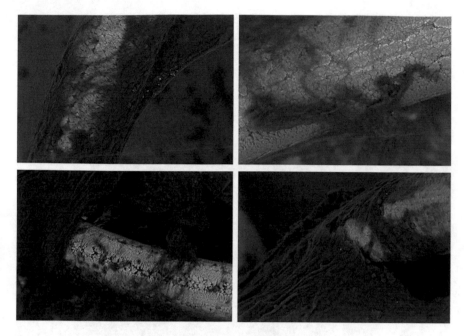

Fig. 9.7 Examples of scanning electron microscope images

As a part of an experimental study using a scanning microscope for processing and analyzing observational data in the process of tissue growth and maturation, the filling of the implant structure with cell elements was observed and evaluated. The data of electron scanning microscopy are represented by monochrome images at different scales (more than 100). Obtained images depending on analyzing part have different resolution. Resolution varies from 500×700 to 5120×3840 pixels.

Examples of raster microscopy images obtained during different days of medical experiment are shown in Fig. 9.7. It also can be seen that brightness and contrast characteristics of the image is varies from image to image.

Taking this into account, it was decided to evaluate the proposed method with and without preprocessing step of brightness correction based on the modified Retinex algorithm.

9.5.2 Evaluation of Brightness Parameters of Medical Images

For the experimental study, 140 images were taken from the entire period of the full-scale medical experiment on rabbits. The results of Retinex algorithm processing for histological images are shown in Fig. 9.8, and for raster microscopy images in Fig. 9.9. As can be seen in the images processed by Retinex algorithm, a

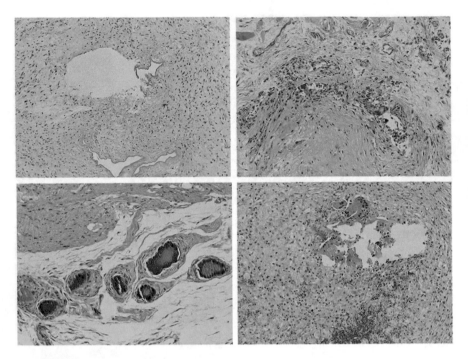

Fig. 9.8 Examples of histological images processed by Retinex algorithm

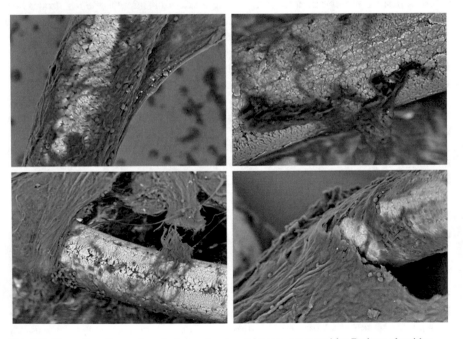

Fig. 9.9 Examples of scanning electron microscope images processed by Retinex algorithm

Table 9.1 Brightness' characteristics of model images

Images	Before brightness correction		After brightness correction	
	Mean (units)	SD (units)	Mean (units)	SD (units)
Histological images				
Image 1	204.22	15.73	190.03	12.43
Image 2	227.91	23.71	190.13	11.20
Image 3	230.95	24.93	189.90	11.02
Image 4	214.67	27.67	189.95	12.20
Scanning electron microscope images				
Image 1	83.26	12.57	127.48	22.14
Image 2	112.89	21.17	127.02	29.53
Image 3	63.25	24.71	126.95	37.20
Image 4	70.10	14.30	127.54	25.42

level of brightness has become approximately the same and local contrast is also changed.

Technical data of the brightness characteristics, such as mean value and Standard Deviation (SD) of medical images before and after brightness correction by Retinex algorithm (Figs. 9.6, 9.7, 9.8 and 9.9), are given in Table 9.1. Tables 9.2 and 9.3 demonstrate statistical data of brightness characteristics for whole set of images obtained in clinical experiments. During research, for the histological images target mean brightness was set as 190 and for scanning electron microscopy images as 127. This will further allow to unify the ranges of the algorithms parameters values for edge detection, color coding, and analysis of objects of interest.

As can be seen from Table 9.3, a range of the mean brightness value for a set of images has been greatly reduced in the images processed by Retinex algorithm, and it is 0.24–0.46 units. A similar result is also observed for images obtained from a

Table 9.2 Medical images brightness' characteristics

Time period (days)	Before brightness correction		After brightness correction	
	Mean (units)	SD (units)	Mean (units)	SD (units)
Histological images				
7	202.12–215.32	14.23–18.32	189.87–190.33	12.29–12.65
14	212.27–245.91	22.70–24.12	189.91–190.15	11.12–11.44
21	218.39–239.45	24.03–26.09	189.95–190.23	10.92–11.14
45	205.87–224.61	26.97–28.17	189.90–190.17	12.03–12.26
Scanning electron microscope images				
7	78.36–88.67	11.39–12.68	126.99–127.12	19.81–24.15
14	69.74–95.82	11.13–34.40	126.17–127.04	21.16–37.15
21	60.39–71.25	25.27–28.31	126.92–127.24	25.48–36.93
45	92.37–116.96	23.85–44.96	126.68–127.98	31.72–39.49

Table 9.3 Evaluation of the brightness' characteristics ranges of medical images

Time period (days)	Before brightness correction		After brightness correction	
	Mean (units)	SD (units)	Mean (units)	SD (units)
Histological images				
7	13.20	4.09	0.46	0.36
14	33.64	1.42	0.24	0.32
21	21.06	2.06	0.28	0.22
45	18.74	1.20	0.27	0.23
Scanning electron microscope images				
7	10.31	1.29	0.13	4.34
14	26.08	23.27	0.87	15.99
21	10.86	3.04	0.32	11.45
45	24.59	21.11	1.3	7.77

scanning electron microscope. In the case of histological images, the standard deviation range after illumination correction decreased.

Although as one can see for the images obtained from a scanning electron microscope, this range has increased. This, in turn, can be characterized by the features of images (different angles and parameters of the observed objects). The influence of brightness correction on edge map and regions map after color coding is demonstrated for histology image (Fig. 9.10) and scanning electron microscope image (Fig. 9.11).

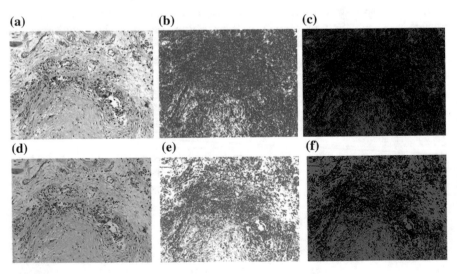

Fig. 9.10 Examples histology image processing: **a** original image, **d** image with brightness correction, **b, e** color coded edge map, **c, f** color coded regions map

(a)   **(b)**   **(c)**

(d)   **(e)**   **(f)**

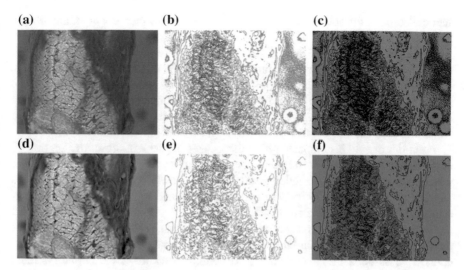

Fig. 9.11 Examples scanning electron microscope image processing: **a** original image, **d** image with brightness correction, **b**, **e** color coded edge map, **c**, **f** color coded regions map

9.5.3   Evaluation of Medical Characteristics

The main computational experiment was carried out using Mathlab software. The following quantitative characteristics were used: the contours and the total area of the implant under study, as well as, the contours and the proportion of tissue germinated on the mesh.

According to the description and estimates based on histological data, the characteristics of the tissue response to the implant mesh are manifested in sufficient form. Thus, in the obtained data on the 7th day a moderate amount of cellular detritus in the form of fine-grained masses, and destroyed cellular elements is observed above the aponeurotic space. Cellular detritus is located in the form of foci. In the vicinity of one area there is a center of white adipose tissue.

On the 14th day in the contact zone cell detritus is weakly expressed. In some places of interaction with the implant there are small focal hemorrhages of red blood cells. Outside the zone of interest a large number of small blood vessels are determined. Collagen fibers are expressed to a moderate degree located near the implant mostly randomly with a tendency to longitudinal course. In the zone of contact with the nickel-titanium implant, the inflammatory response is reduced, fibroblastic cells appear, and actively begin the regeneration process.

On the 21st day, fibromuscular tissue is represented by a network of collagen fibers interlaced into it with diffusely arranged fibroblast forms forming the tissue framework, and there are also many capillary type vessels. In the contact zone around the points of passage of the metal fibers, edema is not detected, there is an accumulation of fibroblasts and a multitude of newly formed collagen and elastic fibers with minor infiltration.

By the end of 45 days there is no edema of fibromuscular tissue, fibrocytes prevail among the cells in the connective tissue. In the surrounding connective tissue, there are thick-walled arteries of small caliber with hypertrophied muscle walls, where in addition collagen fibers form thick bundles located predominantly perpendicular to muscle fibers and concentric around the arteries. Elastic fibers are located around the muscle elements, they form a network of thicker fibers.

A similar picture is demonstrated in the images of raster microscopy. On the 7th day, the titanium nickelide implant is surrounded by granulation tissue, and the mesh cells are filled with cellular elements. Also, newly formed fibrous structures are determined, chaotically located collagen fibrils and multiple erythrocytes presented in the form of "inflated balls" are viewed.

On the 14th day of the medical experiment, the images show the continuation of the formation of tissue around the implant threads, which is a formed regenerate of connective tissue. To a greater extent, the cellular structure of the implant is filled with a tissue matrix, the direction of the fibers is determined equally. On the 21st day, the regeneration of the newly formed tissue continues, the cellular structures of the implant are filled with collagen fibers, connective tissue fibers smoothly pass to the implant, penetrate into its pores.

By 45 days of the experiment, the implant is located in the thickness of the formed tissue, the bunches of fibers are ordered. It is difficult to distinguish, where the implant is, and, where the tissue is. The relief of the regenerate surface becomes smoothed due to compaction and thickening of the structures surrounding the metal threads. The fibrous component of the tissue matrix prevailed over the cellular one, which reflects the processes of connective tissue maturation.

Figures 9.12 and 9.13 show examples of color-coded maps obtained for histological data and scanning electron microscope images. Based on these maps, morphological indicators were calculated and tissue germination was assessed.

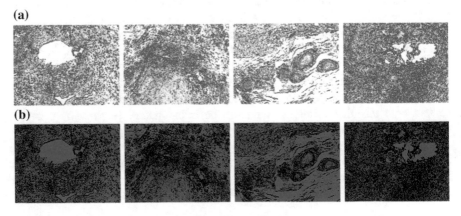

Fig. 9.12 Examples of color coded maps for histological images: **a** edge maps, **b** regions' maps

(a)

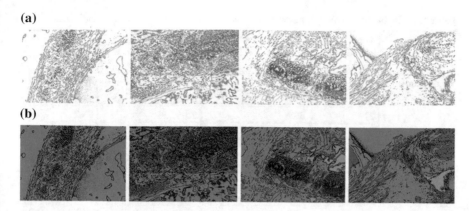

(b)

Fig. 9.13 Examples of color coded maps for scanning electron microscopy images: **a** edge maps, **b** regions' maps

The results of quantitative histological analysis obtained using the developed computational methods are summarized in Tables 9.4 and 9.5. Tables reflect the effect of pre-treatment on the calculated indicators. As an estimate, data on the state of the tissue are used on days 7, 14, 21, and 45 (day 14—the beginning of the active regeneration process), which are obtained on the basis of traditional histological approaches without using computer tools. At the same time, the work of the proposed method was evaluated without and with the implementation of light correction.

Table 9.4 The error estimation of the morphometric parameters in different time moments

Time period (days)	Vessels (%)	Collagen fibers (%)	Elastic fibers (%)
Conventional calculation technique			
7	8.0	2.8	10.3
14	10.5	2.2	8.1
21	13.4	3.6	14.4
45	7.3	1.2	8.5
Calculation with color coding			
7	7.1	2.5	9.6
14	9.6	2.1	7.6
21	12.7	3.4	13.5
45	7.1	1.1	7.9
Calculation with color coding and brightness correction			
7	6.7	2.3	9.2
14	9.2	2.0	7.3
21	12.1	3.2	12.9
45	6.8	1.0	7.5

Table 9.5 Evaluation of the brightness correction influence on the calculated indicators

Indicator	Time period (days)	Without brightness correction (units)	With brightness correction (units)	Range (units)	
				min	max
Amorphous component	7	44.1	44.3	44.0	44.7
	14	27.0	27.2	27.1	27.3
	21	23.3	23.4	23.2	23.4
	45	21.8	21.9	21.6	22.2
Fibroblast	7	6.2	6.3	6.1	6.5
	14	25.7	25.8	25.5	26.1
	21	1.3	1.4	1.2	1.6
	45	0.25	0.26	0.24	0.28
Fibroblasts	7	7.4	7.5	7.3	7.7
	14	26.5	26.7	26.5	26.9
	21	3.8	3.9	3.7	4.1
	45	3.1	3.2	3.0	3.4
Leukocytes	7	9,7	9.8	9.6	10.0
	14	5.7	5.8	5.5	6.1
	21	0.5	0.5	0.4	0.6
	45	0.0	0.0	0.0	0.0
Vessels	7	4.6	4.7	4.4	5.0
	14	14.8	15.0	14.9	15.1
	21	2.3	2.4	2.2	2.6
	45	1.3	1.4	1.3	1.5
Collagen fibers	7	22.3	22.5	22.1	22.9
	14	49.5	49.7	49.3	50.1
	21	17.9	18.0	17.7	18.3
	45	24.1	24.3	24.1	24.5
Elastic fiber	7	11.1	11.2	11.0	11.4
	14	11.3	11.5	11.3	11.7
	21	5.1	5.2	5.0	5.4
	45	9.2	9.3	9.1	9.5

Table 9.4 presents the estimates of the data of morphometric indicators on the calculation error for the main studied parameters. Evaluation conducted for the computational methods: conventional calculation technique, calculation with color coding, and calculation with color coding and brightness correction. Table 9.5 shows estimates of influence of brightness correction on calculated morphometric indicators. In the case of processing without brightness correction, the average values of the indicators are given, and for the results with brightness correction, information on the minimum and maximum values obtained is also provided.

Table 9.6 Evaluation of tissue germination area

Time period (days)	Area (%)		
	Perspective 1	Perspective 2	Perspective 3
Conventional calculation technique			
7	24.4	24.7	24.9
14	47.9	48.1	48.3
21	59.4	59.6	59.8
45	68.3	68.5	68.7
Calculation with color coding			
7	27.1	27.4	27.6
14	53.1	53.3	53.5
21	66.0	66.2	66.4
45	76.0	76.1	76.3
Calculation with color coding and brightness correction			
7	28.1	28.3	28.5
14	55.1	55.3	55.4
21	68.3	68.5	68.7
45	78.6	78.8	78.9

Evaluation of tissue germination was performed on the basis of scanning electron microscopy images. So, more objective data were used for images obtained from different angles (perspectives). As a part of the study, for the evaluation of computer techniques a medical expert specified objects that were defined as tissue, fibers, red blood cells, etc., and the areas with the implant structure were specified separately. For the specified reference samples, parameters were calculated taking into account the indicators of texture characteristics and color code.

With this in mind, the area of germinated tissue share was calculated. It is defined as the ratio of the area of germinated tissue to the area of the corresponding part of the implant mesh. Comparison of tissue germination rates obtained by a medical expert on the basis of the classical assessment approach was carried out with the proposed method in two modes: with brightness correction and without this correction. Table 9.6 present data on the tissues germination, obtained, when measuring the studied geometric texture indicators for the key time stages of medical experiment (7, 14, 21, and 45 days).

According to Table 9.6 one can observe the dependence of the area of germinated tissue in the corresponding pictures (perspectives) on time. In general, similar values of tissue germination area were obtained for different perspectives. Preliminary brightness correction allows to separate tissue and implant features more clearly. In addition, the evaluation of the methodology on the histological data revealed that a brightness correction improves the accuracy of the desired estimates.

9.5.4 Overall Efficiently Evaluation of the Proposed Method

According to the results of the analysis of assessments of histological parameters in the framework of the experiment, we can conclude that the preliminary image processing with regard to the brightness correction improves the accuracy of the estimates, which is important for describing the features of the process under study. Retinex based brightness correction also allows to specify the baseline estimate and significantly reduce the variation of values (error) from 8–12 to 2–3%.

At the same time, the procedure for evaluating the experimental data is significantly simplified due to the fact that the brightness is reduced to the same level, and the parameters of the algorithms do not need to be rearranged. The data presented in Table 9.5 show that the new technique of data processing allows to conduct more accurate assessment of the data during analysis of the studied processes. At the same time, it allow to simplify the work of the expert histologist due to the use of more universal method of processing, which does not require the use of time-consuming traditional procedures of histological analysis.

Analysis of the results of the processing of scanning electron microscope data shows the effectiveness of the developed computational methodology for assessing the main indicators of quantitative histology. This significantly increases the speed of data processing and obtaining updated estimates. Use of the brightness correction mechanism based on Retinex technology makes it possible for medical experts to see the details of the objects, being studied on the images more clearly. During the evaluation of the tissues germination, this is expressed in a clearer definition of regions, on average by 1–3%. This is significant in terms of evaluating medical indicators. The developed algorithmic tools are useful for expert histologists, when processing large amounts of visual data.

9.6 Conclusion

In this research, a modified method of geometric analysis of visual data to solve a problem of quantitative description of the dynamic process on the basis of experimental clinical studies is developed. The technique has shown its effectiveness both in terms of speed of analysis and accuracy (on average by 10–15%) due to the possibility of high-quality preprocessing by brightness correction, as well as, noise reduction. These preliminary computational procedures ensured the efficient execution of the shearlet transform using the modified FFST algorithm. As a result, the application of the developed technique allows to increase the accuracy of selection of linear structures and visual quality of images of the studied clinical objects.

The problem associated with contrast objects under study (tissue and its characteristics) in the images through the procedure of the color coding with the preliminary noise reduction and brightness correction based of Retinex algorithm is successfully solved. Depending on the type of analyzed images (scanning electron microscopy or histological data), the brightness correction allows to reduces the errors of estimation 1-5% and increase the credibility of indicators.

The proposed technique has shown that it significantly reduces the time for processing and analyzing visual data through the use of more unified parameters of key algorithms.

References

1. Khodorenko, V.N., Anikeev, S.G., Kokorev, O.V., Mukhamedov, M.R., Topolnitskiy, E.B., Gunther, V.E.: Structural features of TiNi-based Textile materials and their biocompatibility with cell culture. KnE Mater. Sci. 2(1), 16–24 (2017)
2. Zaworonkow, D., Chekan, V., Kusnierz, K., Lekstan, A., Grajoszek, A., Lekston, Z., Lange, D., Chekalkin, T., Kang, J., Gunther, V. and Lampe P.: Evaluation of TiNi-based wire mesh implant for abdominal wall defect management. Biomed. Phys. Eng. Express 4(2), 027010.1–027010.12 (2018)
3. Muhamedov, M., Kulbakin, D., Gunther, V., Choynzonov, E., Chekalkin, T., Hodorenko, V.: Sparing surgery with the use of TINI-based endografts in larynx cancer patients. J. Surg. Oncol. 111, 231–236 (2015)
4. Iriyanov, Y.M., Chernov, V.F., Radchenko, S.A., Chernov, A.V.: Plastic efficiency of different implants used for repair of soft and bone tissue defects. Bull. Exp. Biol. Med. 155(4), 518–521 (2013)
5. Topolnitskiy, E.B., Dambaev, G.T., Hodorenko, V.N., Fomina, T.I., Shefer, N.A., Gunther, V.E.: Tissue reaction to a titanium-nickelide mesh implant after plasty of postresection defects of anatomic structures of the chest. Bull. Exp. Biol. Med. 153(3), 385–388 (2012)
6. Cobb, W.: A current review of synthetic meshes in abdominal wall reconstruction. Plast. Reconstr. Surg. 142(3S), 64S–71S (2018)
7. Baumann, D.P., Butler, C.E.: Bioprosthetic mesh in abdominal wall reconstruction. Semin. Plast. Surg. 26(1), 18–24 (2012)
8. Lanier, S.T., Fligor, J.E., Miller, K.R., Dumanian, G.A.: Reliable complex abdominal wall hernia repairs with a narrow, well-fixed retrorectus polypropylene mesh: a review of over 100 consecutive cases. Surgery 160(6), 1508–1516 (2016)
9. El-Khatib, H.A., Bener, A.: Abdominal dermolipectomy in an abdomen with pre-existing scars: a different concept. Plast. Reconstr. Surg. 114(4), 992–997 (2004)
10. Vidal, P., Berner, J.E., Will, P.A.: Managing complications in abdominoplasty: a literature review. Arch. Plast. Surg. 44(5), 457–468 (2017)
11. Deeken, C.R., Lake, S.P.: Mechanical properties of the abdominal wall and biomaterials utilized for hernia repair. J. Mech. Behav. Biomed. Mater. 74, 411–427 (2017)
12. Montgomery, A.: The battle between biological and synthetic meshes in ventral hernia repair. Hernia 17, 3–11 (2013)
13. Binnebosel, M., von Trotha, K.T., Jansen, P.L., Conze, J., Neumann, U.P., Junge, K.: Biocompatibility of prosthetic meshes in abdominal surgery. Semin. Immunopathol. 33(3), 235–243 (2011)
14. Dambaev, G.T., Gunther, V.E., Menschikov, A.V., Solovev, M., Avdoshina, E.A., Fatushina, O.A., Kurtseitov, N.E.: Laparascopic hernia repair with the use of TINI-based alloy. KnE Mater. Sci. 2, 193–199 (2017)
15. Radkevich, A.A., Gorbunov, N.A., Khodorenko, V.N., Usoltsev, D.M.: Reparative desmogenez in connective tissue defects after the replacement of NITI implants. Implants Shape Mem. 1, 21–25 (in Russian) (2008)

16. Laschke, M.W., Haufel, J.M., Thorlacius, H., Menger, M.D.: New experimental approach to study host tissue response to surgical mesh materials in vivo. J. Biomed. Mater. Res. A **74**(4), 696–704 (2005)
17. Kolpakov, A.A., Kazantsev, A.A.: Comparative analysis of the results of using titanium silk and polypropylene prostheses in patients with postoperative ventral hernias. Breast Cancer Gastroenterol. Surg. **13**, 774–775 (2015)
18. Ivanov, S.V., Ivanov, I.S., Goryainova, G.N., Tsukanov, A.V., Katunina, T.P.: Comparative tissue morphology in using prostheses from polypropylene and polytetrafluoretilen. Cell Tissue Biol. **6**(3), 309–315 (2012)
19. Roubliova, X.I., Deprest, J.A., Biard, J.M., Ophalvens, L., Gallot, D., Jani, J.C., Van de Ven, C.P., Tibboel, D., Verbeken, E.K.: Morphologic changes and methodological issues in the rabbit experimental model for diaphragmatic hernia. Histol. Histopathol. **25**(9), 1105–1116 (2010)
20. Avtandilov, G.G.: Medical morphometry. Medicine, Moscow (in Russian) (1990)
21. Bourzac, K.: Software: the computer will see you now. Nature **502**(7473), S92–S94 (2013)
22. Howard, C.V., Reed, M.G.: Unbiased stereology. Three-Dimensional Measurement in Microscopy. Bios Scientific Publishers, Oxford (1998)
23. Glaser, J., Greene, G., Hendricks, S.: Stereology for Biological Research with a Focus on Neuroscience. MBF Press, Williston (2007)
24. Hauser, S.: Fast finite shearlet transform: a tutorial. University of Kaiserslautern, Kaiserslautern, Germany (2011)
25. Guo, K., Labate, D., Lim, W.-Q.: Edge analysis and identification using the continuous shearlet transform. Appl. Comput. Harmon. Anal. **27**(1), 24–46 (2009)
26. Kutyniok, G., Sauer, T.: From wavelets to shearlets and back again. In: Neamtu, M., Schumaker, I.I. (eds.) Approximation Theory, vol. XII, pp. 201–209. Hashboro Press, San Antonio, TX, Nachville, TN (2007)
27. Kutyniok, G., Labate, D.: Introduction to shearlets. In: Kutyniok, G., Labate, D. (eds.) Shearlets: Multiscale Analysis for Multivariate Data, pp. 1–38, LLC. Springer Science + Business Media (2012)
28. Lim, W.Q.: The discrete shearlet transform: a new directional transform and compactly supported shearlet frames. IEEE Trans. Image Process. **19**(5), 1166–1180 (2010)
29. Cadena, L., Espinosa, N., Cadena, F., Kirillova, S., Barkova, D., Zotin, A.: Processing medical images by new several mathematics shearlet transform. Int. MultiConf. Eng. Comput. Sci. **I**, 369–371 (2016)
30. Cadena, L., Espinosa, N., Cadena, F., Korneeva, A., Kruglyakov, A., Legalov, A., Romanenko, A., Zotin, A.: Brain's tumor image processing using shearlet transform. In: SPIE 10396, Applications of Digital Image Processing XL, 103961B. https://doi.org/10.1117/12.2272792 (2017)
31. Zotin, A., Simonov, K., Kapsargin, F., Cherepanova, T., Kruglyakov, A., Cadena, L.: Techniques for medical images processing using shearlet transform and color coding. In: Favorskaya, M.N., Jain, L.C. (eds.) Computer Vision in Control Systems-4, ISRL, vol. 136, pp. 223–259. Springer, Cham (2018)
32. Gosset, W.S.: On the error of counting with haemocytometer. Biometrika **5**(3), 351–360 (1907)
33. Noble, D.: Modeling the heart - from genes to cells to the whole organ. Science **295**(5560), 1678–1682 (2002)
34. Setty, Y., Cohen, I.R., Dor, Y., Harel, D.: Four-dimensional realistic modeling of pancreatic organogenesis. PNAS **105**(51), 20374–20379 (2008)
35. Ward, S.T., Rosen, G.D.: Optical disector counting in cryosections and vibratome sections underestimates particle numbers: Effects of tissue quality. Microsc. Res. Tech. **71**(1), 60–81 (2008)
36. Geuna, S., Robecchi, M.G., Raimondo, S.: Morpho-quantitative stereological analysis of peripheral and optic nerve fibers. NeuroQuantol **10**(1), 76–86 (2012)

37. Akazaki, S., Takahashi, T., Nakano, Y., Nishida, T., Mori, H., Takaoka, A., Aoki, H., Chen, H., Kunisada, T., Koike, K.: Three-dimensional analysis of melanosomes isolated from B16 melanoma cells by using ultra high voltage electron microscopy. Microsc. Res. **2**, 1–8 (2014)
38. Takaoka, A., Hasegawa, T., Yoshida, K., Mori, H.: Microscopic tomography with Ultra-HVEM and applications. Ultramicroscopy **108**(3), 230–238 (2008)
39. Kim, Y.J., Jeong, J.Y., Nam, S.Y., Kim, M.J., Oh, J.H., Kim, K.G., Sohn, D.K.: Three dimensional automatic body fat measurement software from CT, and its validation and evaluation. J. Biomed. Sci. Eng. **8**, 665–673 (2015)
40. Shinde, B., Mhaske, D., Dani, A.R.: Study of noise detection and noise removal techniques in medical images. Int. J. Image Graph. Sig. Process. **2**, 51–60 (2012)
41. Sudha, S., Suresh, G.R., Sukanesh, R.: Speckle noise reduction in ultrasound images by wavelet thresholding based on weighted variance. Int. J. Comp. Theory Eng. **1**(1), 1793–8201 (2009)
42. Mamta, J., Mohana, R.: An improved adaptive median filtering method for impulse noise detection. Int. J. Recent Trends Eng. **1**(1), 274–278 (2009)
43. Arin, H.H., Hozheen, O.M., Sardar, P.Y.: Denoising of medical images by using some filters. Int. J. Biotechnol. Res. **3**(1), 10–20 (2015)
44. Kanmani, P., Rajiv, K.A., Deepak, K.P., Ayyappadasan, G.: Performance analysis of noise filters using histopathological tissue images in lung cancer. Int. Res. J. Pharm. **8**(1), 50–54 (2017)
45. Li, Y., Ishitsuka, Y., Hedde, P.N., Nienhaus, G.U.: Fast and efficient molecule detection in localization-based super-resolution microscopy by parallel adaptive histogram equalization. ACS Nano **7**(6), 5207–5214 (2013)
46. Hiremath, P.S., Bannigidad, P., Geeta, S.: Automated identification and classification of white blood cells (leukocytes) in digital microscopicimages. Int. J. Comput. Appl. **2**, 59–63 (2010)
47. Kong, N.S.P., Ibrahim, H., Ooi, C.H., Chieh, D.C.J.: Enhancement of microscopic images using modified self-adaptive plateau histogram equalization. Int. Conf. Comput. Technol. Develop. **2**, 308–310 (2009)
48. Biswajit, B., Pritha, R., Ritamshirsa, C., Biplab, K.S.: Microscopic image contrast and brightness enhancement using multi-scale Retinex and cuckoo search algorithm. Procedia Comput. Sci. **70**, 348–354 (2015)
49. YangDai, T., Zhang, L.: Weighted Retinex algorithm based on histogram for dental CT image enhancement. IEEE Nuclear Science Symposium and Medical Imaging Conference, pp. 1–4 (2014)
50. Weizhen, S., Fei, L., Qinzhen, Z.: The applications of improved Retinex algorithm for X-ray medical image enhancement. IEEE International Conference on Computer Science and Service System, pp. 1655–1658 (2012)
51. Davies, E.: Machine Vision: Theory, Algorithms and Practicalities. Academic (2012)
52. Szeliski, R.: Computer Vision: Algorithms and applications. Springer London Limited, London (2011)
53. Gonzalez, R.C., Woods, R.E.: Digital Image Processing, 3rd edn. Prentice-Hall, Englewood Cliffs (2008)
54. Ahmed, A.S.: Comparative study among Sobel, Prewitt and Canny edge detection operators used in image processing. J. Theor. Appl. Inf. Technol. **96**, 6517–6525 (2018)
55. Stosic, Z., Rutesic, P.: An improved Canny Edge detection algorithm for detecting brain tumors in MRI images. Int. J. Signal Process. **3**, 11–15 (2018)
56. Senthil, K.N., Sathyavathy, S.K.N.: Segmentation of renal calculi from CT abdomen images by incorporating FCM and level set approaches. Int. J. Adv. Res. Comput. Commun. Eng. **5**(7), 132–138 (2016)
57. Cadena, L., Zotin, A., Cadena, F.: Enhancement of medical image using spatial optimized filters and OpenMP technology. Lecture Notes in Engineering and Computer Science. In: International MultiConference *of* Engineers and Computer Scientists, vol. 1, pp. 324–329 (2018)

58. Zotin, A.: Fast algorithm of image enhancement based on multi-scale Retinex. Procedia Comput. Sci. **131**, 6–14 (2018)
59. Gorban, A.N., Kegl, B., Wunsch, D.C., Zinovyev, A.Y.: Principal manifolds for data visualization and dimension reduction. LNCSE, Springer, Berlin-Heidelberg (2008)

Chapter 10
Histological Images Segmentation by Convolutional Neural Network with Morphological Post-filtration

Vladimir Khryashchev, Anton Lebedev, Olga Stepanova and Anastasiya Srednyakova

Abstract An algorithm for segmentation of cell nuclei in histological images is developed and studied. It is based on the implementation of U-Net neural network. As a means of additional processing, the use of morphological filtration is proposed. The training and testing of the algorithm was carried out on NVIDIA DGX-1 supercomputer using a histological image dataset with automatically generated markup. In the course of studies, the values of such metrics as simple match coefficient, Tversky index, and Sørensen coefficient are obtained. According to the results of calculating these metrics, a quality of histological image segmentation developed by the morphological filtering algorithm exceeds the similar performance for the algorithm without using morphological filtering, which allows us to recommend the morphological filter as a means of additional image processing at the output of the neural network algorithm in real medical practice.

Keywords Convolutional neural network · Cell nuclei segmentation · Morphological filtering · Histological image segmentation

V. Khryashchev (✉) · A. Lebedev · O. Stepanova · A. Srednyakova
P.G. Demidov Yaroslavl State University, Sovetskaya 14/2, 150000 Yaroslavl, Russian Federation
e-mail: v.khryashchev@uniyar.ac.ru

A. Lebedev
e-mail: a.lebedev@uniyar.ac.ru

O. Stepanova
e-mail: olga1stepanova@yandex.ru

A. Srednyakova
e-mail: stasya.srednyakova@yandex.ru

M. N. Favorskaya and L. C. Jain (eds.), *Computer Vision in Advanced Control Systems-5*, Intelligent Systems Reference Library 175,
https://doi.org/10.1007/978-3-030-33795-7_10

10.1 Introduction

Cancer has become a real problem for modern society. It occupies the second place after diseases of the cardiovascular system by the number of deaths. This disease ranks second after diseases of the cardiovascular system by the number of deaths [1]. Breast cancer is a malignant tumor of the glandular breast tissue. This type of cancer pathology is the most common type of cancer among women worldwide and the most common disease in the population after lung cancer [2]. An important step towards reducing the mortality of patients with this disease is timely diagnosis, allowing treatment at early stages of the pathological processes development in the body. A number of diagnostic methods are used to detect breast cancer, among which the key role is given to histological examination as the most reliable and accurate approach to the cancer diagnosis.

Histological examination includes a special preparation of histological tumor tissues materials obtained during the operation or by biopsy, as well as, their qualitative and quantitative analysis—morphometry. In morphometry, the specialist records the following data: the general structure of the tissue, the state of the cells and the intercellular substance, the degree of development of the objects present in the visual field, as well as, such quantitative characteristics of the histological slices as the number of objects in the field of view, their size and area, and the ratio between various structures. The number of samples obtained from one patient's tissue fragment reaches 50–100 that makes manual morphometry a rather complicated process requiring a high degree of morphologist's concentration, as well as, considerable time costs for one research [3].

The purpose of this study is to develop algorithms for automatic cell nuclei segmentation on breast tissue histological images based on a convolutional neural network that could provide a segmentation level comparable to the expert one. In addition, it is necessary to achieve high rate of the segmentation system. Such algorithms can be used in decision support systems for early diagnosis of breast cancer by pathologists, as well as, a means of training or control for beginners in the field of breast cancer diagnosis.

The chapter is organized as follows. Section 10.2 presents a short review respect to medical image segmentation using the conventional and deep learning approaches. Histological image dataset is described in Sect. 10.3. Section 10.4 provides a development of algorithms for automatic histological image segmentation. Section 10.5 includes a description of applied morphological filter. Results of nuclei cell segmentation in histological images are reported in Sect. 10.6. Section 10.7 concludes the chapter.

10.2   Related Work

With the development of digital microscopic scanners, it became possible to study histological materials using digital image processing and analysis. An important area is the development of decision support algorithms that provide the quantitative automatic analysis of histological images for morphological specialists. Such algorithms can reduce the number of manual morphometry errors, increase the speed of histological examination, and also these can be used as a training tool for beginners in the field of breast cancer diagnosis. The main difficulty in the development of the automatic histological images analysis algorithms is the need to separate objects with the same features, but belonging to different classes. This separation is difficult to be clearly formalized, and as a result it becomes necessary to search for additional features or to use the methods of fuzzy class separation.

Conventional methods are often used for image segmentation [4, 5]. Despite the fact that they are widely used for solving problems of object detection, their use in medical practice in the task of histological image analysis is difficult for several reasons. Thus, methods based on the binarization process segment images quickly, but do not guarantee high accuracy of the result. Methods based on the search for the boundaries of regions suggest higher accuracy due to low sensitivity to changes in the characteristics of images, but they are processing images slowly. The rate of methods based on region detection is high. However, it can decrease when the algorithms have to process large areas, which is also accompanied by deterioration of results [4, 5]. In addition, some of the existing methods are interactive and require direct user intervention (for example, Flood Fill method, the graph-based segmentation), which also makes it difficult for a specialist to work in real time [5].

A convenient and practical solution of this problem is to use the convolutional neural networks—a tool from the area of deep machine learning, which shows high results of images classification, recognition and segmentation including medical images [6]. The main advantage of using such networks is the resistance to spatial image distortion, zooming, rotations, and shifts. In addition, it is enough to train the network once using a dataset of correctly labeled data so that it can automatically determine the location of elements of interest in images for which no such markup exists [6]. However, a prerequisite for using a trained network in segmentation tasks is the use of images that are similar in type to the images from training base. Thus, a network that recognizes the nuclei of cells of one tissue type can't be used for cells nuclei segmentation of another organ.

Convolutional neural networks underlie a number of studies devoted to the automatic histological images analysis. For example, an ovarian cancer types classifying scheme using cytological images and based on a convolutional neural network with DCNN architecture is presented in [7]. The accuracy of this system is 78.20%. The authors of [8] proposed a convolutional neural network for the diagnosis of osteosarcoma providing a level of accuracy of 92%. The paper [9] presented a deep convolutional neural network for the diagnosis of gastric carcinoma. The accuracy of the proposed approach for detecting cancer was 69% and for detecting mucosal necrosis—81%, which exceeded the level of such detection by traditional machine learning algorithms. The study [10] presents the invasive ductal carcinoma detection system, which shows the following results: the average values of F-measure and the balanced accuracy are 71.80% and 84.23%, respectively. The paper [11] proposed an approach for predicting the prostate cancer recurrence by histological images of prostate tissue, which is based on a convolutional neural network. As a result of the system testing, an AUC value of 0.81 was obtained.

The system of automatic colorectal cancer classification based on a convolutional neural network with VGG16 architecture was described in [12]. The accuracy of the system was about 90%. Authors [13] propose an automatic approach based on high-precision neural networks, which detects prostatectomy whole slide images with high–grade Gleason score is proposed. Such a system is designed for assessing prostate histopathology slides. An accuracy of 78% was obtained in this study.

The study [14] describes a method of epithelial tissue automatic segmentation in histological images of prostate tissue. The method is based on the use of two convolutional neural networks, one of which is U-Net architecture. According to the presented test results, AUC value was 0.97. In the paper [15], a deep learning nuclei segmentation algorithm was developed based on gathering localized information through the generation of superpixels using a simple clustering algorithm and a convolutional neural network. The aim of this system is to squamous epithelium cervical intraepithelial neoplasia classification. The accuracy of this parameter equals 95.97%.

The value of accuracy for the model based on deep networks described in [16] was 99.8%. This paper describes lung cancer classification system. In [17], a computer diagnostic system was introduced for the classification of lung cancer on the basis of CT images. For this, a modified U-Net network was used. The system showed high results: 86.6% accuracy.

The image recognition segmentation system of colon glands [18] showed a result of 98 and 94%, which corresponds to the classes of benign and malignant tissues. The authors of the article [19] decided to use a neural network for segmentation of glands on histological images. The developed system allowed outperforms competing methods by up to 10% in both pixel-level accuracy and object-level Dice.

The accuracy of cervical cancer classification on histopathological images in [20] reached 88.5%. The system is also based on a convolutional neural network. The authors of paper [21] added a convolutional neural network to classify segments as an additional means for processing of histological digital images. This innovation has improved the classification accuracy of uterine cervical cancer by 1.5%. In [22], a system that allows classification of cervical cancer by cytological images is described. The classification accuracy of the model was 93.33% for the original group of images and 89.48% for the group of augmented images.

A large number of studies detected the pathologies indicating the possibility of breast cancer in medical images. The work [23] describes the use of the convolutional neural network architecture for development a breast cancer screening system. The accuracy of the proposed approach on public image bases had reached 96–98%, and the area under ROC curve (AUC) was 0.98–0.99. The value of the metric F1 for the system developed in [24] based on a multilayer convolutional neural network for the classification of invasive ductal carcinoma on medical slides was 89.34%. The paper [25] describes a decision support system based on several segmentation methods, including a convolutional neural network. Development accuracy was approximately 96–100%. The authors of [26] developed a neural network to classify microscopic images of breast tissue into categories: normal, benign, in situ carcinoma, and invasive carcinoma. The accuracy of the proposed method achieved 95%. In work [27], a two-stage model is proposed for classifying histopathological images of the mammary gland, where the convolutional neural network is used to extract traits from images in the learning process at the first stage. At the second stage, classical methods of machine learning are applied. This approach allowed to obtain accuracy up to 99.84%. The authors [28] present an approach to classifying histopathological images into four classes. The following results were obtained: accuracy of 93.8%, AUC 97.3% and sensitivity/specificity of 96.5/88.0% at the high-sensitivity operating point. In study [29], a systemic basis of convolutional neural networks was proposed, which allows the histological images to be classified into four categories (normal tissue, benign lesion, in situ carcinoma, and invasive carcinoma) and non-carcinoma. Accuracy for division into 4 classes was 77.8%, with division into two classes the accuracy was 95.6%. A convolutional neural network in [30], intended for segmentation of histological images, particularly those with Masson's trichrome stain. The accuracy of the algorithm was 0.947.

In [31], an algorithm based on deep convolutional neural networks was presented, allowing diagnosis of pathologies, namely breast cancer and colorectal cancer. F1 metric was 85, 89, and 100%, the accuracy (ACC) was 84, 88, and 100%, and Matthews correlation coefficient was 86, 77, and 100% on different data. In study [32], a method was proposed for the multiclassification of breast cancer using the recently proposed model of deep learning. The model achieved an average accuracy of 93.2%. In [33], the authors propose new methods based on convolutional neural networks, which made it possible to increase the classification accuracy of histological images by 21.54% and 15.07%, compared to their earlier solution. The authors of the article [34] built a classification system of histopathological images based on ResNet neural network architecture. Thus, the resulting algorithm distinguished eight classes of images: sub-classes (adenosis, fibroadenoma, phyllodes tumor, and tubular adenoma) and malignant (ductal carcinoma, lobular carcinoma, mucinous carcinoma, and papillary carcinoma). The accuracy value of the confusion matrix was 95% with a less error rate 0.011. A system based on convolutional neural networks was also presented in [35], which allows the classification of breast tissue histological images: accuracy was 77.8% for four classes and 83.3% for carcinoma/non-carcinoma. The sensitivity of the above method for cancer cases was 95.6%.

A system based on a deep convolutional neural network was used in [36]. This development allows automatic lung cancer classification based on microscopic images. Approximately 71% of all test images are classified correctly. The authors of the article [37] presented a conceptually simple Y-Net network for creating segmentation masks based on histological images of breast tissue. This network identifies important areas for analysis. At the same time, the accuracy of the method (62.5%) turned out to be close to the results of specialists in this field. In the work [38] devoted to the development of an automated system for diagnosing histopathology of the breast, the authors obtained a classification accuracy of about 55%. The system was consisted of several stages: in the first step, convolutional neural networks were used to detect dependencies, and in the second step, a multi-class convolutional network was used. The study showed that the obtained results are comparable with the diagnoses made by pathomorphologists. In [39], the idea of transferring ImageNet knowledge as deep convolutional activation features to the classification and segmentation of histopathology images with little training data was proposed. This approach was applied to histological images with the result that the segmentation accuracy reached 84%, and the classification accuracy was 97.5%.

Thus, the literature review shows the high efficiency of convolutional neural networks in the task of histological images automatic classification and segmentation that confirms the prospects of using this approach to develop medical image analysis systems.

10.3   The Histological Image Dataset

The digitized histological material of the patient is a raster graphic image of the tagged image file format. This format involves storing images with high color depth. Histological images are characterized by high resolution, which allows a specialist to work with several scales without significant loss of quality ($\times 10$, $\times 20$, $\times 30$, and $\times 40$).

The study used data from the medical image dataset, which is freely available [40]. This dataset consists of 143 digitized histological images of patients with ER-positive breast cancer (cancer sensitive to estrogen receptors) stained with hematoxylin and eosin. Image size is 2000×2000 pixels. Images from the dataset contain partial markup: the creators highlighted about 12,000 cell nuclei manually. It is worth noting that it is only a small part of the total number of cell nuclei containing in these histological images. The test dataset was formed by two images belonging to different patients. Each image of the test sample with a resolution of 2000×2000 pixels was divided into 4 fragments. Then 8 test images of 572×572 pixels were formed. After that test images were divided into simple ones (group 1: images 1–1, 1–2, 1–3, 1–4) and complex ones (group 2: images 2–1, 2–2, 2–3, 2–4) depending on the number of selected elements. All images included in the test dataset are shown in Fig. 10.1.

(a)

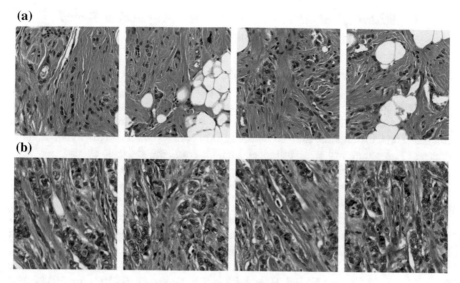

(b)

Fig. 10.1 Examples test images with: **a** simple structure, **b** complex structure

The choice of a small number of test images was made due to the difficulty of expert markup creating: manual marking of histological images is a nontrivial task thanks to the high labor intensity and significant time costs [41].

Diagnosis of different cancer types (the number of the most common types exceeds 10 [16]) requires an individual approach, which consists not only in the biopsy method, but also in the procedure for analyzing the elements depicted on the digitized data. This approach assumes the presence of certain signs (cell nuclei), by which it is possible to detect objects that are important for image analysis among other less important objects (tissue fragments and others).

The cell nuclei were determined in this case according to the recommendations received from specialists in this field. Thus, desired objects have the following features: oval shape, small size compared to other objects (round white objects, for example), dark purple dark pink color (compared to the color of the background), and uneven edges often.

10.4 The Development of Automatic Histological Image Segmentation Algorithms

In this section, the histological image segmentation algorithm based on AlexNet neural network and fast histological image segmentation algorithm based on U-Net neural network are discussed in Sects. 10.4.1–10.4.2, respectively.

10.4.1　Histological Image Segmentation Algorithm Based on AlexNet Neural Network

Training an algorithm based on a convolutional neural network involves the creation of data markup, i.e. binary masks of the corresponding images, where the pixels with the value "1" characterize the location of the desired nuclei of cells. The used image dataset was characterized by the presence of partial markup of some images. Thus, the creators marked about 12,000 nuclei of cells on the average in the dataset. The examples of histological images from the dataset, as well as, their corresponding binary masks containing partial markings are shown in Fig. 10.2. This number of cells in partial markup is small compared to the total number of cells in each image (Fig. 10.3).

To solve a problem of automatic segmentation an algorithm based on the use of the AlexNet [42] convolutional neural network architecture (hereinafter, Algorithm 1) was developed. AlexNet architecture proposed by Alex Krizhevesky and others, won the most difficult ImageNet challenge for visual object recognition called the ImageNet Large Scale Visual Recognition Challenge (ILSVRC) in 2012. The choice was made in favor of AlexNet architecture, since this architecture is still successfully used in solving problems of detecting objects of a particular class in images [43–45]. This architecture consists of five convolutional layers and three

(a)

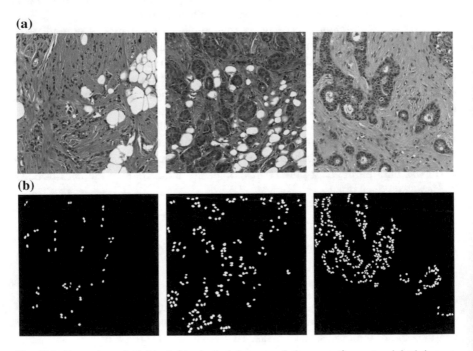

(b)

Fig. 10.2 Examples of digitized histological images of tissue sections: **a** original images, **b** corresponding binary masks containing marking

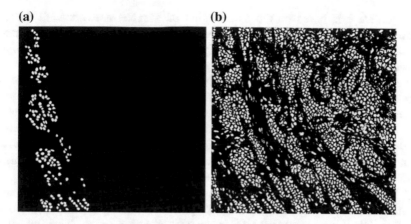

Fig. 10.3 Example of markup for image from a training sample: **a** partial, **b** full

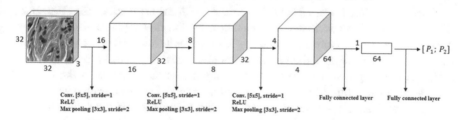

Fig. 10.4 The architecture of the convolutional neural network AlexNet

fully connected ones. The activation function used in this architecture is rectified linear unit (ReLU) (see Fig. 10.4). The aim of this algorithm is to create a segmentation binary mask dataset for histological images. The scheme of training and testing of Algorithm 1 is shown in Fig. 10.5.

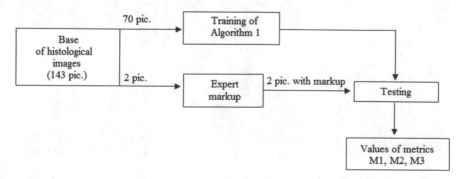

Fig. 10.5 The scheme of training and testing of Algorithm 1

Algorithm 1 is based on the modification of AlexNet network with the input layer of 32 × 32 pixels, which is due to the small size of fragments of training histological images with markup. More than that, this allows reducing the number of computational operations.

The network consists of successively alternating layers of convolution and pooling, the last two layers are fully connected. A vector of two P_i values is formed at the output of the network: P_1 is the probability that the pixel in question corresponds to the cell nucleus, P_2 is the probability that the pixel does not belong to the cell nucleus. Thus, to process a histological image in the 2000 × 2000 format, it's necessary first add auxiliary tools to process the extreme pixels, and then run the network 2,000 × 2,000 = 4,000,000 times.

A half of the histological images dataset was used for training of Algorithm 1. From these images fragments of 32 × 32 pixels containing the markup were cut out. Then they were multiplied by geometric transformations. The size of the resulting dataset was 816,500 image fragments. The training and testing of the algorithm was carried out on a supercomputer for deep learning NVIDIA DGX-1.

Then Algorithm 1 was tested using images with expert markup (see Fig. 10.6). The result of automatic image segmentation by the developed algorithm is presented in Fig. 10.6c. Noticeable that Algorithm 1 quite successfully copes with the task of cell nuclei segmentation in histological images, due to that a fragment of the image at the output of the network has a significant similarity with the result of manual segmentation (Fig. 10.6b). Visually, several "missing elements" can be noted, as well as, the discrepancy between the boundaries of the same elements. Nevertheless, conclusion making about the correctness of the obtained results is possible only through an objective evaluation using specialized metrics.

(a)   **(b)**   **(c)**

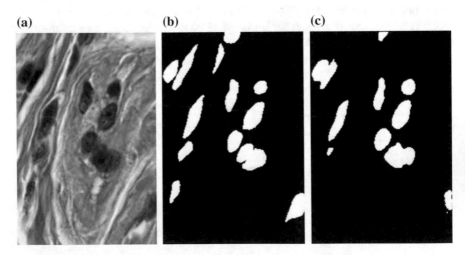

Fig. 10.6 The comparison of expert markup with the automatic segmentation result: **a** original image, **b** reference markup, **c** markup on the output of Algorithm 1

Despite a high quality of segmentation, it is worth noting that the processing of one histological image by Algorithm 1 takes about 3 h. Such time costs impose restrictions on the use of such a network for the direct segmentation of histological images in real medical practice and necessitate to search faster approaches. In this regard, it was decided to use Algorithm 1 as an auxiliary tool for creating automatic markup of the histological image dataset. As a result, 141 images of the test dataset were re-marked.

The creation of automatic markup allowed to reduce significantly the time required to create the data markup, as well as, to evaluate the feasibility of automatic markup using convolutional neural networks. In addition, this stage provides the possibility of learning the neural networks using images obtained from the output of other networks ("unsupervised learning").

10.4.2 Fast Histological Image Segmentation Algorithm Based on U-Net Neural Network

To solve the problem of cell nuclei segmentation in the real time mode, an algorithm based on the convolutional neural network U-Net [46] has been developed (hereinafter—Algorithm 2). The choice of this architecture was made for several reasons. First, it allows to receive a large number of images for training automatically. This aspect is important, because when medical image analysis algorithms are developing, the amount of input data is very limited, and that does not contribute to effective training of networks. Second, U-Net architecture allows to achieve clearer separation in the case, when objects of the same class on the image are in contact. The third crucial quality in favor of the chosen architecture is its speed.

A schematic representation of the network architecture is presented in Fig. 10.7.

The network architecture consists of a contracting path (left side) and an expanding path (right side). The contracting path represents the typical convolutional neural network architecture and consists of several blocks (4 blocks in the original version). Each such block consists of repeated applications of two 3×3 convolutions, each of that is followed by an activation function ReLU, and a pooling operation with a 2×2 filter size with stride 2 for downsampling. The number of channels is doubled at each step of downsampling. The expanding path includes the same number of blocks as the contracting path. Each of its blocks consists of an upsampling of the feature map with a reduction the number of channels by two times followed by 2×2 convolution ("deconvolution"), a concatenation with the correspondingly cropped feature map from the contracting path and two 3×3 convolutions, each followed by ReLU. The cropping is necessary due to the loss of border pixels at each convolution. At the last layer, a 1×1 convolution is used to map each 64-component feature vector to the desired number of classes (getting a flat image). On the whole, the network has 23 convolutional layers.

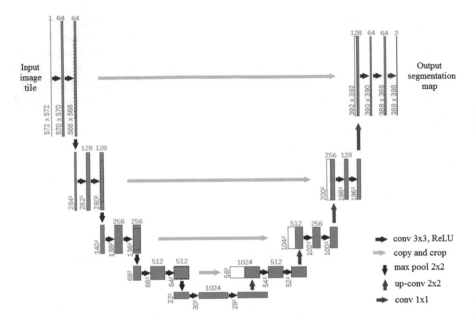

Fig. 10.7 The architecture of the convolutional neural network U-Net

The training and testing scheme of Algorithm 2 is shown in Fig. 10.8. Fragments of images with automatic markup of 572 × 572 pixels were used for training, and then were multiplied using geometric transformations. The total size of the training dataset was 1,920,000 images. The training took place in parallel on four video cards on the supercomputer NVIDIA DGX-1, the size of the batch was 128 images. Learning was stopped after completing about 390,000 iterations.

The results of processing the test images using such a network are shown in Fig. 10.9. Visual subjective assessment shows that the algorithm successfully recognizes the objects corresponding to the cell nuclei description. It also shows that the result of the segmentation contains "false positives" and "goal pass".

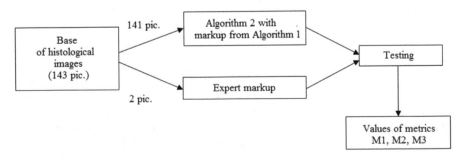

Fig. 10.8 Training and testing of Algorithm 2

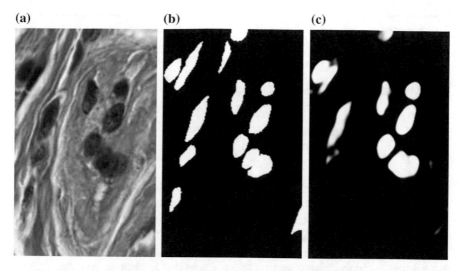

Fig. 10.9 The comparison of expert markup with the automatic segmentation result: **a** original image, **b** reference markup, **c** markup on the output of Algorithm 2

10.5 Morphological Filter

This chapter proposes the introduction of the stage of morphological filtering of the histological images binary masks at the output of the neural network. It is used as an additional means to increase the reliability of the results of histological image processing by the algorithm based on the network with U-Net architecture considered earlier.

The application of mathematical morphology to image analysis is a perspective approach, which is used in materials science, as well as, in the analysis of medical images including histological preparations [47–49]. The main advantage of this type of image processing is its simplicity: both the input and output of the morphological filter produce a binary image, and the use of morphological operations usually contributes minimally to the overall computational complexity of the algorithm.

Erosion and dilatation are the basic morphological operations [50]. The erosion of the set A by the primitive B is a set of all image points z, when shifted into which the set B is entirely contained in A:

$$A \ominus B = \{z | (B)_z \subseteq A\}, \tag{10.1}$$

where $A \in Z^2$, $B \in Z^2$.

The dilatation of the set A by the primitive B is the set of all image points z, when shifted into which the sets B and A coincide in at least one element:

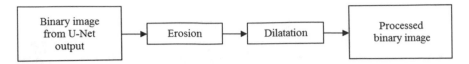

Fig. 10.10 Block diagram of the morphological filter

$$A \oplus B = \left\{ z | \left(\widehat{B} \right)_z \cap A \neq \emptyset \right\}, \qquad (10.2)$$

where $\widehat{B}$ is the central reflection of set B relative to its origin, $A \in Z^2$, $B \in Z^2$.

The morphological filter developed for solving the created objectives consists of successive application of erosion and dilation to a binary image using a mask (structure-forming set) in the form of a circle with a radius of 5 pixels (Fig. 10.10). The choice of primitive in favor of the circle was made due to the fact that the desired objects in the binary image have a shape that is close to rounded.

The result of applying the morphological filter to the test image is shown in Fig. 10.11.

The main objectives of the morphological filtering stage in the context of this study are to eliminate small, compared to the cell size, objects found as a result of segmentation by the network with U-Net architecture, as well as, to restore the clear contours of the detected cell nuclei. However, there is an increase in the number of "goal pass".

(a)   **(b)**   **(c)**

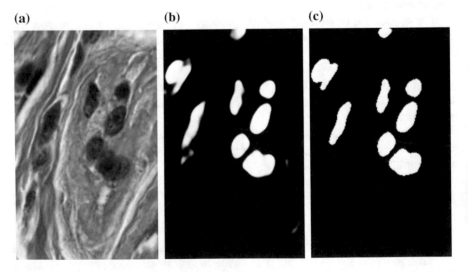

Fig. 10.11 The comparison of expert markup with the automatic segmentation result: **a** original image, **b** U-Net markup, **c** markup on the output of morphological filter

10.6 Results of Nuclei Cell Segmentation in Histological Images

In this section, the comparative segmentation results provided by Algorithm 1 and Algorithm 2 are represented in Sect. 10.6.1. Section 10.6.2 provides the use of morphological filtration as a means of additional image processing.

10.6.1 Comparison of the Segmentation Results for Histological Images at the Output of Algorithm 1 and Algorithm 2

Images at the output of Algorithm 1 and Algorithm 2 with expert marking were compared to assess the quality of segmentation. The output images of the algorithms are shown in Fig. 10.12.

The following metrics were used for the analysis: simple match coefficient (M1), Tversky index (M2), Sørensen-Dice coefficient (M3) [51]. The measurement results of the metrics data for eight images of varying complexity from the test sample are presented in Figs. 10.13, 10.14 and 10.15, as well as in Table 10.1. The choice of several metrics is due to the achievement of greater objectivity of the presented results.

Let us consider in more detail the data obtained using each of the metrics. We introduce the necessary notation: N_{00} is the total number of pixels, where the reference and segmentation result of algorithm both have a value of "0" (there is a background), N_{11} is the total number of pixels, where the reference and segmentation result of algorithm both have a value of "1" (presence of the object), N_{10} is the total number of pixels, for which the reference value is "1", and algorithm

(a) (b) (c) (d)

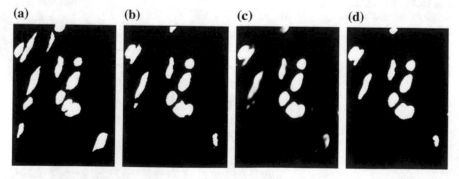

Fig. 10.12 Comparison of expert markup with the automatic segmentation result: **a** reference markup, **b** markup on the output of Algorithm 1, **c** markup on the output of Algorithm 2, **d** markup on the output of morphological filter

Fig. 10.13 The results of M1 metric calculation for the developed algorithms

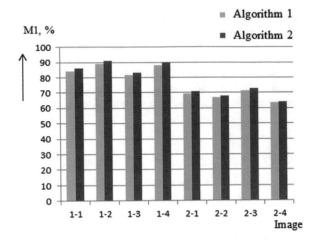

Fig. 10.14 The results of M2 metric calculation for the developed algorithms

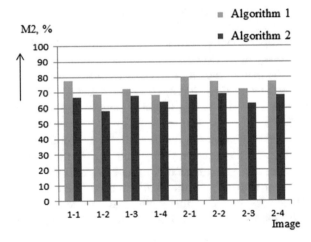

Fig. 10.15 The results of M3 metric calculation for the developed algorithms

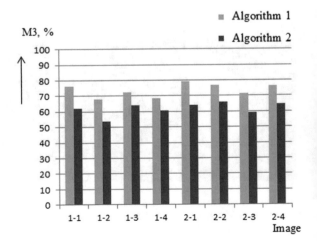

Algorithm	Image complexity	Metric		
		M1, %	M2, %	M3, %
Algorithm 1	Simple	84.22	77.89	76.41
		89.13	69.20	67.99
		81.45	72.25	72.42
		87.98	68.54	68.62
	Complex	69.65	80.14	79.32
		67.01	77.43	76.99
		71.30	72.57	71.55
		63.71	77.04	76.29
Algorithm 2	Simple	85.90	66.90	62.30
		90.80	58.10	53.70
		83.40	67.80	64.20
		89.80	64.00	60.40
	Complex	71.10	68.60	64.30
		68.00	69.60	66.00
		72.80	63.00	59.10
		64.20	68.30	64.50

Table 10.1 M1, M2, M3 metric values for test dataset consisting of 8 images with different complexity

segmentation result is "0" ("false negative"), N_{01} is the total number of pixels, for which the reference value is "0", and the algorithm markup is "1" ("false positive").

The results of calculating the simple match coefficient measured by Eq. 10.3 for the segmentation masks at the output of the algorithms based on U-Net and AlexNet networks are presented in Fig. 10.13.

$$M1 = \frac{N_{00} + N_{11}}{N_{00} + N_{01} + N_{10} + N_{11}} * 100\% \qquad (10.3)$$

According to M1, a quality of cell nuclei segmentation performed by Algorithm 2 outperforms the results obtained at the output of Algorithm 1. In addition, the results of evaluating M1 metric show that segmentation of simple images with fewer elements is performed more successfully.

Tversky index (M2 metric) allows adjusting the value of the coefficients α and β, which control the magnitude of penalties for "false positives" and "false negatives", respectively:

$$M2 = \frac{N_{11}}{N_{11} + \alpha N_{01} + \beta N_{10}} * 100\%. \qquad (10.4)$$

In this chapter, the following values of parameters for calculating Tversky index were chosen: $\alpha = 0.4$ and $\beta = 0.6$. The results of calculating M2 metric are presented in Fig. 10.14.

According to M2 metric, the segmentation similarity performed by Algorithm 2 with the reference markup is from 58.1 to 69.6%. M2 values for Algorithm 1 are slightly higher and fall in the range of 68.54–80.14%, which is an acceptable result for using this algorithm in order to create automatic markup.

M3 metric (Sørensen-Dice coefficient) is a variation of Tversky index obtained with the coefficients $\alpha = \beta = 0.5$. In this case, the penalties for "false positives" and "false negatives" are the same. The results characterizing the values of M3 metric are presented in Fig. 10.15.

From this we can see that the similarity of segmentation performed by Algorithm 2 to the reference markup is in the range from 53.7 to 66.0%, while the result for Algorithm 1 is 73.6% on average

One of the central issues in the field of medical image analysis is the trade-off between quality and computational complexity of algorithms. Modern approaches to diagnostics including screening studies require the creation of algorithms with reasonable processing time for a single clinical case. Thus, the speed of the algorithm, which determines the number of histological images that can be processed per unit of time, plays an important role. The results of performance evaluating of the developed algorithms obtained on NVIDIA DGX-1 supercomputer are presented in Table 10.2. As can be seen from Table 10.2, Algorithm 1 takes about 3 h to process one histological image with the resolution of 2000×2000 pixels, which is connected with the need to run the neural network 4 million times for classifying each pixel of the image. Algorithm 2 takes about 4 s to process one image because it process image by one pass.

According to the results of the metric calculation, as well as, visual analysis, Algorithm 1 shows a high quality of histological image segmentation. However, the use of this algorithm in practice is associated with significant time costs. Thus, Algorithm 1 can be recommended as an auxiliary tool for creating automatic markup of medical images. Algorithm 2 with an almost similar level of segmentation provides significantly better performance, which allows us to recommend it for use in real medical practice.

The results of the research lead to the intermediate conclusions that are the following:

- Algorithm 1 developed on the base of AlexNet convolutional neural network can be used to automatically create the markup of the training dataset, but if there are a large number of objects in the image (complex image), it is better to mark up images manually (according to the measurement results of such a metric as simple match coefficient). In addition, despite the high quality of histological image segmentation, this approach cannot be used for direct

Algorithm	Time per one processed image, sec
Algorithm 1	10 800
Algorithm 2	4

Table 10.2 Computational complexity of the developed algorithms

analysis of medical images in real time due to the too long processing time of a single image.

- The developed Algorithm 2 based on the network with U-Net architecture can be successfully used to implement the segmentation of histological images on the base of automatically obtained markup in real medical practice, as evidenced by the high level of similarity of the obtained markup with the reference one. In addition, Algorithm 2 allows to process histological images in 2,700 times faster than Algorithm 1.

10.6.2   The Use of Morphological Filtration as a Means of Additional Processing

Analysis of the results of histological images segmentation by the developed algorithms showed that Algorithm 2 based on U-Net neural network turned out to be more preferable for use in real medical practice. As a consequence, it was decided to apply a morphological filter to the images at the output of Algorithm 2.

The results of comparing the quality of histological image segmentation at the output of Algorithm 2 and the system [Algorithm 2 + morphological filter] are presented in Table 10.3 and in Figs. 10.16, 10.17 and 10.18. The output images of the algorithms are shown in Fig. 10.12.

Table 10.3 M1, M2, M3 metric values for test dataset consisting of 8 images with different complexity

Algorithm	Image complexity	Metric		
		M1, %	M2, %	M3, %
Algorithm 2	Simple	85.9	66.9	62.3
		90.8	58.1	53.7
		83.4	67.8	64.2
		89.8	64.0	60.4
	Complex	71.1	68.6	64.3
		68.0	69.6	66.0
		72.8	63.0	59.1
		64.2	68.3	64.5
Algorithm 2 + Morphological filter	Simple	84.2	76.8	75.4
		89.1	67.4	66.1
		84.2	71.2	71.5
		87.8	68.0	68.3
	Complex	69.5	80.7	80.0
		66.7	77.8	77.6
		71.1	72.8	72.0
		63.4	77.5	77.0

Fig. 10.16 The results of M1 metric calculation for U-Net based algorithm with and without using the morphological filter

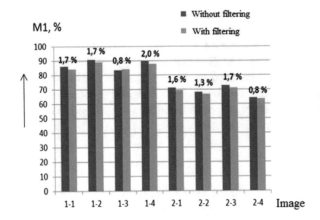

Fig. 10.17 The results of M2 metric calculation for U-Net based algorithm with and without using the morphological filter

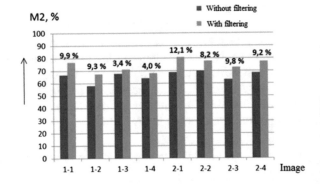

Fig. 10.18 The results of M3 metric calculation for U-Net based algorithm with and without using the morphological filter

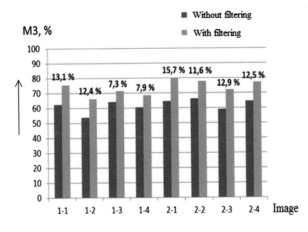

The difference between the values of the metric in question for the algorithm without filtering and the algorithm with filtering is indicated above the histogram columns relating to the image in question.

From Table 10.3 and Fig. 10.16, it can be seen that, according to the measurement results of M1 metric, the use of a morphological filter leads to a slight decrease in the quality of the resulting markup from 78.25 to 77% (average values of M1 for algorithm with filtering and without filtering).

Figure 10.17 shows that the difference between the values of M2 metric for algorithms with filtering and without it increased in comparison with the indicators for M1 metric. The average value of this difference for a group of test images is 8.24%, while the morphological filtering gives an improvement in the quality of the obtained binary segmentation masks. The similarity with the reference markup of data ranges from 58.1 to 69.6% with no filtering and from 67.4 to 80.7% if it present. Therefore, according to this metric, the use of a morphological filter is recommended for optimal results.

The results characterizing the values of M3 metric for the developed algorithms are shown in Fig. 10.18. According to the obtained values, the best segmentation results are observed with the additional use of the morphological filter. Improving the similarity of the reference markup and the image from the output of the filter averages 11.68% for a group of images of varying complexity. Thus, according to the indicators of this metric, the use of a morphological filter is necessary to improve the quality of the obtained segmentation masks.

In addition, it is clear from the histograms that, according to M1 metric, the use of a morphological filter leads to the best results for a group of simple images, while according to M2 and M3 metrics, morphological filtering leads to better results for a group of complex images.

The performance of the developed algorithm was also evaluated when it was launched on NVIDIA DGX-1 computer. According to the assessment, the processing time of a single histological image by a neural network is approximately 3–4 s, depending on the complexity of the image, and the processing time of one frame with a morphological filter is, on average, 0.74 s. Thus, processing of one image is carried out in the mode close to real time, which will allow to reduce significantly the time spent on histological studies in real practice.

10.7   Conclusions

This chapter presents an algorithm for cell nuclei segmentation of histological images based on U-Net convolutional neural network with subsequent morphological filtering. The results of the presented algorithm were evaluated using such segmentation quality assessment metrics as a simple match coefficient, Tversky index, and Sørensen coefficient. The results of the study show that Algorithm 2 developed on the basis of U-Net network can be successfully used to implement the segmentation of histological images based on automatically obtained markup in real medical practice, as evidenced by the high level of similarity of the resulting markup to the reference one. In addition, Algorithm 2 allows histological images to be processed 2,700 times faster than Algorithm 1.

Algorithm 1 based on AlexNet neural network can be used to automatically create the markup of the training dataset, but if there are a large number of objects in the image (complex image), it is better to mark up the image manually (according to the results of calculating such metrics as simple match coefficient and Hausdorff distance). In addition, despite the high quality of the segmentation of histological images, this approach cannot be used for direct analysis of medical images in real time due to significant time costs. According to simple match coefficient, the quality of segmentation of histological images using a morphological filter worsens by 1.25%, while the data obtained as a result of calculating Tversky index and Sørensen coefficient indicates an improvement in the segmentation quality by 8.24% and 11.68%, respectively. Thus, the analysis of the results of numerical experiments confirms the need to use morphological filtering as a means of additional processing of histological images binary masks obtained at the output of the neural network algorithm.

References

1. World Health Organization: Cancer. http://www.who.int/cancer/en. Accessed 11 Aug 2019
2. Mytsik, A.V.: Using the ImageJ program for automatic morphometry in histological studies. Omsk Sci. Herald **2**(100), 187–189 (2011) (in Russian)
3. Wu, H.-S., Xu, R., Harpaz, N., Burstein, D., Gil, J.: Segmentation of intestinal gland images with iterative region growing. J. Microscopy **220**(3), 190–204 (2005)
4. Priorov, A.L., Khryashchev, V.V., Stepanova, O.A., Srednyakova, A.S.: Development and research of the algorithm of segmentation of cell nuclei on histological images. Biomed. Radioelectron. **11**, 13–20 (2018) (in Russian)
5. Rother, C., Kolmogorov, V., Blake, A.: Grabcut—interactive foreground extraction using iterated graph cuts. Microsoft Technical Report, MSRTR-2011 (2004)
6. Kieffer, B., Babaie, M., Kalra, S., Tizhoosh, H.: Convolutional neural networks for histopathology image classification: training vs. using pre-trained networks. 1–6 (2017). arXiv:1710.05726
7. Wu, M., Yan, C., Liu, H., Liu, Q.: Automatic classification of ovarian cancer types from cytological images using deep convolutional neural networks. Biosci. Rep. **38**(3), 1–11 (2018)
8. Mishra, R., Daescu, O., Leavey, P., Rakheja, D., Sengupta, A.: Convolutional neural network for histopathological analysis of osteosarcoma. J. Comput. Biol. **25**(3), 313–325 (2018)
9. Sharma, H., Zerbe, N., Klempert, I., Hellwich, O., Hufnagl, P.: Deep convolutional neural networks for automatic classification of gastric carcinoma using whole slide images in digital histopathology. Comput. Med. Imaging Graph. **61**, 2–13 (2017)
10. Cruz-Roaa, A., Basavanhallyb, A., Gonzáleza, F.: Automatic detection of invasive ductal carcinoma in whole slide images with convolutional neural networks. In: Proceedings of SPIE 9041, Medical Imaging 2014: Digital Pathology, vol. 9041, pp. 904103-1–904103-15 (2014)
11. Kumar, N., Verma, R., Arora, A., Kumar, A., Gupta, S., Sethi, A., Gann, P.H.: Convolutional neural networks for prostate cancer recurrence prediction. In: Proceedings of SPIE 10140, Medical Imaging 2017: Digital Pathology, vol. 101400H, pp. 1–12 (2017)

12. Ponzio, F., Macii, E., Ficarra, E., Di, Cataldo, S.: Colorectal cancer classification using deep convolutional networks—an experimental study. In: 11th International Joint Conference on Biomedical Engineering Systems and Technologies, vol. 2, pp. 58–66 (2018)
13. Jimenez–del–Toroab, O., Atzoria, M., Otaloraab, S., Anderssonc, M., Eurenc, K., Hedlundc, M., Ronnquistc, P., Mullerabd, H.: Convolutional neural networks for an automatic classification of prostate tissue slides with high–grade Gleason score. In: Proceedings of SPIE 10140, Medical Imaging 2017: Digital Pathology, vol. 101400O, pp. 1–10 (2017)
14. Bulten, W., Litjens, G.J.S., Hulsbergen-van de Kaa, C.A., van der Laak, J.: Automated segmentation of epithelial tissue in prostatectomy slides using deep learning. In: Proceedings of Medical Imaging 2018: Digital Pathology, vol. 105810S, pp. 1–7 (2018)
15. Sornapudi, S., Stanley, R.J., Stoecker, W.V., Almubarak, H., Long, R., Antani, S., Thoma. G., Zuna, R., Frazier, S.R.: Deep learning nuclei detection in digitized histology images by superpixels. J. Pathol. Inform. **9**, 5.1–5.10 (2018)
16. Nogay, H.S.: Deep convolutional neural networks to detect lung cancer stage. J. Cogn. Syst. **2** (2), 33–36 (2017)
17. Alakwaa, W., Nassef, M., Badr, A.: Lung cancer detection and classification with 3D convolutional neural network (3D-CNN). Int. J. Adv. Comput. Sci. Appl. **8**(8), 409–417 (2017)
18. Kainz, P., Pfeiffer, M., Urschler, M.: Semantic segmentation of colon glands with deep convolutional neural networks and total variation segmentation (2015). arXiv:1511.06919
19. Taieb, A.B., Hamarneh, G.: Topology aware fully convolutional networks for histology gland segmentation. In: International Conference on Medical Image Computing and Computer-Assisted Intervention, pp. 460–468 (2016)
20. Guo, P., Banerjee, K., Joe Stanley, R., Long, R., Antani, S., Thoma, G., Zuna, R., Frazier, S. R., Moss, R.H., Stoecker, W.V.: Nuclei-based features for uterine cervical cancer histology image analysis with fusion-based classification. IEEE J. Biomed Health Inform. **20**(6), 1595–1607 (2016)
21. Almubaraka, H.A., Stanleya, R.J., Longb, R., Antanib, S., Thomab, G., Zunac, R., Frazierd, S.R.: Convolutional neural network based localized classification of uterine cervical cancer digital histology images. Comput. Sci. **114**, 281–287 (2017)
22. Wu, M., Yan, C., Liu, H., Liu, Q., Yin, Y.: Automatic classification of cervical cancer from cytological images by using convolutional neural network. Biosci. Rep. **38**(6), BSR20181769 (2018)
23. Chougrada, H., Zouakia, H., Alheyane, O.: Deep convolutional neural networks for breast cancer screening. Comput. Methods Program. Biomed. **157**, 19–30 (2018)
24. Jamil-Ur Rahman, Md., Mahmud, F., Sultan, R., Ahsan, S.A.: Automatic system for detecting invasive ductal carcinoma using convolutional neural networks. In: 2018 IEEE Region 10 Conference, TENCON 2018, pp. 673–678 (2018)
25. Kowal, M., Obuchowicz, A., Filipczuk, P., Korbicz, J.: Computer-aided diagnosis of breast cancer based on fine needle biopsy microscopic images. Comput. Biol. Med. **43**(10), 1563–1572 (2013)
26. Nazeri, K., Ebrahimi, M., Aminpour, A.: Two-stage convolutional neural network for breast cancer histology image classification. In: Campilho, A., Karray, F., ter Haar Romeny, B. (eds) 15th International Conference on Image Analysis and Recognition, LNCS, vol. 10882, pp. 717–726. Springer International Publishing (2018)
27. Kiambe, K.: Breast histopathological image feature extraction with convolutional neural networks for classification. Trans. Image Process. Pattern Recognit. **4**(2), 4–12 (2018)
28. Rakhlin, A., Shvets. A., Iglovikov, V., Kalinin, A.: Deep convolutional neural networks for breast cancer histology image analysis. In: Campilho, A., Karray, F., ter Haar Romeny, B. (eds) 15th International Conference on Image Analysis and Recognition, LNCS, vol. 10882, pp. 737–744. Springer International Publishing (2018)
29. Araújo, T., Aresta, G., Castro, E., Rouco, J., Aguiar, P., Eloy, C., Polónia, A., Campilho, A.: Classification of breast cancer histology images using convolutional neural networks. PLoS ONE **12**(6), e0177544 (2017). https://doi.org/10.1371/journal.pone.0177544

30. Fu, X., Liu, T., Xiong, Z., Smaill, B.H., Stiles, M.K., Zhao, J.: Segmentation of histological images and fibrosis identification with a convolutional neural network. Comput. Biol. Med. **98**, 147–158 (2018)
31. Xu, J., Luo, X., Wang, G., Gilmore, H., Madabhushi, A.: A deep convolutional neural network for segmenting and classifying 12 epithelial and stromal regions in histopathological images. Eurocomputing **191**, 214–223 (2016)
32. Han, Z., Benzheng, W., Yuanjie, Z., Yilong, Y., Kejian, L., Shuo, L.: Breast cancer multi-classification from histopathological images with structured deep learning model. Sci. Rep. **7**(1), 4172.1–4172.6 (2017)
33. Murthy, V., Hou, L., Samaras, D., Kurc, T.M., Saltz, J.H.: Center-focusing multi-task CNN with injected features for classification of glioma nuclear images. In: IEEE Winter Conference on Applications of Computer Vision, pp. 834–841 (2017)
34. Nawaz, M.A., Sewissy, A.A., Soliman, T.H.A.: Automated classification of breast cancer histology images using deep learning based convolutional neural networks. Int. J. Comput. Sci. Netw. Secur. **4**, 152–160 (2018)
35. Aresta, G., Castro, E., Rouco, J., Aguiar, P., Eloy, C., Polónia, A., Campilho, A.: Classification of breast cancer histology images using convolutional neural networks. PLoS One, **12**(6), e0177544 (2017)
36. Teramoto, A., Tsukamoto, T., Kiriyama, Y., Fujita, H.: Automated classification of lung cancer types from cytological images using deep convolutional neural networks. BioMed. Res. Int. **2017**, 1, 4067832.1–4067832.6 (2017)
37. Mehta, S., Mercan, E., Bartlett, J., Weaver, D., Shapiro, L.: Y-Net: joint segmentation and classification for diagnosis of breast biopsy images. In: Proceedings of Medical Image Computing and Computer Assisted Intervention, pp. 893–901 (2018)
38. Gecer, B., Aksoy, S., Mercan, E., Shapiro, L.G., Weaver, D.L., Elmore, J.G.: Detection and classification of cancer in whole slide breast histopathology images using deep convolutional networks. Pattern Recogn. **84**, 345–356 (2018)
39. Xu, Y., Jia, Z., Wang, L., Ai, Y., Zhang, F., Lai, M., Chang, E.: Large scale tissue histopathology image classification, segmentation, and visualization via deep convolutional activation features. BMC Bioinform. **18**(1), 281.1–281.17 (2017)
40. Janowczyk, A., Madabhushi, A.: Deep learning for digital pathology image analysis: A comprehensive tutorial with selected use cases. J. Pathol. Inform. **7**, 29.1–29.18 (2016)
41. Khryashchev, V., Lebedev, A., Stepanova, O., Srednyakova, A.: Using convolutional neural networks in the problem of cell nuclei segmentation on histological images. In: Dolinina, O., Brovko, A., Pechenkin, V., Lvov, A., Zhmud, V., Kreinovich, V. (eds) Proceedings of Recent Research in Control Engineering and Decision Making, ICIT 2019, SSDC, vol. 199, pp. 149–161. Springer, Cham (2019)
42. Krizhevsky, A., Sutskever, I., Hinton, G.: ImageNet classification with deep convolutional neural networks. In: 26th Annual Conference on Neural Information Processing Systems, vol. 1, pp. 1097–1105 (2012)
43. Wang, S.H., Xie, S., Chen, X., Guttery, D.S., Tang, C., Sun, J., Zhang, Y.D.: Alcoholism identification based on an AlexNet transfer learning model. Front. Psychiat. **10**, 205.1–205.13 (2019)
44. Lu, S., Lu, Z., Zhang, Y.D.: Pathological brain detection based on AlexNet and transfer learning. J. Comput. Sci. **30**, 41–47 (2019)
45. Fairuz, S., Habaebi, M., Elsheikh, E.: Finger vein identification based on transfer learning of AlexNet. In: 7th International Conference on Computer and Communication Engineering, pp. 465–469 (2018)
46. Fischer, P., Ronneberger, O., Brox, T.: U-net: convolutional networks for biomedical image segmentation. In: Navab, N., Hornegger, J. (eds) Medical Image Computing and Computer-Assisted Intervention (MICCAI) 2015, LNCS, vol. 9351, pp. 234–341. Springer, Munich, Germany (2015)

47. Pingel, T.J., Clarke, K.C., McBride, W.A.: An improved simple morphological filter for the terrain classification of airborne LiDAR data. ISPRS J. Photogr. Remote Sens. **77**, 21–30 (2013)
48. Wang, Q., Wu, L., Xu, Z., Tang, H., Wang, R., Li, F.: A progressive morphological filter for point cloud extracted from UAV images. In: International Geoscience and Remote Sensing Symposium, pp. 2023–2026 (2014)
49. Zhao, Y., Gui, W., Chen, Z.: Edge detection based on multi-structure elements morphology. In: 6th World Congress Intelligent Control and Automation, vol. 2, pp. 9795–9798 (2006)
50. Soille, P.: Morphological Image Analysis. Springer, Berlin, Heidelberg, New York (1999)
51. Minervini, M., Rusu, C., Tsaftaris, S.A.: Learning computationally efficient approximations of complex image segmentation metrics. In: 8th International Symposium on Image and Signal Processing and Analysis, pp. 60–65 (2013)

Author Index

Printed in the United States
By Bookmasters